GAME THEORY AND BUSINESS APPLICATIONS

INTERNATIONAL SERIES IN OPERATIONS RESEARCH & MANAGEMENT SCIENCE

Frederick S. Hillier, Series Editor
Stanford University

Saigal, R. / *LINEAR PROGRAMMING: A Modern Integrated Analysis*

Nagurney, A. & Zhang, D. / *PROJECTED DYNAMICAL SYSTEMS AND VARIATIONAL INEQUALITIES WITH APPLICATIONS*

Padberg, M. & Rijal, M. / *LOCATION, SCHEDULING, DESIGN AND INTEGER PROGRAMMING*

Vanderbei, R. / *LINEAR PROGRAMMING: Foundations and Extensions*

Jaiswal, N.K. / *MILITARY OPERATIONS RESEARCH: Quantitative Decision Making*

Gal, T. & Greenberg, H. / *ADVANCES IN SENSITIVITY ANALYSIS AND PARAMETRIC PROGRAMMING*

Prabhu, N.U. / *FOUNDATIONS OF QUEUEING THEORY*

Fang, S.-C., Rajasekera, J.R. & Tsao, H.-S.J. / *ENTROPY OPTIMIZATION AND MATHEMATICAL PROGRAMMING*

Yu, G. / *OPERATIONS RESEARCH IN THE AIRLINE INDUSTRY*

Ho, T.-H. & Tang, C. S. / *PRODUCT VARIETY MANAGEMENT*

El-Taha, M. & Stidham, S. / *SAMPLE-PATH ANALYSIS OF QUEUEING SYSTEMS*

Miettinen, K. M. / *NONLINEAR MULTIOBJECTIVE OPTIMIZATION*

Chao, H. & Huntington, H. G. / *DESIGNING COMPETITIVE ELECTRICITY MARKETS*

Weglarz, J. / *PROJECT SCHEDULING: Recent Models, Algorithms & Applications*

Sahin, I. & Polatoglu, H. / *QUALITY, WARRANTY AND PREVENTIVE MAINTENANCE*

Tavares, L. V. / *ADVANCED MODELS FOR PROJECT MANAGEMENT*

Tayur, S., Ganeshan, R. & Magazine, M. / *QUANTITATIVE MODELING FOR SUPPLY CHAIN MANAGEMENT*

Weyant, J./ *ENERGY AND ENVIRONMENTAL POLICY MODELING*

Shanthikumar, J.G. & Sumita, U./*APPLIED PROBABILITY AND STOCHASTIC PROCESSES*

Liu, B. & Esogbue, A.O. / *DECISION CRITERIA AND OPTIMAL INVENTORY PROCESSES*

Gal, T., Stewart, T.J., Hanne, T./ *MULTICRITERIA DECISION MAKING: Advances in MCDM Models, Algorithms, Theory, and Applications*

Fox, B. L./ *STRATEGIES FOR QUASI-MONTE CARLO*

Hall, R.W. / *HANDBOOK OF TRANSPORTATION SCIENCE*

Grassman, W.K./ *COMPUTATIONAL PROBABILITY*

Pomerol, J-C. & Barba-Romero, S. / *MULTICRITERION DECISION IN MANAGEMENT*

Axsäter, S. / *INVENTORY CONTROL*

Wolkowicz, H., Saigal, R., Vandenberghe, L./ *HANDBOOK OF SEMI-DEFINITE PROGRAMMING: Theory, Algorithms, and Applications*

Hobbs, B. F. & Meier, P. / *ENERGY DECISIONS AND THE ENVIRONMENT: A Guide to the Use of Multicriteria Methods*

Dar-El, E./ *HUMAN LEARNING: From Learning Curves to Learning Organizations*

Armstrong, J. S./ *PRINCIPLES OF FORECASTING: A Handbook for Researchers and Practitioners*

Balsamo, S., Personé, V., Onvural, R./ *ANALYSIS OF QUEUEING NETWORKS WITH BLOCKING*

Bouyssou, D. et al/ *EVALUATION AND DECISION MODELS: A Critical Perspective*

Hanne, T./ *INTELLIGENT STRATEGIES FOR META MULTIPLE CRITERIA DECISION MAKING*

Saaty, T. & Vargas, L./ *MODELS, METHODS, CONCEPTS & APPLICATIONS OF THE ANALYTIC HIERARCHY PROCESS*

GAME THEORY AND BUSINESS APPLICATIONS

edited by
Kalyan Chatterjee
Pennsylvania State University

and

William F. Samuelson
Boston University

Kluwer Academic Publishers
Boston/Dordrecht/London

Distributors for North, Central and South America:
Kluwer Academic Publishers
101 Philip Drive
Assinippi Park
Norwell, Massachusetts 02061 USA
Telephone (781) 871-6600
Fax (781) 871-6528
E-Mail <kluwer@wkap.com>

Distributors for all other countries:
Kluwer Academic Publishers Group
Distribution Centre
Post Office Box 322
3300 AH Dordrecht, THE NETHERLANDS
Telephone 31 78 6392 392
Fax 31 78 6546 474
E-Mail <orderdept@wkap.nl>

 Electronic Services <http://www.wkap.nl>

Library of Congress Cataloging-in-Publication

Chatterjee, Kalyan.
    Game theory and business applications / edited by Kalyan Chatterjee and William F. Samuelson.
        p. cm. -- (International series in operations research & management science; 35)
    Includes bibliographical references and index.
    ISBN 0-7923-7332-4 (alk. paper)
    1. Game theory. 2. Decision making. 3. Management. I. Samuelson, William F. II. Title. III. Series

HB144 .G372 2001
658.4'0353--dc21
                                                                    2001022503

Copyright © 2001 by Kluwer Academic Publishers.

All rights reserved. No part of this publication may be reproduced, stored in a retrieval system or transmitted in any form or by any means, mechanical, photocopying, recording, or otherwise, without the prior written permission of the publisher, Kluwer Academic Publishers, 101 Philip Drive, Assinippi Park, Norwell, Massachusetts 02061

Printed on acid-free paper.

Printed in the United States of America

Table of Contents

List of Contributors	vii
Acknowledgements	x

1.	Introduction *Kalyan Chatterjee and William Samuelson*	1
2.	Game Theory Models in Finance *Franklin Allen and Stephen Morris*	17
3.	Game Theory Models in Accounting *Chandra Kanodia*	49
4.	Game Theory Models in Operations Management and Information Systems *Lode Li and Seungjin Whang*	95
5.	Incentive Contracting and the Franchise Decision *Francine Lafontaine and Margaret E. Slade*	133
6.	Cooperative Games and Business Strategy *H.W. Stuart, Jr.*	189
7.	Renegotiation in the Repeated Amnesty Dilemma, with Economic Applications *Joseph Farrell and Georg Weizsäcker*	213
8.	Reputation and Signalling Quality Through Price Choice *Taradas Bandyopadhyay, Kalyan Chatterjee, and Navendu Vasavada*	247
9.	Game Theory and The Practice of Bargaining *Kalyan Chatterjee*	273

10.	Auctions in Theory and Practice *William F. Samuelson*	295
11.	The Economics of Auctions and Bidder Collusion *Robert C. Marshall and Michael J. Meurer*	339
12.	Activity Rules for an Iterated Double Auction *Robert Wilson*	371
	Index	**387**

CONTRIBUTORS

Franklin Allen is the Nippon Life Professor of Finance and Economics at the Wharton School of the University of Pennsylvania. From 1990-1993 he was Vice Dean and Director of Wharton Doctoral Programs and from 1993-1996 he was Executive Editor of the *Review of Financial Studies*. In 1999 he served as President of the Western Finance Association and President of the Society for Financial Studies. In 2000, he is serving as President of the American Finance Association. Dr. Allen's main areas of interest include corporate finance, asset pricing, financial innovation and comparative financial systems.

Taradas Banyopadhyay is Professor of Economics at the University of California, Riverside. His current research interests include individual decision making and social choice, and his work has been published widely in academic journals.

Kalyan Chatterjee is the Distinguished Professor of Management Science and Economics at Pennsylvania State University. His research applies game theory to a variety of incomplete information models: arbitration and bargaining, coalition formation, and market competition. For many years, he has taught graduate courses in game theory and negotiation.

Joseph Farrell is Professor of Economics and Affiliate Professor of Business at the University of California, Berkeley, where he is the founding Director of the Competition Policy Center. In 1996-97 he was Chief Economist at the Federal Communications Commission. He is Editor of the *Journal of Industrial Economics*. His work has addressed game theory (particularly renegotiation and communication), standard-setting, industrial organization, and telecommunications.

Chandra Kanodia is the Honeywell Professor of Accounting at the University of Minnesota. He serves on the editorial boards of the *Journal of Accounting Research*, the *Review of Accounting Studies*, and the *Asia Pacific Journal of Accounting and Economics*. His published research has appeared in such journals as *Econometrica*, the *Accounting Review*, and *Management Science*.

Francine Lafontaine is Professor of Business Economics and Public Policy at the University of Michigan Business School, and a faculty research fellow at the National Bureau of Economic Research. Her research focuses on vertical relationships and contracting, with a special emphasis on the application of recent advances in contract theory to the analysis of franchising arrangements. She has published her research in such journals as the *Journal of Political Economy*, the *RAND Journal of Economics*, and the *Journal of Industrial Economics*.

Lode Li is a Professor at Yale School of Management. His research addresses a broad range of issues in managerial economics and operations management. These include operations strategies in competitive, time-sensitive markets, the coordination of global operations, and the value of information in a competitive environment. His recent work focuses on incentives for vertical chain coordination and information sharing in the presence of horizontal competition.

Robert C. Marshall is Professor and Head of the Economics Department at Pennsylvania State University. His research has primarily focused on bidder collusion in auctions and procurements. His papers in this area have appeared in the *Journal of Political Economy*, the *American Economic Review*, and the *Quarterly Journal of Economics*.

Michael J. Meurer is Associate Professsor at Boston University School of Law. He previously taught in the law school at the University at Buffalo and in the economics department and law school at Duke University. He has published in economics and law journals on topics including antitrust law, intellectual property law, and industrial organization. He has consulted with the Federal Trade Commission on merger issues relating to patent licensing, and with AID on antitrust issues in Mongolia.

Stephen Morris is Professor of Economics at Yale University. His recent research in theoretical and applied game theory has emphasized the importance of players' uncertainty about others' beliefs.

William F. Samuelson is Professor of Economics and Finance at Boston University School of Management. His research on decision making under uncertainty, competitive bidding, and bargaining has been published in economics, finance, and management science journals. His teaching centers in the areas of microeconomics, decision making, and game theory. Currently, he is working on the fourth edition of his textbook in managerial economics.

Margaret Slade is a professor of economics at the University of British Columbia in Vancouver, Canada and a member of GREQAM in Marseille, France. She works in the area of applied industrial organization, which includes econometric modeling of the games that firms play when they interact in markets as well as the games that agents play when they interact inside firms.

Harborne W. Stuart, Jr. is an Associate Professor in the Management Science Department at Columbia Business School. In his research, he studies business decisions using game-theoretic approaches. His research includes the further development of "interactive decision theory," which takes strategic uncertainty as the primary focus, and "added value theory," which uses cooperative game theory to study businesses as the central players in economic value creation. Application of this research is principally to the fields of strategy, negotiation, and operations.

Navendu Vasavada co-founded Lumin Asset Management in 1998 and manages its $50 million hedge fund. Before this, he held academic appointments in Finance at Pennsylvania State University, the Wharton School, and the Wharton Center for Applied Research. Early in his professional career, he worked at the Indian Petrochemicals Corporation and advised the Government of India on investments in the petrochemical and pharmaceutical industries.

Georg Weizsäcker grew up in Kaiserslautern, Germany, studied economics in Berlin and Berkeley, and is currently enrolled in the Ph.D. program at Harvard University. His research interests lie in the areas of game theory and behavioral/experimental economics.

Seungjin Whang is Associate Professor of Operations, Information and Technology at the Graduate School of Business, Stanford University. He also serves as Associate Director of Global Supply Chain Forum at Stanford University. His research interests focus on supply chain management and information technology, including incentive issues for supply chain management and pricing for congestion-prone services.

Robert Wilson is a professor at the Stanford Business School where he teaches courses on game theory and market design. His theoretical research on auctions and related markets over thirty-five years stems from practical work in applied contexts. In recent years he has worked on the design of the FCC spectrum auctions and the California electricity auctions, among others. His chapter in this book is based on a project sponsored by the California Power Exchange.

Acknowledgements

We would like to thank everyone who made this book possible. We owe a special debt of gratitude to those colleagues who gave up their time to referee the papers, among others Sudipto Bhattacharya, Sugato Bhattacharyya, Steve Chiu, Tom Gresik, Manish Kacker, Vijay Krishna, Darwin Neher, Mike Pangburn, John Riley, Phillip Stocken and Shankar Sundaresan.

Nicole Duffala provided invaluable help in retyping and formatting papers submitted in Scientific Workplace and in proofreading what she had typed. Siddhartha Bandyopadhyay read through several chapters for the keywords. The editors from Kluwer, especially Gary Folven whose idea it was to publish such a book, were patient as we moved along.

1 INTRODUCTION

Kalyan Chatterjee and William Samuelson

The aim of this volume is to provide the interested reader a broad and (we hope) deep view of the way business decisions can be modelled and analyzed using game theory. Indeed, the chapter contents embrace a wide variety of business functions – from accounting to finance to operations to strategy to organizational design. Moreover, specific application areas include numerous kinds of market competition, bargaining, auctions and competitive bidding. All of these applications involve competitive decision settings – that is to say, situations where a number of economic agents in pursuit of their respective self-interests take actions that together affect all of their fortunes. In the language of game theory, players take actions consistent with the given "rules of the game," and these joint actions determine final outcomes and payoffs. Besides providing a structure in which to model competitive settings, game theory goes a long way toward answering the key question: What optimal decision or action should a competitor take, all the while anticipating optimal actions from one's rivals? Provided the game-theoretic description faithfully captures the real-world competitive situation at hand, game theory provides a compelling guide for business strategy.

From its birth with von Neumann and Morgenstern's, *Theory of Games and Economic Behavior* (1944), and for its next twenty-five years, game theory was principally a mathematical discipline. To be sure, elegant solution concepts were developed – but for a relatively limited class of settings (such as zero-sum games). In the last twenty-five years, however, game theory has been applied to a growing number of practical problems: from antitrust analysis to monetary policy; from the design of auction institutions to the structuring of incentives within firms; from patent races to dispute resolution. Game theory is a staple of doctoral studies in economics, finance, and management science. Indeed, there are many excellent texts in the subject: Fudenberg and Tirole (1991), Gibbons (1992), Osborne and Rubinstein (1994), Rasmusen (1994), or (Myerson (1997), to name a few. Perhaps, a better barometer of the growing importance of game-theoretic ideas in business and management is the growing place of the discipline in mainstream texts in strategy, microeconomics, and managerial economics. Notable among these are Milgrom and Roberts (1992), Oster (1999), and Besanko, Dranove and Shanley (1999). Besides economics, strategy, and finance, the marketing area has also been a fertile area for game-theoretic applications. See, for example, Chatterjee and Lilien, (1986), Eliashberg and Lilien (1993), and the perceptive article by Wernerfelt (1995). In a personal communication, a leading researcher in marketing estimates that some 3 of 4 doctoral theses in marketing use some game theory. By contrast, twenty-five years ago, game theory was absent from most textbooks and most research work.

The chapters of this volume outline some of the most interesting applications of game theory in *non-technical terms*. We presuppose that the reader has a working knowledge of basic game-theoretic concepts and results: from Nash and Bayesian equilibrium, to moral hazard and adverse selection. (Any of the game theory and economics texts cited above are good places to turn for re-enforcing the basic tools.)

The first section of this volume discusses game-theoretic applications in four functional areas of business: finance, accounting, operations management and information systems, and organization design. The second section considers competitive strategies in "imperfect" markets. Using cooperative and non-cooperative game-theoretic approaches, these four chapters consider various topics: spatial competition, signaling of product quality, trust and cooperation in ongoing relationships, strategic behavior in bargaining, and the "balance of power" between the firm and its buyers and suppliers. The last section of the book deals in detail with auctions and competitive bidding institutions. The emphasis is on the contributions of game theory to both auction theory and

Introduction 3

practice. Topics considered include optimal auctions, bidder collusion, and the design of institutions for selling the radio spectrum and trading electrical power.

It goes without saying that no single book can include the contributions of game theory in all aspects of business. Omitted from this volume are such important topic areas as: research and development, law and economics, and regulation (including antitrust practices).

1. Functional Business Areas

The opening chapter in this section, by Franklin Allen and Stephen Morris, deals with finance, the area in business that has historically drawn most on economics for its theoretical development. Allen and Morris consider two broad fields in finance, asset pricing and corporate finance. The first is concerned primarily with the decisions of investors and the second with the decisions of firms. The pre-game-theoretic approach to these fields relied heavily on the assumptions of perfect information and perfect markets. While these early models were successful in explaining many empirical phenomena, sophisticated empirical work uncovered the existence of "anomalies" that the theory could not explain. This was especially the case in corporate finance, where the puzzle of why firms pay dividends and how firms choose levels of debt and equity resisted simple explanations based, for example, on incorporating tax effects into the models of Miller and Modigliani. This gave rise to the first wave of models that incorporated asymmetric information and signalling. Allen and Morris discuss these in some detail, noting their insights and also their limitations. In addition, strategic interaction plays an important role when analyzing the links between product markets and financial decisions – whether these be models of debt levels, takeover bids and the market for corporate control, or models of financial intermediation. The main features of these models are also considered in this chapter. While the area of asset pricing has seen less game-theoretic scholarly work, the chapter does examine the best-known models of market microstructure.

Allen and Morris also survey the "second wave" of game-theoretic modeling in finance. This second wave reflects recent work on higher order beliefs and explores the consequences of departing from the common knowledge assumptions typical of traditional models. The authors take up the recent work on information cascades, applied to such phenomena as bank runs (also the subject of influential "first wave" models) and the effect of different priors versus different information. While the extensive use of game theory and contract theory models has brought theory more in line with observation, much more remains to be done. Allen and Morris provide pointers to where this research is heading in the future.

The field of accounting, though linked with finance in the popular mind, has traditionally pursued somewhat different research methods. In Chapter 3, Chandra Kanodia, discusses the changes in accounting research that have necessitated the use of game theory and the methods of mechanism design. The focus is now not so much on *ex post* analysis of accounting data as on the strategic interactions among individuals and the *ex ante* incentives to engage in various kinds of behavior. Kanodia considers new approaches to management accounting, with the focus on three main research avenues: (i) The design of performance incentives (including the role of stochastic monitoring) to alleviate the problems of "moral hazard" in principal-agent problems; (ii) The problems of coordination in organizations, especially the use of transfer pricing and budgeting to induce efficient internal decisions; and (iii) Strategies for auditor hiring and auditor pricing.

Concerning the first research topic, moral hazard occurs when an agent takes actions (in its own self-interest) that are unobservable to the principal and which may conflict with the principal's interests. In Kanodia's model, the principal cannot observe the agent's action, but can observe two separate signals – performance measures, if you will – correlated with his action. The first signal is costless, but the second is costly. Having observed the first signal, when should the principal pay to observe the second? Should the investigation of the second performance measure be triggered by seemingly poor performance or by good performance? Kanodia addresses these important questions in a number of different settings. He also shows how to determine the optimal weights when combining multiple performance measures. (A particular performance measure should have a larger relative weight, the lower its variance and the greater its sensitivity to the agent's action.)

Transfer pricing and budgeting are alternative means to coordinate the modern firm's many activities. Transfer pricing schemes (based on *ex ante* measures of marginal cost) set internal prices for "upstream" activities to supply "downstream" activities. By contrast, budget schemes achieve coordination by assigning targets – cost targets, revenue targets, production targets – to managers. Kanodia shows how the firm can construct a cost target for the upstream activity and revenue target for the downstream activity to induce both divisions to i) report truthful information, ii) take profit-maximizing actions, and iii) coordinate on a profit-maximizing level of output. The same allocation could be achieved by a suitably designed transfer pricing scheme (though such a scheme is somewhat more round-about). Kanodia contends that transfer pricing (and therefore, delegated decision making) tends to be most advantageous when there is imperfect communication between divisional managers and the firm's central authority.

Kanodia goes on to consider strategies for hiring and compensating auditors, when auditing costs are uncertain *ex ante* but are revealed over time with audit experience in a multi-period model. The firm seeks to minimize auditing costs, recognizing that the incumbent auditor will have an informational advantage. In a variety of settings Kanodia shows that the firm's optimal multi-period payment schedule marshals competition among potential auditors to squeeze the informational rents of the incumbent. Kanodia concludes the chapter by discussing the contributions and limitations of game-theoretic models with respect to accounting practice. A major success is the literature's focus on the practical virtues of self-selected contracts. The most important limitation involves the informational demands of these contracts. The theory would be enriched by including notions of robustness, simplicity and bounded rationality.

Research in operations management and information systems, the third functional area considered, has frequently drawn on the techniques of operations research and related disciplines on the fringes of economics. By and large, the thrust and stress of these research programs have focused on improving system performance as measured by the objectives of a single decision maker. Nonetheless, operations managers have always acknowledged the importance of designing internal incentive schemes to coordinate the differing objectives of manufacturing and marketing. More recently has come an emphasis on manufacturing strategy – that is, the role that manufacturing plays in creating and sustaining the competitive advantage of the firm. Game theory is ideally suited to model these questions.

In Chapter 4, Lode Li and Seungjin Whang identify and discuss a number of important issues in the areas of operations management and information systems. These topics include time-based competition, the coordination of internal incentives, and the effects of network externalities, especially in the software market.

The chapter's first model focuses on competition among firms offering customers a choice of processing priorities and service quality at various prices. In characterizing the Nash equilibrium among firms, the authors show that faster processing (and lower variance) firms command higher prices and greater market shares. A second version of this model incorporates heterogeneous customers and produces an interesting set of equilibrium results. Firms give service priority (and charge higher prices) to impatient customers. In equilibrium, a firm that is both faster and lower cost uses a two-pronged strategy. It competes for impatient customers via faster delivery and for time-insensitive customers via lower prices. In a third version of the model, the focus is on pricing and overall system performance, recognizing that in the aggregate customer jobs impose externalities in the form of congestion. The authors characterize an optimal priority-pricing scheme (i.e. one that maximizes the net value of the system as a whole). In equilibrium, heterogeneous customers self-select into different pricing and priority categories. One interesting feature of the optimal pricing system is that longer jobs (regardless of priority) are penalized more than proportionately to time taken. In addition, higher priority jobs pay a surcharge that is proportional to service duration.

Echoing the coordination and internal incentive issues considered by Kanodia, Li and Whang's second main model looks at the coordination between production and marketing. A manufacturing manager expends effort and resources to expand overall productive capacity. Separate product managers expend resources to increase product demand while "competing" for the firm's limited manufacturing capacity. When all managers are risk neutral, the authors characterize an optimal incentive scheme that implements a first-best (value-maximizing) solution for the firm as a whole. The decentralized pricing mechanism has each product group receive 100 percent of its revenues and pay a capacity fee equal to the shadow price of capacity. The manufacturing manager is paid the expected shadow price of capacity.

A third operations management model considers information sharing in an oligopolistic market. Against the backdrop of Cournot equilibrium among quantity-setting firms, the authors examine the incentives for competing firms to share information about individual firm costs and about market demand. A two-stage game with information exchange in the first stage and Cournot

Introduction 7

competition in the second is analyzed under the assumption of normally distributed demand. There turns out to be an equilibrium in which cost data is exchanged but not demand data.

Li and Whang conclude by considering three explanations for high market concentration (the evolution of a few dominant firms) in the software industry. One cause is the existence of network externalities. A second stems from learning effects both on the buyers' side (consumers prefer familiar software) and the seller's side (large suppliers learn through experience how to deliver higher quality and lower costs). A third source of concentration particular to software is its enormous development cost and very low production cost.

In Chapter 5, Francine LaFontaine and Margaret Slade survey the empirical research on franchising contracts, concentrating on the choice a firm makes between in-house, "integrated" retail operations and "separated" franchised sales. The focus is on "business format" franchising, where the franchisor provides know-how and quality control to the franchisee in return for royalty payments. (The franchisee is responsible for production.) Thus, the thrust of the chapter is the tradeoff between centralized control and decentralized decisions in the modern, large-scale corporation. The authors' monumental review of the empirical research on franchise contracts is felicitously organized using a rich and flexible principal-agent model. Of particular concern is the agent's incentive compensation, that is, the degree to which the agent is paid based on results (so-called high-powered incentives). In the empirical studies of firm contracting, the authors note that franchising – where vertical separation is the norm and the franchisee's net income depends directly on its decisions and efforts – embodies high-powered incentives. By contrast, under vertical integration, the principal typically sets a tighter rein on the agent's behavior and the latter's compensation is much less sensitive to the business outcomes resulting from its actions and behavior.

The principal-agent model generates a rich set of predictions. Other things equal, franchising (vertical separation and high-powered incentives) should predominate:
i) when the agent's actions and efforts significantly affect profitability (i.e. the agent's job is entrepreneurial in nature) or create positive externalities (spillovers) for the business as a whole;
ii) in settings where the agent's risk is low and in markets where products are close substitutes;

iii) when the agent's *outputs* can be measured precisely and/or at low cost. By contrast, if the principal can monitor and measure the agent's *effort* precisely and at low cost, vertical integration, not franchising, is the favored form of organization.

The empirical studies of these factors are remarkably consistent with one another and with these model predictions. For instance, franchising is more likely in settings where business outcomes are highly sensitive to agent actions and where these outcomes can be well-measured. In reviewing scores of empirical studies addressing the hypotheses above, LaFontaine and Slade discuss the data used to capture the explanatory factors in question and interpret the study results. They also offer explanations when the empirical pattern is inconsistent with the theoretical prediction. Thus, they note an important and persistent anomaly. Counter to the model prediction, the empirical studies show that franchising becomes more likely in high-risk business settings. The prevalence of contracts where the royalty is a percentage of revenues and not profits is another puzzle for which the authors offer possible explanations.

The chapter therefore serves several purposes. It gives a detailed description of theory and empirical work in an area of great importance in marketing and industrial organization. In addition, it ties the empirical findings to the principal-agent model, demonstrating the prevalence of similar structures across different functional areas of business. As we noted above, important models in accounting and in operations management (not to mention finance) also rely on the principal-agent framework.

2. Competitive Strategies in Imperfect Markets

The second section of the book collects a number of game-theoretic studies of strategic behavior in imperfect markets.

In Chapter 6, the first contribution in this section, Harborne Stuart analyzes business strategy through the lens of cooperative game theory. While the mainstream treatments of business strategy take a non-cooperative point of view, Stuart's analysis underscores the additional insights provided by the cooperative approach. Accordingly, the intent is to present the collection of ideas he has developed with Adam Brandenburger, and by Brandenburger

Introduction 9

and Nalebuff (1997), rather than to survey the substantial existing literature in the field.

Cooperative game theory emphasizes the structure of the value-creating relationships among players. In other words, the cooperative model specifies the set of players and the value each group of players can obtain by themselves, without needing to specify the exact rules of the interaction, or the procedure, by which the actual game proceeds. By contrast, the extensive or strategic form models of non-cooperative game theory rely on precisely specified rules of the game and posit equilibrium behavior as the norm. Stuart focuses on two (of many) solution concepts that have been suggested for cooperative games: the "added-value principle" and the core. The first principle states that no player gets more than the difference in values between the coalition of all players and the coalition of all but the one player in question. The familiar concept of the core posits that the allocation of player payoffs should be such that no coalition of players can secure better payoffs for its members simply by "going it alone."

Stuart applies the added-value principle in a basic "supplier-firm-buyer" game. (Here, the firm acquires inputs from a supplier and produces a final good to sell to a buyer.) By varying the degree of competition at each of the three market levels (firms, suppliers, buyers), Stuart demonstrates how the added-value principle can be used to establish definite payoffs for the interacting parties (without recourse to specific bargaining or pricing procedures). Stuart goes on to extend his framework to analyze so-called "biform" games, settings that combine aspects of cooperative and non-cooperative games. In these settings, players make strategic, non-cooperative choices in the first stage of play – choices that determine the nature of the game in the second stage. Ultimate outcomes in the second stage are identified using the cooperative game-theoretic principles noted above. This framework is used to consider two interesting settings: a monopolist's choice of production capacity and the location choices of firms competing for the business of a dispersed set of customers. This second model is particularly intriguing. In the second stage of the game, Stuart shows how the location choices of firms determine payoffs to firms and customers using the added-value principle. He then identifies equilibrium location choices in the first stage of the game and shows that self-interested behavior leads to socially-efficient spatial differentiation of firms. By contrast, in the usual non-cooperative treatments of spatial competition, there is no guarantee that an equilibrium exists.

In Chapter 7, Joseph Farrell and Georg Weizsäcker model cooperation in repeated economic interactions between two parties. The framework embraces a number of important business and economic applications. One example is repeated transactions between buyer and seller where the latter determines the quality and price of the item for sale. A second example is designing a contract between principal and agent to induce optimal work effort by the latter. A third example is a loan agreement between lender and borrower, where period by period the latter chooses to repay or default and the former chooses whether or not to renew the loan.

As the authors note, all of these are examples of *the repeated amnesty dilemma*. In each period, each player has two possible actions. Both sides profit if the first player takes the "trusting" action and the second takes the "honest" action. Unfortunately, this is not an equilibrium in the single stage of the game. (The second player can profitably deviate to its "cheating" action.) This cooperative outcome can be sustained in the infinitely repeated game if players are sufficiently patient (don't discount future payoffs too highly). But cooperation can be sustained only by the first player threatening not to trust the other – an action that hurts both players. Farrell and Weizsäcker note that carrying out these punishments is not plausible; the players would prefer to renegotiate to Pareto-superior continuation equilibria should a deviation ever occur. Of course, such renegotiations would obviate the threat of punishment, thereby failing to sustain the cooperative equilibrium in the first place.

Accordingly, the authors consider only equilibrium behavior that is renegotiation proof. This new requirement constrains the degree of cooperation achievable in the infinitely repeated game. In either the normal form or extensive form representations, the second player cheats with a certain positive probability. In general the first player takes the trusting action also with positive probability. Thus, some degree of cooperation can be achieved. In accord with one's intuition, raising the second player's payoff from cheating in the one-stage game reduces the equilibrium level of cooperation in repeated play. (In the repeated extensive form, if the short-term incentive to cheat is too large, no cooperation is possible in the repeated game.) Finally, the authors show that cooperation is enhanced, if the players are free to make the current terms of trade (the price, the wage, or the interest rate in the examples above) depend on the past history of the game.

Introduction

In Chapter 8, Taradas Bandyopadhyay, Kalyan Chatterjee, and Navendu Vasavada analyze a related buyer-seller model. Here, sellers can choose "high-quality" production methods or "low quality" methods and set high or low prices. (The high-quality production method produces a more favorable probability distribution of quality for the good in question.) After making a purchase in the first period, the buyer experiences the actual quality of the good and reports this realized quality by "word of mouth" to a new potential buyer in the second period. As in the Farrell and Weizsäcker model, the one-period version of the model yields a negative result; the unique outcome is a pooling "lemons" equilibrium, where all types of firms choose low-quality production methods. However, the two-period model sustains multiple equilibria which are of some interest. Besides the lemons equilibrium, there is i) a "sorting" equilibrium in which the more-productive sellers choose high-quality production methods and all sellers charge a common price and ii) a partial signalling equilibrium, with high-quality sellers charging high prices and low-quality sellers randomizing between high and low prices. Buyer experience in period one, passed on via word of mouth to buyers in period two, is the key to these equilibria. Intuitively, more-productive sellers (but not their less-productive counterparts) choose high-quality technology in period one and signal this via high first-period prices, precisely because they are then able to reap the ensuing second-period benefits. In short, the authors' analysis provides an important foundation for price signalling.

In Chapter 9, Kalyan Chatterjee examines the contributions of game theory to the study of bargaining and dispute resolution. Chatterjee begins by showing how the game-theoretic point of view influences bargaining behavior and strategy. Starting with Rubinstein's (1982) classic analysis of alternating offers, he shows how factors such as the power of commitment, patience, and outside options influence bargaining outcomes. In turn, the presence of private information fundamentally changes the bargaining problem. As a general rule, equilibrium behavior means making aggressive offers – offers that risk disagreement even when a zone of agreement exists. Sophisticated game-theoretic models lead to the "negotiation dance," where concessions (or the lack thereof) reveal information about player types. To sum up, the aggressive self-interested use of private information can lead to disagreement and inefficiency. Thus, there is no easy way out of the "bargainer's dilemma." Typically, it will not be possible to maximize the players' joint negotiation gains and then to bargain over the split. Self-interested behavior inevitably compromises value maximization. Finally, Chatterjee discusses a number of current topics in bargaining: the use of incomplete contracts, the sequencing of negotiations, and the different means of communication.

Chatterjee acknowledges that game-theoretic models can be criticized on several grounds. They presume a relentless degree of rationality and an unrealistic amount of information on the part of bargainers. They frequently suffer from a multiplicity of possible equilibrium outcomes or from outcomes that are highly sensitive to the details and parameters of the model. Finally, the model predictions are often difficult to put to empirical test. Nonetheless, he sees substantial value in the game-theoretic method. Most important, the approach explains and highlights important features of bargaining behavior. The models succeed in offering "contingent guidance" rather than general advice. Continuing research is aimed at remedying many of the deficiencies of the current generation of bargaining models.

3. Auctions

The final three chapters share a common focus on strategic bidding behavior and auction performance. Auctions and competitive bidding institutions are important for both theoretical and practical reasons. Today, auctions are used in an increasing range of transactions – from the sale of oil leases and treasury securities to online sales via the internet. Auction theory provides rich and flexible models of price formation in the absence of competitive markets. This theory lends valuable insight into optimal bidding strategies under different auction institutions. Moreover, it provides normative guidelines concerning "market" performance – whether performance is measured by market efficiency or maximum seller revenue.

In Chapter 10, William Samuelson compares the theoretical and empirical evidence on auction behavior and performance. While theory points to equilibrium bidding as a benchmark, there is considerable empirical evidence (from controlled experiments and field data) that actual bidding behavior only loosely follows this normative prescription. First, with respect to the two most common auction methods, the English and sealed-bid auctions, his analysis delivers a mixed message. While the English auction can be expected to outperform its sealed-bid counterpart on efficiency grounds, the efficiency advantage of the English auction in a number of representative settings is relatively small. Furthermore, neither auction method can claim high efficiency marks in complex auction settings, for instance, when multiple items are to be allocated and buyer values are non-additive. In addition, the preponderance of practical evidence favors the sealed-bid auction as a revenue generator, contrary to the benchmark prediction of auction theory. Thus, one implication of his analysis is to redirect attention to the practical virtues of sealed-bid procedures.

In addition, the relative performance of various auction institutions in complex environments – when multiple items are for sale and values are non-additive – remains an open question. Only recently have theoretical and empirical investigations focused on alternative auction institutions. For instance, the simultaneous ascending auction has performed well in experiments and in practice. (This method was used in the FCC's multi-billion-dollar spectrum auctions.) Serious attention is also being paid to the sealed-bid combinatorial auction, which allows buyers to bid for individual items and combinations of items. In combination with a Vickrey payment scheme, the combinatorial auction promises favorable performance in theory and practice.

In Chapter 11, Robert Marshall and Michael Meurer consider the important problem of bidder collusion in auctions. As the authors note, in recent years the vast majority of criminal cases under Section One of the Sherman Act have been brought for bid-rigging and price fixing in road construction and government procurement. Thus, this is but one sign of the potential severity of the problem. Taking a theoretical lens to the problem, the authors examine the strategic behavior of bidders and the susceptibility of alternative auction institutions to collusion. The analysis underscores the fundamental point that the English oral auction is more susceptible to collusion than its sealed-bid counterpart.

One way of making this argument intuitively is to consider the behavior of a ring of bidders who agree to depress the winning bid and share the additional profit gained. Implementation of a collusive agreement is straightforward in an English auction. The designated bidder for the ring follows the same strategy as he would absent collusion – that is, he stands willing to bid up to his value of the item if necessary. All other ring bidders suppress their bids, so the bidding stops at a depressed price (to the collective gain of the ring). Under a sealed-bid auction, there is a subtle difference. To secure a lower price for the bidder ring, the designated winning bidder must reduce his own sealed bid. This leaves the designated winner vulnerable to a ring member who can bid slightly above the collusive bid and win the item for his own personal gain. The authors note that to deter such bidding, would be cheaters must receive a large share of the collusive gain, and this itself makes collusion problematic. (There is not enough gain to go around.) Moreover, they show that the English auction continues to be susceptible to collusion in more general settings: when there are potential bidders outside the ring, and when multiple objects are for sale. In short, the sealed-bid auction is attractive to the seller (and as a bidding institution in general) precisely because it makes bidder collusion more difficult.

The usual presumption is that collusive agreements, by distorting competitive prices, lead to allocative inefficiencies and deadweight losses. (If collusive buyers depress auction prices for fish, the returns to fishing suffer and ultimately the supply of fresh fish is curtailed.) In addition, measures that a seller might take – higher reserve prices, for instance – to protect itself against collusive behavior by buyers can also lead to inefficiencies. However, in the latter part of their analysis, Marshall and Meurer point out two exceptions to this presumption. First, collusive bidding behavior can act as a countervailing force against a seller with some degree of monopoly power in auctioning a unique or differentiated item. In a wonderfully simple example, the authors show that bidder collusion can induce the seller to increase the quantity of items to be auctioned. Though the seller's welfare declines (justifying the seller's concern about collusion), the collusive bidders' collective welfare increases by a greater amount, therefore, raising efficiency. Second, the elevation of bidder profits via collusion has the potential advantage of inducing bidders to make informational investments that have private and social value. (These investments would not be made in the absence of the profit incentive afforded by collusion.)

In Chapter 12, Robert Wilson provides a fitting closing chapter that investigates the implementation of a new exchange mechanism. The application is to the design of a wholesale market for forward trades (one day ahead) of electrical power between power producers and large customers in California (the California Power Exchange). Wilson describes an iterated double auction that allows sellers and buyers to revise bids as the auction proceeds. Final prices are determined and trade takes place only at the close of the double auction. The goal of the iterated auction is to ensure early and reliable price discovery and to deliver efficient prices and exchanges. Reliable price discovery means that interim prices in the early stages of the auction should be good predictors of the direction of final prices. Early price discovery is essential to power generators who must decide which plants to

Introduction

start and operate for consecutive hours during the next day. (Discovery is also important for large buyers who must plan their power needs over the different hours of the day.) However, as Wilson emphasizes, the rules of the iterated auction must be carefully designed to deter "gaming" behavior. Specifically, suppliers and buyers may have incentives to defer serious bidding to the close of the auction. (That way a player can observe the exact pattern of prices revealed by the demand and supply of other traders.) However, the strategy of delaying serious bids (free riding) defeats the goal of early and reliable price discovery.

Wilson proposes a simple but ingenious set of activity rules to spur serious bids. All new bids must be submitted in the first iteration of the auction. (This rule prevents any trader from waiting to submit new bids at the last iteration, and it ensures that the maximum volumes of supply and demand are revealed immediately.) At any iteration, the intersection of supply and demand determines the current (interim) market-clearing price. Suppose that a player's current bid is excluded from trade in the current iteration. For instance, the interim clearing price is $23 but a seller bid seeks to sell a specified number of units at $25. Wilson's revision rule is as follows. Any such excluded bid must be improved immediately, i.e. in the next iteration or else it is "frozen." Here, the seller would have to lower its bid below $23 or forfeit the chance to do so later. This revision rule is based on the principle of revealed preference: a bidder's refusal to improve a previous clearing price is presumptive evidence that it cannot do so profitably. (Note that if the interim clearing price increases later to $24, the seller would then have the chance to beat this price (his bid would be partially unfrozen). By preventing free riding, this revision rule induces timely revelation of demands and costs. As Wilson notes, the iterated double auction, guided by a careful choice of activity rules, performs well in experimental tests.

Together, the eleven chapters in this volume apply game-theoretic thinking to a wide variety of business functions within the firm – from accounting to organization design – and to numerous realms of market competition – between buyers and sellers, bargaining parties, or competing bidders. Along the way, the reader will gain insight into the power and subtlety of game-theoretic methods: the identification of equilibria (simple, sequential, and Bayesian) and the design of optimal mechanisms. To the reader, enjoy!

References

Besanko, D., D. Dranove, and M. Shanley (1999), *Economics of Strategy*, John Wiley and Sons, New York.

Brandenburger, A.M. and B.J. Nalebuff (1997), *Co-opetition*, Doubleday Publishers, New York.

Chatterjee, K. and G.L. Lilian (1986), "Game Theory in Marketing Science: Uses and Limiatations," *International Journal of Research in Marketing*, 3, 73-93.

Eliashberg, J. and G.L. Lilien (Eds) (1993), *Handbooks in Operations Research and Management Science, 5: Marketing* Elsevier-North Holland.

Fudenberg, D. and J. Tirole (1991), *Game Theory*, MIT Press, Cambridge, MA.

Gibbons, R. (1992), *Game Theory for Applied Economists*, Princeton University Press, Princeton, NJ.

Myerson, R.B. (1997), *Game Theory: Analysis of Conflict*, Harvard University Press, Cambridge, MA.

Milgrom, P. and J. Roberts (1992), *Economics, Organization, and Management*, Prentice Hall, Englewood Cliffs, NJ.

Osborne, M. and A. Rubinbstein (1994), *A Course in Game Theory*, MIT Press, Cambridge, MA.

Oster, S. (1999), *Modern Competitive Analysis*, Oxford University Press, New York.

Rasmusen, E. (1994), *Games and Information*, Blackwell Publishers, Oxford, UK.

Rubinstein, A. (1982), "Perfect Equilibrium in a Bargaining Model," *Econometrica*, 50, 97-109.

von Neumann, J. and O. Morgenstern (1944*), Theory of Games and Economic Behavior*, John Wiley and Sons, New York.

Wernerfelt, B. (1995), "An Efficiency Criterion for Marketing Design," *Journal of Marketing Research*, 31, 462-470.

2 GAME THEORY MODELS IN FINANCE

Franklin Allen and Stephen Morris

Finance is concerned with how the savings of investors are allocated through financial markets and intermediaries to firms, which use them to fund their activities. Finance can be broadly divided into two fields. The first is *asset pricing*, which is concerned with the decisions of investors. The second is *corporate finance*, which is concerned with the decisions of firms. Traditional neoclassical economics did not attach much importance to either kind of finance. It was more concerned with the production, pricing and allocation of inputs and outputs and the operation of the markets for these. Models assumed certainty and in this context financial decisions are relatively straightforward. However, even with this simple methodology, important concepts such as the time value of money and discounting were developed.

Finance developed as a field in its own right with the introduction of uncertainty into asset pricing and the recognition that classical analysis failed to explain many aspects of corporate finance. In Section 1, we review the set of issues raised and some of the remaining problems with the pre-game-theoretic literature. In Section 2, we recount how a first generation of game theory models tackled those problems, and discuss the successes and failures. Our purpose in this section is to point to some of the main themes in the various sub-fields. We do not attempt to provide an introduction to game theory. See Gibbons (1992) for a general introduction to applied game theory and Thakor (1991) for a survey of game theory in finance including an introduction to game theory. Nor do we attempt to be encyclopedic.

This first generation of game-theoretic models revolutionized finance but much remains to be explained. Game-theoretic methods continue to develop and we believe that extensions involving richer informational models are especially relevant for finance. In Section 3, we review recent work concerning higher-order beliefs and informational cascades and discuss its relevance for finance. We also review work that entails differences in beliefs not explained by differences in information.

1. The Main Issues in Finance

Asset Pricing

The focus of Keynesian macroeconomics on uncertainty and the operation of financial markets led to the development of frameworks for analyzing risk. Keynes (1936) and Hicks (1939) took account of risk by adding a risk premium to the interest rate. However, there was no systematic theory underlying this risk premium. The key theoretical development which eventually lead to such a theory was von Neumann and Morgenstern's (1947) axiomatic approach to choice under uncertainty. Their notion of expected utility, developed originally for use in game theory, underlies the vast majority of theories of asset pricing.

<u>The Capital Asset Pricing Model.</u> Markowitz (1952; 1959) utilized a special case of von Neumann and Morgenstern's expected utility to develop a theory of portfolio choice. He considered the case where investors are only concerned with the mean and variance of the payoffs of the portfolios they are choosing. This is a special case of expected utility provided the investor's utility of consumption is quadratic and/or asset returns are multinormally distributed. Markowitz's main result was to show that diversifying holdings is optimal and the benefit that can be obtained depends on the covariances of asset returns. Tobin's (1958) work on liquidity preference helped to establish the mean-variance framework as the standard approach to portfolio choice problems. Subsequent authors have made extensive contributions to portfolio theory. See Constantinides and Malliaris (1995).

It was not until some time after Markowitz's original contribution that his framework of individual portfolio choice was used as the basis for an equilibrium theory, namely the capital asset pricing model (CAPM). Brennan (1989) has argued that the reason for the delay was the boldness of the assumption that all investors have the same beliefs about the means and variances of all assets. Sharpe (1964) and Lintner (1965) showed that in equilibrium

$$Er_i = r_f + \beta_i(Er_M - r_F),$$

where Er_i is the expected return on asset i, r_f is the return on the risk free asset, Er_M is the expected return on the market portfolio (i.e. a value weighted portfolio of all assets in the market) and $\beta_i = cov(r_i, r_M)/var(r_M)$. Black (1972) demonstrated that the same relationship held even if no risk free asset existed provided r_F was replaced by the expected return on a portfolio or asset with $\beta = 0$. The model formalizes the risk premium of Keynes and Hicks and shows that it depends on the covariance of returns with other assets.

Despite being based on the very strong assumptions of mean-variance preferences and homogeneity of investor beliefs, the CAPM was an extremely important development in finance. It not only provided key theoretical insights concerning the pricing of stocks, but also led to a great deal of empirical work testing whether these predictions held in practice. Early tests such as Fama and Macbeth (1973) provided some support for the model. Subsequent tests using more sophisticated econometric techniques have not been so encouraging. Ferson (1995) contains a review of these tests.

The CAPM is only one of many asset-pricing models that have been developed. Other models include the Arbitrage Pricing Theory (APT) of Ross (1977a) and the representative agent asset-pricing model of Lucas (1978). However, the CAPM was the most important not only because it was useful in its own right for such things as deriving discount rates for capital budgeting but also because it allowed investigators to easily adjust for risk when considering a variety of topics. We turn next to one of the most important hypotheses that resulted from this ability to adjust for risk.

Market Efficiency. In models involving competitive markets, symmetric information and no frictions such as transaction costs, the only variations in returns across assets are due to differences in risk. All information that is available to investors becomes reflected in stock prices and no investor can earn higher returns except by bearing more risk. In the CAPM, for example, it is only differences in β's that cause differences in returns. The idea that the differences in returns are due to differences in risk came to be known as the *Efficient Markets Hypothesis*. During the 1960's a considerable amount of research was undertaken to see whether U.S. stock markets were in fact efficient. In a well-known survey, Fama (1970) argued that the balance of the evidence suggested markets were efficient. In a follow up piece, Fama (1991) continued to argue that by and large markets were efficient despite the continued documentation of numerous anomalies.

Standard tests of market efficiency involve a joint test of efficiency and the equilibrium asset-pricing model that is used in the analysis. Hence a rejection of the joint hypothesis can either be a rejection of market efficiency or the asset-pricing model used or both. Hawawini and Keim (1995) survey these "anomalies." Basu (1977) discovered one of the first. He pointed out that price to earnings (P/E) ratios provided more explanatory power than β's. Firms with low P/E ratios (value stocks) tend to outperform stocks with high P/E ratios (growth stocks). Banz (1981) showed that there was a significant relationship between the market value of common equity and returns (the size effect). Stattman (1980) and others have demonstrated the significant predictive ability of price per share to book value per share (P/B) ratios for returns. In an influential paper, Fama and French (1993) have documented that firm size and the ratio of book to market equity are important factors in explaining average stock returns. In addition to these cross-sectional effects there are also a number of significant time-series anomalies. Perhaps the best known of these is the January effect. Rozeff and Kinney (1976) found that returns on an equal weighted index of NYSE stocks were much higher in January than in the other months of the year. Keim (1983) demonstrated that the size effect was concentrated in January. Cross (1973) and French (1980) pointed out that the returns on the S&P composite index are negative on Mondays. Numerous other studies have confirmed this weekend effect in a wide variety of circumstances.

These anomalies are difficult to reconcile with models of asset pricing such as the CAPM. Most of them are poorly understood. Attempts have been made to explain the January effect by tax loss selling at the end of the year. Even this is problematic because in countries such as the U.K. and Australia where the tax year does not end in December there is still a January effect. It would seem that the simple frameworks most asset pricing models adopt are not sufficient to capture the richness of the processes underlying stock price formation.

Instead of trying to reconcile these anomalies with asset pricing theories based on rational behavior, a number of authors have sought to explain them using behavioral theories based on foundations taken from the psychology literature. For example, Dreman (1982) argues that the P/E effect can be explained by investors' tendency to make extreme forecasts. High (low) P/E ratio stocks correspond to a forecast of high (low) growth by the market. If investors predict too high (low) growth, high P/E stocks will underperform (overperform). De Bondt and Thaler (1995) surveys behavioral explanations for this and other anomalies.

Continuous Time Models. Perhaps the most significant advance in asset pricing theory since the early models were formulated was the extension of the paradigm to allow for continuous trading. This approach was developed in a series of papers by Merton (1969; 1971; 1973a) and culminated in his development of the intertemporal capital asset pricing model (ICAPM). The assumptions of expected utility maximization, symmetric information and frictionless markets are maintained. By analyzing both the consumption and portfolio decisions of an investor through time and assuming prices per share are generated by Ito processes, greater realism and tractability compared to the mean-variance approach is achieved. In particular, it is not necessary to assume quadratic utility or normally distributed returns. Other important contributions that were developed using this framework were Breeden's (1979) Consumption CAPM and Cox, Ingersoll and Ross's (1985) modeling of the term structure of interest rates.

The relationship between continuous time models and the Arrow–Debreu general equilibrium model was considered by Harrison and Kreps (1979) and Duffie and Huang (1985). Repeated trading allows markets to be made effectively complete even though there are only a few securities.

One of the most important uses of continuous time techniques is for the pricing of derivative securities such as options. This was pioneered by Merton (1973b) and Black and Scholes (1973) and led to the development of a large literature that is surveyed in Ross (1992). Not only has this work provided great theoretical insight but it has also proved to be empirically implementable and of great practical use.

Corporate Finance

The second important area considered by finance is concerned with the financial decisions made by firms. These include the choice between debt and equity and the amount to pay out in dividends. The seminal work in this area was Modigliani and Miller (1958) and Miller and Modigliani (1961). They showed that with perfect markets (i.e., no frictions and symmetric information) and no taxes the total value of a firm is independent of its debt/equity ratio. Similarly they demonstrated that the value of the firm is independent of the level of dividends. In their framework it is the investment decisions of the firm that are important in determining its total value.

The importance of the Modigliani and Miller theorems was not as a description of reality. Instead it was to stress the importance of taxes and capital market imperfections in determining corporate financial policies. Incorporating the tax deductibility of interest but not dividends and bankruptcy costs lead to the trade-off theory of capital structure. Some debt is desirable because of the tax shield arising from interest deductibility but the costs of bankruptcy and financial distress limit the amount that should be used. With regard to dividend policy, incorporating the fact that capital gains are taxed less at the personal level than dividends into the Modigliani and Miller framework gives the result that all payouts should be made by repurchasing shares rather than by paying dividends.

The trade-off theory of capital structure does not provide a satisfactory explanation of what firms do in practice. The tax advantage of debt relative to the magnitude of expected bankruptcy costs would seem to be such that firms should use more debt than is actually observed. Attempts to explain this, such as M. Miller (1977), that incorporate personal as well as corporate taxes into the theory of capital structure, have not been successful. In the Miller model, there is a personal tax advantage to equity because capital gains are only taxed on realization and a corporate tax advantage to debt because interest is tax deductible. In equilibrium, people with personal tax rates above the corporate tax rate hold equity while those with rates below hold debt. This prediction is not consistent with what occurred in the U.S. in the late 1980's and early 1990's when there were no personal tax rates above the corporate rate. The Miller model suggests that there should have been a very large increase in the amount of debt used by corporations but there was only a small change.

The tax-augmented theory of dividends also does not provide a good explanation of what actually happens. Firms have paid out a substantial amount of their earnings as dividends for many decades. Attempts to explain the puzzle using tax based theories such as the clientele model have not been found convincing. They are difficult to reconcile with the fact that many people in high tax brackets hold large amounts of dividend paying stocks and on the margin pay significant taxes on the dividends.

Within the Modigliani and Miller framework other corporate financial decisions also do not create value except through tax effects and reductions in frictions such as transaction costs. Although theoretical insights are provided, the theories are not consistent with what is observed in practice. As with the asset pricing models discussed above this is perhaps not surprising given their simplicity. In particular, the assumptions of perfect information and perfect markets are very strong.

2. The Game-Theory Approach

The inability of standard finance theories to provide satisfactory explanations for observed phenomena lead to a search for theories using new methodologies. This was particularly true in corporate finance where the existing models were so clearly unsatisfactory. Game theory has provided a methodology that has brought insights into many previously unexplained phenomena by allowing asymmetric information and strategic interaction to be incorporated into the analysis. We start with a discussion of the use of game theory in corporate finance where to date it has been most successfully applied. We subsequently consider its role in asset pricing.

Corporate Finance

<u>Dividends as Signals</u>. The thorniest issue in finance has been what Black (1976) termed "the dividend puzzle." Firms have historically paid out about a half of their earnings as dividends. Many of these dividends were received by investors in high tax brackets who, on the margin, paid substantial amounts of taxes on them. In addition, in a classic study Lintner (1956) demonstrated that managers "smooth" dividends in the sense that they are less variable than earnings. This finding was confirmed by Fama and Babiak (1968) and numerous other authors. The puzzle has been to explain these observations. See Allen and Michaely (1995) for a survey of this literature.

In their original article on dividends, Miller and Modigliani (1961) had suggested that dividends might convey significant information about a firm's prospects. However, it was not until game-theoretic methods were applied that any progress was made in understanding this issue. Bhattacharya's (1979) model of dividends as a signal was one of the first papers in finance to use these tools. His contribution started a large literature.

Bhattacharya assumes that managers have superior information about the profitability of their firm's investment. They can signal this to the capital market by "committing" to a sufficiently high level of dividends. If it turns out the project is profitable these dividends can be paid from earnings without a problem. If the project is unprofitable then the firm has to resort to outside finance and incur deadweight transaction costs. The firm will therefore only find it worthwhile to commit to a high dividend level if in fact its prospects are good. Subsequent authors like Miller and Rock (1985) and John and Williams (1985) developed models which did not require committing to a certain level of dividends and where the deadweight costs required to make the signal credible were plausible.

One of the problems with signaling models of dividends is that they typically suggest that dividends will be paid to signal new information. Unless new information is continually arriving there is no need to keep paying them. But in that case the level of dividends should be varying to reflect the new information. This feature of dividend signaling models is difficult to reconcile with smoothing. In an important piece, Kumar (1988) develops a 'coarse signaling' theory that is consistent with the fact that firms smooth dividends. Firms within a range of productivity all pay the same level of dividends. It is only when they move outside this range that they will alter their dividend level.

Another problem in many dividend signaling models (including Kumar (1988)) is that they do not explain why firms use dividends rather than share repurchases. In most models the two are essentially equivalent except for the way that they are taxed since both involve transferring cash from the firm to the owners. Dividends are typically treated as ordinary income and taxed at high rates whereas repurchases involve price appreciations being taxed at low capital gains rates. Building on work by Ofer and Thakor (1987) and Barclay and Smith (1988), Brennan and Thakor (1990) suggest that repurchases have a disadvantage in that informed investors are able to bid for undervalued stocks and avoid overvalued ones. There is thus an adverse selection problem. Dividends do not suffer from this problem because they are pro rata.

Some progress in understanding the dividend puzzle has been made in recent years. This is one of the finance applications of game theory that has been somewhat successful.

Capital Structure. The trade-off theory of capital structure mentioned above has been a textbook staple for many years. Even though it had provided a better explanation of firms' choices than the initial dividend models, the theory is not entirely satisfactory because the empirical magnitudes of bankruptcy costs and interest tax shields do not seem to match observed capital structures. The use of game-theoretic techniques in this field has allowed it to move ahead significantly. Harris and Raviv (1991) survey the area.

The first contributions in a game-theoretic vein were signaling models. Ross (1977b) develops a model where managers signal the prospects of the firm to the capital markets by choosing an appropriate level of debt. The reason this acts as a signal is that bankruptcy is costly. A high debt firm with good prospects will only incur these costs occasionally while a similarly levered firm with poor prospects will incur them often. Leland and Pyle (1977) consider a situation where entrepreneurs use their retained share of ownership in a firm to signal its value. Owners of high-value firms retain a high share of the firm to signal their type. Their high retention means they don't get to diversify as much as they would if there was symmetric information, and it is this that makes it unattractive for low value firms to mimic them.

Two influential papers based on asymmetric information are Myers (1984) and Myers and Majluf (1984). If managers are better informed about the prospects of the firm than the capital markets, they will be unwilling to issue equity to finance investment projects if the equity is undervalued. Instead they will have a preference for using equity when it is overvalued. Thus equity is regarded as a bad signal. Myers (1984) uses this kind of reasoning to develop the "pecking order" theory of financing. Instead of using equity to finance investment projects, it will be better to use less information sensitive sources of funds. Retained earnings are the most preferred, with debt coming next and finally equity. The results of these papers and the subsequent literature such as Stein (1992) and Nyborg (1995) are consistent with a number of stylized facts concerning the effect of issuing different types of security on stock price and the financing choices of firms. However, in order to derive them, strong assumptions such as overwhelming bankruptcy aversion of managers are often necessary. Moreover, as Dybvig and Zender (1991) and others have stressed, they often assume sub-optimal managerial incentive schemes. Dybvig and Zender show that if managerial incentive schemes are chosen optimally, the Modigliani and Miller irrelevance results can hold even with asymmetric information.

A second contribution of game theory to understanding capital structure lies in the study of agency costs. Jensen and Meckling (1976) pointed to two kinds of agency problems in corporations. One is between equity holders and bondholders and the other is between equity holders and managers. The first arises because the owners of a levered firm have an incentive to take risks; they receive the surplus when returns are high but the bondholders bear the cost when default occurs. Diamond (1989) has shown how reputation considerations can ameliorate this risk shifting incentive when there is a long time horizon. The second conflict arises when equity holders cannot fully control the actions of managers. This means that managers have an incentive to pursue their own interests rather than those of the equity holders. Grossman and Hart (1982) and Jensen (1986) among others have shown how debt can be used to help overcome this problem. Myers (1977) has pointed to a third agency problem. If there is a large amount of debt outstanding which is not backed by cash flows from the firm's assets, i.e. a "debt overhang," equity holders may be reluctant to take on safe, profitable projects because the bondholders will have claim to a large part of the cash flows from these.

The agency perspective has also lead to a series of important papers by Hart and Moore and others on financial contracts. These use game-theoretic techniques to shed light on the role of incomplete contracting possibilities in determining financial contracts and in particular debt. Hart and Moore (1989) consider an entrepreneur who wishes to raise funds to undertake a project. Both the entrepreneur and the outside investor can observe the project payoffs at each date, but they cannot write explicit contracts based on these payoffs because third parties such as courts cannot observe them. The focus of their analysis is the problem of providing an incentive for the entrepreneur to repay the borrowed funds. Among other things, it is shown that the optimal contract is a debt contract and incentives to repay are provided by the ability of the creditor to seize the entrepreneur's assets. Subsequent contributions include Hart and Moore (1994; 1998), Aghion and Bolton (1992), Berglof and von Thadden (1994) and von Thadden (1995). Hart (1995) contains an excellent account of the main ideas in this literature.

The Modigliani and Miller (1958) theory of capital structure is such that the product market decisions of firms are separated from financial market decisions. Essentially this is achieved by assuming there is perfect competition in product markets. In an oligopolistic industry where there are strategic interactions between firms in the product market, financial decisions are also likely to play an important role. Allen (1986), Brander and Lewis (1986) and Maksimovic (1986) and a growing subsequent literature (see Maksimovic (1995) for a survey) have considered different aspects of these interactions between financing and product markets. Allen (1986) considers a duopoly model where a bankrupt firm is at a strategic disadvantage in choosing its investment because the bankruptcy process forces it to delay its decision. The bankrupt firm becomes a follower in a Stackelberg investment game instead of a simultaneous mover in a Nash-Cournot game. Brander and Lewis (1986) and Maksimovic (1986) analyze the role of debt as a precommitment device in oligopoly models. By taking on a large amount of debt a firm effectively precommits to a higher level of output. Titman (1984) and Maksimovic and Titman (1993) have considered the interaction between financial decisions and customers' decisions. Titman (1984) looks at the effect of an increased probability of bankruptcy on product price because, for example, of the difficulties of obtaining spare parts and service should the firm cease to exist. Maksimovic and Titman (1993) consider the relationship between capital structure and a firm's reputational incentives to maintain high product quality.

A significant component of the trade-off theory is the bankruptcy costs that limit the use of debt. An important issue concerns the nature of these bankruptcy costs. Haugen and Senbet (1978) argued that the extent of bankruptcy costs was limited because firms could simply renegotiate the terms of the debt and avoid bankruptcy and its associated costs. The literature on strategic behavior around and within bankruptcy relies extensively on game-theoretic techniques. See Webb (1987), Giammarino (1988), Brown (1989) and, for a survey, Senbet and Seward (1995). This work shows that Haugen and Senbet's argument depends on the absence of frictions. With asymmetric information or other frictions, bankruptcy costs can occur in equilibrium.

<u>The Market for Corporate Control</u>. The concept of the market for corporate control was first developed by Manne (1965). He argued that in order for resources to be used efficiently, it is necessary that firms be run by the most able and competent managers. Manne suggests that the way in which modern

capitalist economies achieve this is through the market for corporate control. There are several ways in which this operates including tender offers, mergers and proxy fights.

Traditional finance theory with its assumptions of symmetric information and perfectly competitive frictionless capital markets had very little to offer in terms of insights into the market for corporate control. In fact the large premiums over initial stock market valuations paid for targets appeared to be at variance with market efficiency and posed something of a puzzle. Again it was not until the advent of game-theoretic concepts and techniques that much progress was made in this area.

The paper that provided a formal model of the takeover process and renewed interest in the area was Grossman and Hart (1980). They pointed out that the tender offer mechanism involved a free rider problem. If a firm makes a bid for a target in order to replace its management and run it more efficiently then each of the target's shareholders has an incentive to hold out and say no to the bid. The reason is that they will then be able to benefit from the improvements implemented by the new management. They will only be willing to tender if the offer price fully reflects the value under the new management. Hence a bidding firm cannot make a profit from tendering for the target. In fact if there are costs of acquiring information in preparation for the bid or other bidding costs, the firm will make a loss. The free rider problem thus appears to exclude the possibility of takeovers. Grossman and Hart's solution to this dilemma was that a firm's corporate charter should allow acquirors to obtain benefits unavailable to other shareholders after the acquisition. They term this process "dilution."

Another solution to the free rider problem, pointed out by Shleifer and Vishny (1986a), is for bidders to be shareholders in the target before making any formal tender offer. In this way they can benefit from the price appreciation in the "toehold" of shares they already own even if they pay full price for the remaining shares they need to acquire. The empirical evidence is not consistent with this argument, however. Bradley, Desai and Kim (1988) find that the majority of bidders own no shares prior to the tender offer.

A second puzzle that the empirical literature has documented is the fact that bidding in takeover contests occurs through several large jumps rather than many small ones. For example, Jennings and Mazzeo (1993) found that the majority of the initial bid premiums exceed 20% of the market value of the target 10 days before the offer. This evidence conflicts with the standard solution of the English auction model that suggests there should be many small bid increments. Fishman (1988) argues that the reason for the large initial premium is to deter potential competitors. In his model, observing a bid alerts the market to the potential desirability of the target. If the initial bid is low a second bidder will find it worthwhile to spend the cost to investigate the target. This second firm may then bid for the target and push out the first bidder or force a higher price to be paid. By starting with a sufficiently high bid the initial bidder can reduce the likelihood of this competition.

Much of the theoretical literature has attempted to explain why the defensive measures that many targets adopt may be optimal for their shareholders. Typically the defensive measures are designed to ensure that the bidder that values the company the most ends up buying it. For example, Shleifer and Vishny (1986b) develop a model where the payment of greenmail to a bidder, signals to other interested parties that no "white knight" is waiting to buy the firm. This puts the firm in play and can lead to a higher price being paid for it than initially would have been the case.

A survey of the literature on takeovers is contained in Hirshleifer (1995). Since strategic interaction and asymmetric information are the essence of takeover contests, game theory has been central to the literature.

Initial Public Offerings (IPOs). In 1963 the U.S. Securities and Exchange Commission undertook a study of IPOs and found that the initial short-run return on these stocks was significantly positive. Logue (1973), Ibbotson (1975) and numerous subsequent academic studies have found a similar result. In a survey of the literature on IPOs, Ibbotson and Ritter (1995) give a figure of 15.3% for the average increase in the stock price during the first day of trading based on data from 1960-1992. The large short-run return on IPOs was for many years one of the most glaring challenges to market efficiency. The standard symmetric information models that existed in the 1960s and 1970s were not at all consistent with this observation.

The first paper to provide an appealing explanation of this phenomenon was Rock (1986). In his model the under-pricing occurs because of adverse selection. There are two groups of buyers for the shares, one is informed about the true value of the stock while the other is uninformed. The informed group will only buy when the offering price is at or below the true value. This implies that the uninformed will receive a high allocation of overpriced stocks since they will be the only people in the market when the offering price is above the true value. Rock suggested that in order to induce the uninformed to participate they must be compensated for the overpriced stock they ended up buying. Under-pricing on average is one way of doing this.

Many other theories of under-pricing followed. These include under-pricing as a signal (Allen and Faulhaber (1989); Grinblatt and Hwang (1989) and Welch (1989)), as an inducement for investors to truthfully reveal their valuations (Benveniste and Spindt (1989)), to deter lawsuits (Hughes and Thakor (1992)), and to stabilize prices (Ruud (1993)), among others.

In addition to the short run under-pricing puzzle, there is another anomaly associated with IPOs. Ritter (1991) documents significant long-run under-performance of newly issued stocks. During 1975-1984, he finds a cumulative average under-performance of around 15% from the offer price relative to the matching firm-adjusted return. Loughran (1993) and Loughran and Ritter (1995) confirmed this long run under-performance in subsequent studies.

Several behavioral theories have also been put forward to explain long-run under-performance. E. Miller (1977) argues that there is a wide range of opinion concerning IPOs and the initial price will reflect the most optimistic opinion. As information is revealed through time, the most optimistic investors will gradually adjust their beliefs and the price of the stock will fall. Shiller (1990) argues that the market for IPOs is subject to an 'impresario' effect. Investment banks will try to create the appearance of excess demand and this will lead to a high price initially but subsequently to underperformance. Finally, Ritter (1991) and Loughran and Ritter (1995) suggest that there are swings of investor sentiment in the IPO market and firms use the "window of opportunity" created by overpricing to issue equity.

Although IPOs represent a relatively small part of financing activity, they have received a great deal of attention in the academic literature. The reason perhaps is the extent to which underpricing and overpricing represent a violation of market efficiency. It is interesting to note that while game-theoretic techniques have provided many explanations of underpricing they have not been utilized to explain overpricing. Instead the explanations presented have relied on relaxing the assumption of rational behavior by investors.

Game Theory Models in Finance 31

Intermediation. A second area that has been significantly changed by game-theoretic models is intermediation. Traditionally, banks and other financial intermediaries were regarded as vehicles for reducing transaction costs (Gurley and Shaw (1960)). The initial descriptions of bank behavior were relatively limited. Indeed, the field was dramatically changed by the modeling techniques introduced in Diamond and Dybvig (1983). This paper develops a simple model where a bank provides insurance to depositors against liquidity shocks. At an intermediate date customers find out whether they require liquidity then or at the final date. There is a cost to liquidating long term assets at the intermediate date. A deposit contract is used where customers who withdraw first get the promised amount until resources are exhausted after which nothing is received (i.e., the first come first served constraint). These assumptions result in two self-fulfilling equilibria. In the good equilibrium, everybody believes only those who have liquidity needs at the intermediate date will withdraw their funds and this outcome is optimal for both types of depositor. In the bad equilibrium, everybody believes everybody else will withdraw. Given the assumptions of first come first served and costly liquidating of long-term assets, it is optimal for early and late consumers to withdraw and there is a run on the bank. Diamond and Dybvig argue the bad equilibrium can be eliminated by deposit insurance. In addition to being important as a theory of runs, the paper was also instrumental in modeling liquidity needs. Similar approaches have been adopted in the investigation of many topics.

Diamond and Dybvig (1983) together with an earlier paper by Bryant (1980) led to a large literature on bank runs and panics. For example, Chari and Jagannathan (1988) consider the role of aggregate risk in causing bank runs. They focus on a signal extraction problem where part of the population observes a signal about the future returns of bank assets. Others must then try to deduce from observed withdrawals whether an unfavorable signal was received by this group or whether liquidity needs happen to be high. The authors are able to show that panics occur not only when the economic outlook is poor but also when liquidity needs turn out to be high. Jacklin and Bhattacharya (1988) compare what happens with bank deposits to what happens when securities are held directly so runs are not possible. In their model some depositors receive a signal about the risky investment. They show that either bank deposits or directly held securities can be optimal depending on the characteristics of the risky investment. The comparison of bank-based and stock market-based financial systems has become a widely considered topic in recent years. See Thakor (1996) and Allen and Gale (1999).

Other important papers in the banking and intermediation literature are Stiglitz and Weiss (1981) and Diamond (1984). The former paper developed an adverse selection model in which rationing credit is optimal. The latter paper considers a model of delegated monitoring where banks have an incentive to monitor borrowers because otherwise they will be unable to pay off depositors. A full account of the recent literature on banking is contained in Bhattacharya and Thakor (1993).

Asset Pricing

Early work incorporating asymmetric information into the asset pricing literature employed the (non-strategic) concept of rational expectations equilibrium as in Grossman and Stiglitz (1980). Each market participant is assumed to learn from market prices but still believes that he does not influence market prices. This literature helped address a number of novel issues, for example, free riding in the acquisition of information. But a number of conceptual problems arose in attempting to reconcile asymmetric information with competitive analysis, and an explicitly strategic analysis seemed to be called for as in Dubey, Geanakoplos and Shubik (1987).

This provided one motive for the recent literature on market microstructure. Whereas general equilibrium theory simply assumes an abstract price formation mechanism, the market microstructure literature seeks to model the process of price formation in financial markets under explicit trading rules. The papers that contained the initial important contributions are Kyle (1985) and Glosten and Milgrom (1985). O'Hara (1995) provides an excellent survey of the extensive literature that builds on these two papers.

Kyle (1985) develops a model with a single risk-neutral market maker, a group of noise traders who buy or sell for exogenous reasons such as liquidity needs, and a risk-neutral informed trader. The market maker selects efficient prices, and the noise traders simply submit orders. The informed trader chooses a quantity to maximize his expected profit. In Glosten and Milgrom (1985) there are also a risk-neutral market maker, noise traders, and informed traders. In contrast to Kyle's model, Glosten and Milgrom treat trading quantities as fixed and instead focus on the setting of bid and ask prices. The market maker sets the bid-ask spread to take into account the possibility that the trader may be informed and have a better estimate of the true value of the security. As orders are received, the bid and ask prices change to reflect the trader's informational advantage. In addition, the model is competitive in the sense that the market maker is constrained to make zero expected profits.

Besides the field of market microstructure, a number of other asset-pricing topics have been influenced by game theory. These include market manipulation models. See Cherian and Jarrow (1995) for a survey. Many financial innovation models, for instance Allen and Gale (1994) and Duffie and Rahi (1995), also use game-theoretic techniques. However, these areas do not as yet have the visibility of other areas in asset pricing.

Pricing anomalies such as those associated with P/E or P/B ratios that have received so much attention in recent years are intimately associated with accounting numbers. Since these numbers are to some extent the outcome of strategic decisions, analysis of these phenomena using game-theoretic techniques seems likely to be a fruitful area of research.

3. Richer Models of Information and Beliefs

Despite the great progress in finance using game-theoretic techniques, many phenomena remain unexplained. One reaction to this has been to move away from models based on rational behavior and develop behavioral models. We argue that it is premature to abandon rationality. Recent developments in game theory have provided powerful new techniques that explain many important financial phenomena. In this section, we review three lines of research and consider their implications for finance.

Higher Order Beliefs

Conventional wisdom in financial markets holds that participants are concerned not just about fundamentals, but also about what others believe about fundamentals, what others believe about others' beliefs, and so on. Remarkably, the mainstream finance literature largely ignores such issues. When such concerns are introduced and discussed, it is usually in the context of models with irrational actors. Yet the game-theory literature tells us that when there are coordination aspects to a strategic situation, such higher order beliefs are crucially important for fully rational actors.

How do these issues come to be bypassed? In our view, this happens because models of asymmetric information to date – though tractable and successful in examining many finance questions -- are not rich enough to address issues of higher-order beliefs. If it is assumed that players' types, or signals, are independent, it is (implicitly) assumed that there is common knowledge of players' beliefs about other players' beliefs. If it is assumed that each signal implies a different belief about fundamentals, it is (implicitly) assumed that a player's belief about others' beliefs is uniquely determined by his belief about fundamentals. Modeling choices made for "tractability" often have the effect of ruling out an interesting role for higher order beliefs.

We will discuss one example illustrating how higher order beliefs about fundamentals determine outcomes in a version of Diamond and Dybvig's (1983) model of intermediation and bank runs. In the environment described, there is a unique equilibrium. Thus for each possible "state of the world", we can determine whether there is a run, or not. But the "state of the world" is not determined only by the "fundamentals," i.e., the amount of money in the bank. Nor is the state determined by "sunspots," i.e., some payoff irrelevant variable that has nothing to do with fundamentals. Rather, what matters is depositors' higher order beliefs: what they believe about fundamentals, what they believe others believe, and so on. Our example illustrates why game theory confirms the common intuition that such higher order beliefs matter and determine outcomes. After the example, we will review a few attempts to incorporate this type of argument in models of financial markets.

The Example. There are two depositors in a bank. Depositor i's type is ξ_i. If ξ_i is less than 1, then depositor i has liquidity needs that require him to withdraw money from the bank; if ξ_i is greater than or equal to 1, he has no liquidity needs and acts to maximize his expected return. If a depositor withdraws his money from the bank, he obtains a guaranteed payoff of $r > 0$. If he keeps his money on deposit and the other depositor does likewise, he gets a payoff of R, where $r < R < 2r$. Finally, if he keeps his money in the bank and the other depositor withdraws, he gets a payoff of zero.

Notice that there are four states of "fundamentals": both have liquidity needs, depositor 1 only has liquidity needs, depositor 2 only has liquidity needs, and neither has liquidity needs. If there was common knowledge of fundamentals,

Game Theory Models in Finance

and at least one depositor had liquidity needs, the unique equilibrium has both depositors withdrawing. But if it were common knowledge that neither depositor has liquidity needs, they are playing a coordination game with the following payoffs:

	Remain	Withdraw
Remain	R, R	0, r
Withdraw	r, 0	r, r

With common knowledge that neither investor has liquidity needs, this game has two equilibria: both remain and both withdraw. We will be interested in a scenario where neither depositor has liquidity needs, both know that no one has liquidity needs, both know that both know this, and so on up to any large number of levels, but nonetheless it is not common knowledge that no one has liquidity needs. We will show that in this scenario, the unique equilibrium has both depositors withdrawing. Clearly, higher-order beliefs, in addition to fundamentals, determine the outcome.

Here is the scenario. The depositors' types, ξ_1 and ξ_2, are highly correlated; in particular suppose that a random variable T is drawn from a smooth distribution on the non-negative numbers and each ξ_i is distributed uniformly on the interval $[T - \varepsilon, T + \varepsilon]$, for some small $\varepsilon > 0$. Given this probability distribution over types, types differ not only in fundamentals, but also in beliefs about the other depositor's fundamentals, and so on. To see why, recall that a depositor has liquidity needs exactly if ξ_i is less than 1. But when do both depositors *know* that both ξ_i are greater than or equal to 1? Only if both ξ_i are greater than $1 + 2\varepsilon$ (since each player knows only that the other's signal is within 2ε of his own)? When do both depositors know that both know that both ξ_i are greater than 1? Only if both ξ_i are greater than $1 + 4\varepsilon$. To see this, suppose that $\varepsilon = .1$ and depositor 1 receives the signal $\xi_1 = 1.3$. She can deduce that T is within the range 1.2 to 1.4 and hence that depositor 2's signal is within the range 1.1 to 1.5. However, if depositor 2 received the signal $\xi_2 = 1.1$, then he sets a positive probability of depositor 1 having ξ_1

smaller than 1. Only if depositor 1's signal is greater or equal to $1 + 4\varepsilon = 1.4$ would this possibility be avoided. By iterating this argument, we see that it can never be common knowledge that both players are free of liquidity needs.

What do these higher order beliefs imply? In fact, for small enough ε, the unique equilibrium of this game has both depositors always withdrawing, whatever signals they observe. Observe first that by assumption each depositor must withdraw if ξ_i is smaller than 1, i.e., if she or he has liquidity needs. But suppose depositor 1's strategy is to remain only if ξ_1 is greater than some k, for k > 1. Further, consider the case that depositor 2 observes signal, $\xi_2 = k$. For small ε, he would attach probability about ½ to depositor 1 observing a lower signal, and therefore withdrawing. Therefore depositor 2 would have an expected payoff of about ½R for remaining and r for withdrawing. Since r > ½R by assumption, he would have a strict best response to withdraw if he observed k. In fact, his unique best response is to withdraw if his signal is less than some cutoff point strictly larger than k. But this implies that each depositor must have a higher cutoff for remaining than the other. This is a contradiction. So the unique equilibrium has both depositors always withdrawing.

This argument may sound paradoxical. After all, we know that if there was common knowledge that payoffs were given by the above matrix (i.e., both ξ_i were above 1), then there would be an equilibrium where both depositors remained. The key feature of the incomplete information environment is that while there are only four states of fundamentals, there is a continuum of states corresponding to different higher order beliefs. In all of them, there is a lack of common knowledge that both depositors do not have liquidity needs. Given our assumptions on payoffs, this is enough to guarantee withdrawal.

We do not intend to imply by the above argument that depositors are able to reason to very high levels about the beliefs and knowledge of other depositors. The point is simply that some information structures fail to generate sufficient common knowledge to support coordination on risky outcomes. How much common knowledge is "sufficient" is documented in the game-theory literature: what is required is the existence of "almost public" events, i.e., events that everyone believes very likely whenever they are true. See Monderer and Samet (1989) and Morris, Rob and Shin (1995). While participants in financial markets may be unable to reason to very high levels of beliefs and knowledge, they should be able to recognize the existence or non-existence of almost public events.

The above example is a version of one introduced by Carlsson and van Damme (1993). Earlier work by Halpern (1986) and Rubinstein (1989) developed the link between coordination and common knowledge. See Morris and Shin (1997) for a survey of these developments. Morris and Shin (1998) generalize the logic of the above example to a model with a continuum of investors deciding whether or not to attack a currency with a fixed peg. Higher-order beliefs are crucial to the ability of investors to coordinate their behavior, and thus a key factor in determining when currency attacks occur.

A number of other models have explored the role of higher order beliefs in finance. In Abel and Mailath (1994), risk-neutral investors subscribe to securities paid from a new project's revenues. They note that it is possible that all investors subscribe to the new securities even though all investors' expected returns are negative. This could not happen if it was common knowledge that all expected returns were negative.

Allen, Morris and Postlewaite (1993) consider a rational expectations equilibrium of a dynamic asset trading economy with a finite horizon, asymmetric information and short sales constraints. They note that an asset may trade at a positive price, even though every trader knows that the asset is worthless. Even though each trader knows that the asset is worthless, he attaches positive probability to some other trader assigning positive expected value to the asset in some future contingency. It is worth holding the asset for that reason. Again, this could not occur if it was common knowledge that the asset was worthless.

Kraus and Smith (1989) describe a model where the arrival of information about others' information (not new information about fundamentals) drives the market. Kraus and Smith (1998) consider a model where multiple self-fulfilling equilibria arise because of uncertainty about other investors' beliefs. They term this "endogenous sunspots". They show that such sunspots can produce "pseudo-bubbles" where asset prices are higher than in the equilibrium with common knowledge.

Shin (1996) compares the performance of decentralized markets with dealership markets. While both perform the same in a complete information environment, he notes that the decentralized market performs worse in the presence of higher order uncertainty about endowments. The intuition is that a decentralized market requires coordination that is sensitive to a lack of common knowledge, whereas the dealership requires less coordination.

Information Cascades

There is an extensive literature concerned with informational cascades. Welch (1992) is an early example. A group of potential investors must decide whether to invest in an initial public offering (IPO) sequentially. Each investor has some private information about the IPO. Suppose that the first few investors happen to observe bad signals and choose not to invest. Later investors, even if they observed good signals, would ignore their own private information and not invest on the basis of the (public) information implicit in others' decisions not to invest. But now even if the majority of late moving investors has good information, their good information is never revealed to the market. Thus inefficiencies arise in the aggregation of private information because the investors' actions provide only a coarse signal of their private information. This type of phenomenon has been analyzed more generally by Banerjee (1992) and Bikhchandani, Hirshleifer and Welch (1992). Finance applications are surveyed in Devenow and Welch (1996).

It is important to note that informational cascades occur even in the absence of any payoff interaction between decision makers. In the Welch (1992) account of initial public offerings, investors do not care whether others invest or not; they merely care about the information implicit in others' decisions whether to invest. But the argument does rely on decisions being made sequentially and publicly. Thus an informational cascades account of bank runs would go as follows. Either the bank is going to collapse or it will not, *independent of the actions of depositors*. Depositors decide whether to withdraw sequentially. If the first few investors happened to have good news, the bank would survive; if they happened to have bad news, the bank would not survive. By contrast, in the previous section, we described a scenario where despite the fact that all investors knew for sure that there was no need for the bank to collapse, it had to collapse because of a lack of common knowledge that the bank was viable. That scenario arose only because of payoff interaction (each depositor's payoff depends on other depositors' actions, because they influence the probability of collapse); but it occurred even when all decisions were made simultaneously.

One major weakness of the informational cascade argument is that it relies on action sets being too coarse to reveal private information (see Lee (1993)). There are some contexts where this assumption is natural: for example, investors' decisions whether to subscribe to initial public offerings at a fixed offer price (although even then the volume demanded might reveal information continuously). But once prices are endogenized, the (continuum) set of possible prices will tend to reveal information. Researchers have identified two natural reasons why informational cascades might nonetheless occur in markets with endogenous price formation. If investors face transaction costs, they may tend not to trade on the basis of small pieces of information (Lee (1997)). In this case, market crashes might occur when a large number of investors, who have observed bad news but not acted on it, observe a (small) public signal that pushes them into trading despite transaction costs. Avery and Zemsky (1996) exploit the fact that although prices may provide rich signals about private information, if private information is rich enough (and, in particular, multi-dimensional), the market will not be able to infer private information from prices.

Heterogeneous Prior Beliefs

Each of the two previous topics we reviewed concerned richer models of asymmetric information. We conclude by discussing the more basic question as to how differences in beliefs are modeled. A conventional modeling assumption in economics and finance is the "common prior" assumption: rational agents may observe different signals (i.e., there may be asymmetric information) but it is assumed that their posterior beliefs could have been derived by updating a common prior belief on some state space. Put differently, it is assumed that all differences in beliefs are the result of differences in information, not differences in prior beliefs.

For some purposes, it does not matter if differences in beliefs are explained by different information or differences in priors. For example, Lintner (1969) derived a CAPM with heterogeneous beliefs and – assuming, as he did, that investors do not learn from prices – the origin of their differences in beliefs did not matter. It is only once it is assumed that individuals learn from others' actions (or prices that depend on others' actions) that the difference becomes important. Thus the distinction began to be emphasized in finance exactly when game-theoretic and information-theoretic issues were introduced. Most importantly, "no trade" theorems, such as that of Milgrom and Stokey

(1982), established that differences in beliefs based on differences in information alone could not lead to trade.

But while the distinction is important, this does not justify a claim that heterogeneous prior beliefs are inconsistent with rationality. See Morris (1995) for a review of attempts to justify this claim and also Gul (1998) and Aumann (1998). In any case, there is undoubtedly a significant middle ground between the extreme assumptions that (1) participants in financial markets are irrational; and (2) all differences in beliefs are explained by differences in information. We will briefly review some work in finance within this middle ground.

Harrison and Kreps (1978) considered a dynamic model where traders were risk neutral, had heterogeneous prior beliefs (*not* explained by differences in information) about the dividend process of a risky asset, and were short sales constrained in that asset. They observed that the price of an asset would typically be more than any trader's fundamental value of the asset (the discounted expected dividend) because of the option value of being able to sell the asset to some other trader with a higher valuation in the future. Morris (1996) examined a version of the Harrison and Kreps model where although traders start out with heterogeneous prior beliefs, they are able to learn the true dividend process through time; a re-sale premium nonetheless arises, one that reflects the divergence of opinion before learning has occurred. Thus this model provides a formalization of E. Miller's (1977) explanation of the opening market overvaluation of initial public offerings: lack of learning opportunities implies greater heterogeneity of beliefs implies higher prices.

The above results concerned competitive models and were, therefore, non-strategic. But heterogeneous prior beliefs play a similar role in strategic models of trading volume. Trading volume has remained a basic puzzle in the finance literature. It is hard to justify the absolute volume of trade using standard models where trade is generated by optimal diversification with common prior beliefs. Empirically relevant models thus resort to modeling shortcuts, such as the existence of noise traders. But ultimately the sources of speculative trades must be modeled and differences of opinion (heterogeneous prior beliefs) are surely an important source of trade.

In Harris and Raviv (1993), traders disagree about the likelihood of alternative public signals conditional on payoff relevant events. They present a simple model incorporating this feature that naturally explains the positive autocorrelation of trading volume and the correlation between absolute price changes and volume as well as a number of other features of financial market data. A number of other authors, Varian (1989) and Biais and Bossaerts (1998), have derived similar results. The intuition for these findings is similar to that of noise trader models. In our view, however, explicitly modeling the rational differences in beliefs leading to trade will ultimately deepen our understanding of financial markets.

References

Abel, A. and G. Mailath (1994), "Financing Losers in Financial Markets," *Journal of Financial Intermediation*, 3, 139-165.

Aghion, P. and P. Bolton (1992), "An 'Incomplete Contracts' Approach to Financial Contracting," *Review of Economic Studies*, 59, 473-494.

Allen, F. (1986), "Capital Structure and Imperfect Competition in Product Markets," working paper, University of Pennsylvania.

Allen, F. and G. Faulhaber (1989), "Signalling by Underpricing in the IPO Market," *Journal of Financial Economics*, 23, 303-323.

Allen, F. and D. Gale (1999), *Comparing Financial Systems*, MIT Press, Cambridge, Massachusetts (forthcoming).

Allen, F. and D. Gale (1994), *Financial Innovation and Risk Sharing*, MIT Press, Cambridge, Massachusetts.

Allen, F. and R. Michaely (1995), "Dividend Policy," in Jarrow, Maksimovic, Ziemba (1995), 793-837.

Allen, F., S. Morris and A. Postlewaite (1993), "Finite Bubbles with Short Sales Constraints and Asymmetric Information," *Journal of Economic Theory*, 61, 209-229.

Aumann, R. (1998), "Common Priors: A Reply to Gul," *Econometrica*, 66, 929-938.

Avery, C. and P. Zemsky (1996), "Multi-Dimensional Uncertainty and Herd Behavior in Financial Markets," forthcoming in *American Economic Review*.

Banerjee, A. (1992), "A Simple Model of Herd Behavior," *Quarterly Journal of Economics*, 107, 797-817.

Banz, R. (1981), "The Relationship Between Return and Market Value of Common Stock," *Journal of Financial Economics*, 9, 3-18.

Barclay, M. and C. Smith, Jr. (1988), "Corporate Payout Policy: Cash Dividends Versus Open-Market Repurchases," *Journal of Financial Economics*, 22, 61-82.

Basu, S. (1977), "Investment Performance of Common Stocks in Relation to their Price-Earnings Ratio: A Test of the Efficient Market Hypothesis," *Journal of Finance*, 32, 663-682.

Benveniste, L. and P. Spindt (1989), "How Investment Bankers Determine the Offer Price and Allocation of New Issues," *Journal of Financial Economics*, 24, 343-361.

Berglof, E. and E. von Thadden (1994), "Short-Term versus Long-Term Interests: Capital Structure with Multiple Investors," *Quarterly Journal of Economics*, 109, 1055-1084.

Bhattacharya, S. (1979), "Imperfect Information, Dividend Policy, and the 'Bird in the Hand' Fallacy," *Bell Journal of Economics*, 10, 259-270.

Bhattacharya, S. and A. Thakor (1993), "Contemporary Banking Theory," *Journal of Financial Intermediation*, 3, 2-50.

Biais, B. and P. Bossaerts (1998), "Asset Prices and Trading Volumes in a Beauty Contest," *Review of Economic Studies*, 65, 307-340.

Bikhchandani, S., D. Hirshleifer, and I. Welch (1992), "A Theory of Fads, Fashions, Customs and Cultural Change as Informational Cascades," *Journal of Political Economy*, 100, 992-1026.

Black, F. (1972), "Capital Market Equilibrium with Restricted Borrowing," *Journal of Business*, 45, 444-455.

Black, F. (1976), "The Dividend Puzzle," *Journal of Portfolio Management*, 2, 5-8.

Black, F. and M. Scholes (1973), "The Pricing of Options and Corporate Liabilities," *Journal of Political Economy*, 81, 637-659.

Bradley, M., A. Desai and E. Kim (1988), "Synergistic Gains from Corporate Acquisitions and Their Division Between the Stockholders of Target and Acquiring Firms," *Journal of Financial Economics*, 21, 3-40.

Brander, J. and T. Lewis (1986), "Oligopoly and Financial Structure: The Limited Liability Effect," *American Economic Review*, 76, 956-970.

Breeden, D. (1979), "An Intertemporal Asset Pricing Model with Stochastic Consumption and Investment Opportunities," *Journal of Financial Economics*, 7, 265-296.

Brennan, M. (1989), "Capital Asset Pricing Model," in J. Eatwell, M. Milgate and P. Newman (Eds.), *The New Palgrave Dictionary of Economics*, Stockton Press, New York.

Brennan, M. and A. Thakor (1990), "Shareholder Preferences and Dividend Policy," *Journal of Finance*, 45, 993-1019.

Brown, D. (1989), Claimholder Incentive Conflicts in Reorganization: The Role of Bankruptcy Law," *Review of Financial Studies*, 2, 109-123.

Bryant, J. (1980), "A Model of Reserves, Bank Runs, and Deposit Insurance," *Journal of Banking and Finance*, 4, 335-344.

Chari, V. and R. Jagannathan (1988), "Banking Panics, Information, and Rational Expectations Equilibirum," *Journal of Finance*, 43, 749-760.

Cherian, J. and R. Jarrow (1995), "Market Manipulation," in Jarrow, Maksimovic and Ziemba (1995), 611-630.

Carlsson, H. and E. van Damme (1993), "Global Games and Equilibrium Selection," *Econometrica*, 61, 989-1018.

Constantinides, G. and A. Malliaris (1995), in Jarrow, Maksimovic and Ziemba (1995), 1-30.

Cox, J., J. Ingersoll, J. and S. Ross (1985), "A Theory of the Term Structure of Interest Rates," *Econometrica*, 53, 385-407.

Cross, F. (1973), "The Behavior of Stock Prices on Fridays and Mondays," *Financial Analysts Journal*, 29, 67-69.

De Bondt, W. and R. Thaler (1995), "Financial Decision Making in Markets and Firms: A Behavioral Perspective," in Jarrow, Maksimovic and Ziemba (1995), 385-410.

Devenow, A. and I. Welch (1996), "Rational Herding in Financial Economics," *European Economic Review*, 40, 603-615.

Diamond, D. (1984), "Financial Intermediation and Delegated Monitoring," *Review of Economic Studies*, 51, 393-414.

Diamond, D. (1989), "Reputation Acquisition in Debt Markets," *Journal of Political Economy*, 97, 828-862.

Diamond, D. and P. Dybvig (1983), "Bank Runs, Deposit Insurance and Liquidity," *Journal of Political Economy*. 91, 401-419.

Dreman, D. (1982), *The New Contrarian Investment Strategy*, Random House, New York.

Dubey, P. J., J. Geanakoplos and M. Shubik (1987). "The Revelation of Information in Strategic Market Games: A Critique of Rational Expectations Equilibrium," *Journal of Mathematical Economics*, 16, 105-137.

Duffie, D. and C. Huang (1985), "Implementing Arrow-Debreu Equilibria by Continuous Trading of Few Long-Lived Securities," *Econometrica*, 53, 1337-1356.

Duffie, D. and R. Rahi (1995), "Financial Market Innovation and Security Design: An Introduction," *Journal of Economic Theory*, 65, 1-42.

Dybvig, P. and J. Zender (1991), "Capital Structure and Dividend Irrelevance with Asymmetric Information," *Review of Financial Studies*, 4, 201-219.

Fama, E. (1970), "Efficient Capital Market: A Review of Theory and Empirical Work," *Journal of Finance*, 25, 382-417.

Fama, E. (1991), "Efficient Capital Market, II," *Journal of Finance*, 46, 1575-1617.

Fama, E. and H. Babiak (1968), "Dividend Policy: An Empirical Analysis," *Journal of the American Statistical Association*, 63, 1132-1161.

Fama, E. and J. Macbeth (1973), "Risk, Return and Equilibrium: Empirical Tests," *Journal of Political Economy*, 71, 607-636.

Fama, E. and K. French (1993), "Common Risk Factors in the Returns on Stocks and Bonds," *Journal of Financial Economics*, 33, 3-56.

Ferson, W. (1995), "Theory and Empirical Testing of Asset Pricing Models," in Jarrow, Maksimovic and Ziemba (1995), 145-200.

Fishman, M. (1988), "Theory of Pre-Emptive Takeover Bidding," *Rand Journal of Economics*, 19, 88-101.
French, K. (1980), "Stock Returns and the Weekend Effect," *Journal of Financial Economics*, 8, 55-69.

Giammarino, R. (1988), "The Resolution of Financial Distress," *Review of Financial Studies*, 2, 25-47.

Gibbons, R. (1992), *Game Theory for Applied Economists*, Princeton University Press, Princeton, New Jersey.

Glosten, L. and P. Milgrom (1985), "Bid, Ask, and Transaction Prices in a Specialist Market with Heterogeneously Informed Traders," *Journal of Financial Economics*, 13, 71-100.

Grinblatt, M. and C. Hwang (1989), "Signalling and the Pricing of New Issues," *Journal of Finance*, 44, 393-420.

Grossman, S and O. Hart (1980), "Takeover Bids, the Free-Rider Problem and the Theory of the Corporation," *Bell Journal of Economics*, 11, 42-64.

Grossman, S. and O. Hart (1982), "Corporate Financial Structure and Managerial Incentives," in J. McCall (Ed.), *The Economics of Information and Uncertainty*, University of Chicago Press, Chicago, Illinois.

Grossman, S. and J. Stiglitz (1980), "On the Impossibility of Informationally Efficient Markets," *American Economic Review*, 70, 393-408.

Gul, F. (1998), "A Comment on Aumann's Bayesian View," *Econometrica*, 66, 923-928.

Gurley, J. and E. Shaw (1960), *Money in a Theory of Finance*, The Brookings Institution, Washington, D.C.

Halpern, J. (1986), "Reasoning about Knowledge: An Overview," in J. Halpern (Ed.), *Theoretical Aspects of Reasoning about Knowledge*, Morgan Kaufmann, Los Altos, California.

Harris, M. and A. Raviv (1991), "The Theory of Capital Structure," *Journal of Finance*, 46, 297-355.

Harris, M. and A. Raviv (1993), "Differences of Opinion Make a Horse Race," *Review of Financial Studies*, 6, 473-506.

Harrison, M. and D. Kreps (1978), "Speculative Investor Behavior in a Stock Market with Heterogeneous Expectations," *Quarterly Journal of Economics*, 92, 323-336.

Harrison, M. and D. Kreps (1979), "Martingales and Arbitrage in Multiperiod Securities Markets," *Journal of Economic Theory*, 12, 381-408.

Hart, O. (1995), *Firms, Contracts and Financial Structure*, Oxford University Press, New York, New York.

Hart, O. and J. Moore (1989), "Default and Renegotitation: A Dynamic Model of Debt," MIT Working Paper 520.

Hart, O. and J. Moore (1994), "A Theory of Debt Based on the Inalienability of Human Capital," *Quarterly Journal of Economics*, 109, 841-879.

Hart, O. and J. Moore (1998), "Default and Renegotiation: A Dynamic Model of Debt," *Quarterly Journal of Economics*, 113, 1-41.

Haugen, R. and L. Senbet (1978), "The Insignificance of Bankruptcy Costs to the Theory of Optimal Capital Structure," *Journal of Finance*, 33, 383-392.

Hawawini, G. and D. Keim (1995), "On the Predictability of Common Stock Returns: World-Wide Evidence," in Jarrow, Maksimovic and Ziemba (1995), 497-544.

Hicks, J. (1939), *Value and Capital*, Oxford University Press, New York.

Hirshleifer, D. (1995), "Mergers and Acquisitions: Strategic and Informational Issues," in Jarrow, Maksimovic and Ziemba (1995), 839-885.

Hughes, P. and A. Thakor (1992), "Litigation Risk, Intermediation, and the Underpricing of Initial Public Offerings," *Review of Financial Studies*, 5, 709-742.

Ibbotson, R. (1975), "Price Performance of Common Stock New Issues," *Journal of Financial Economics*, 2, 235-272.

Ibbotson, R. and J. Ritter (1995), "Initial Public Offerings," in Jarrow, Maksimovic and Ziemba (1995), 993-1016.

Jacklin, C. and S. Bhattacharya (1988), "Distinguishing Panics and Information-Based Bank Runs: Welfare and Policy Implications," *Journal of Political Economy*, 96, 568-592.

Jarrow, R., V. Maksimovic and W. Ziemba (Eds.), *Handbooks in Operations Research and Management Science, Volume 9, Finance*, North-Holland Elsevier, Amsterdam, The Netherlands.

Jennings, R. and M. Mazzeo (1993), "Competing Bids, Target Management Resistance and the Structure of Takeover Bids," *Review of Financial Studies*, 6, 883-910.

Jensen, M. (1986), "Agency Costs of Free Cash Flow, Corporate Finance and Takeovers," *American Economic Review*, 76, 323-339.

Jensen, M. and W. Meckling (1976), "Theory of the Firm: Managerial Behavior, Agency Costs, and Capital Structure," *Journal of Financial Economics*, 3, 305-360.

John, K. and J. Williams (1985), "Dividends, Dilution and Taxes: A Signaling Equilibrium," *Journal of Finance*, 40, 1053-1070.

Keim, D. (1983), "Size-Related Anomalies and Stock Return Seasonality: Further Empirical Evidence," *Journal of Financial Economics*, 12, 13-32.

Keynes, J. (1936), *The General Theory of Employment, Interest and Money*," Harcourt Brace and Company, New York.

Kraus, A. and M. Smith (1989), "Market Created Risk," *Journal of Finance*, 44, 557-569.

Kraus, A. and M. Smith (1998), "Endogenous Sunspots, Pseudo-bubbles, and Beliefs about Beliefs," *Journal of Financial Markets* 1, 151-174.

Kumar, P. (1988), "Shareholder-Manager Conflict and the Information Content of Dividends," *Review of Financial Studies*, 1, 111-136.

Kyle, A. (1985), "Continuous Auctions and Insider Trading," *Econometrica*, 53, 1315-1336.

Lee, I. (1993), "On the Convergence of Informational Cascades," *Journal of Economic Theory*, 61, 395-411.

Lee, I. (1997), "Market Crashes and Informational Avalanches," forthcoming in the *Review of Economic Studies*.

Leland, H. and D. Pyle (1977), "Information Asymmetries, Financial Structure, and Financial Intermediation," *Journal of Finance*, 32, 371-388.

Lintner, J. (1956), "Distribution of Incomes of Corporations among Dividends, Retained Earnings, and Taxes," *American Economic Review*, 46, 97-113.

Lintner, J. (1965), "The Valuation of Risk Assets and the Selection of Risky Investments in Stock Portfolios and Capital Assets," *Review of Economics and Statistics*, 47, 13-37.

Lintner, J. (1969), "The Aggregation of Investors' Diverse Judgements and Preferences in Pure Competitive Markets," *Journal of Financial and Quantitative Analysis*, 4, 347-400.

Logue, D. (1973), "On the Pricing of Unseasoned Equity Issues: 1965-69," *Journal of Financial and Quantitative Analysis*, 8, 91-103.

Loughran, T. (1993), "NYSE vs. NASDAQ Returns: Market Microstructure or the Poor Performance of IPO's?" *Journal of Financial Economics*, 33, 241-260.

Loughran, T. and J. Ritter (1995), "The New Issues Puzzle," *Journal of Finance*, 50, 23-51.

Lucas, R., Jr. (1978), "Asset Prices in an Exchange Economy," *Econometrica*, 46, 1429-1445.

Maksimovic, V. (1986), "Optimal Capital Structure in Oligopolies," unpublished Ph. D. dissertation, Harvard University.

Maksimovic, V. (1995), "Financial Structure and Product Market Competition," in Jarrow, Maksimovic and Ziemba (1995), 887-920.

Maksimovic, V. and S. Titman (1991), "Financial Reputation and Reputation for Product Quality," *Review of Financial Studies*, 2, 175-200.

Manne, H. (1965), "Mergers and the Market for Corporate Control," *Journal of Political Economy*, 73, 110-120.

Markowitz, H. (1952), "Portfolio Selection," *Journal of Finance*, 7, 77-91.

Markowitz, H. (1959), "Portfolio Selection: Efficient Diversification of Investments, Wiley, New York.

Merton, R. (1969), "Lifetime Portfolio Selection: The Continuous Time Case," *Review of Economics and Statistics*, 51, 247-257.

Merton, R. (1971), "Optimum Consumption and Portfolio Rules in a Continuous Time Model," *Journal of Economic Theory*, 3, 373-413.

Merton, R. (1973a), "An Intertemporal Capital Asset Pricing Model," *Econometrica*, 41, 867-887.

Merton, R. (1973b), "Theory of Rational Option Pricing," *Bell Journal of Economics and Management Science*, 4, 141-183.

Milgrom, P. and N. Stokey (1982), "Information, Trade and Common Knowledge," *Journal of Economic Theory*, 26, 17-27.

Miller, E. (1977), "Risk, Uncertainty and Divergence of Opinion," *Journal of Finance*, 32, 1151-1168.

Miller, M. (1977), "Debt and Taxes," *Journal of Finance*, 32, 261-275.

Miller, M. and F. Modigliani (1961), "Dividend Policy, Growth and the Valuation of Shares," *Journal of Business*, 34, 411-433.

Miller, M. and K. Rock (1985), "Dividend Policy under Asymmetric Information," *Journal of Finance*, 40, 1031-1051.

Modigliani, F. and M. Miller (1958), "The Cost of Capital, Corporation Finance and the Theory of Investment," *American Economic Review*, 48, 261-297.

Monderer, D. and D. Samet (1989), "Approximating Common Knowledge with Common Beliefs," *Games and Economic Behavior*, 1, 170-190.

Morris, S. (1995), "The Common Prior Assumption in Economic Theory," *Economics and Philosophy*, 11, 227-253.

Morris, S. (1996), "Speculative Investor Behavior and Learning," *Quarterly Journal of Economics*, 111, 1111-1133.

Morris, S., R. Rob and H. Shin (1995), "p-Dominance and Belief Potential," *Econometrica*, 63, 145-157.

Morris, S. and H. Shin (1997), "Approximate Common Knowledge and Coordination: Recent Lessons from Game Theory," *Journal of Logic, Language and Information*, 6, 171-190.

Morris, S. and H. Shin (1998), "Unique Equilibrium in a Model of Self-Fulfilling Attacks," *American Economic Review*, 88, 587-597.

Myers, S. (1977), "Determinants of Corporate Borrowing," *Journal of Financial Economics*, 5, 147-175.

Myers, S. (1984), "The Capital Structure Puzzle," *Journal of Finance*, 39, 575-592.

Myers, S. and N. Majluf (1984), "Corporate Financing and Investment Decisions When Firms Have Information that Investors do not Have," *Journal of Financial Economics*, 13, 187-221.

Nyborg, K. (1995), "Convertible Debt as Delayed Equity: Forced versus Voluntary Conversion and the Information Role of Call Policy," *Journal of Financial Intermdiation*, 4, 358-395.

Ofer, A. and A. Thakor (1987), "A Theory of Stock Price Responses to Alternative Corporate Cash Disbursement Methods: Stock Repurchases and Dividends," *Journal of Finance*, 42, 365-394.

O'Hara, M. (1995), *Market Microstructure Theory*, Blackwell, Cambridge, Massachussetts.

Ritter, J. (1991), "The Long Run Performance of Initial Public Offerings," *Journal of Finance*, 46, 3-28.

Rock, K. (1986), "Why New Issues are Underpriced," *Journal of Financial Economics*, 15, 187-212.

Ross, S. (1977a), "The Arbitrage Theory of Capital Asset Pricing," *Journal of Economic Theory*, 13, 341-360.

Ross, S. (1977b), "The Determination of Financial Structure: The Incentive Signalling Approach," *Bell Journal of Economics*, 8, 23-40.

Ross, S. (1992), "Finance," in J. Eatwell, M. Milgate, and P. Newman (eds.), *The New Palgrave Dictionary of Money and Finance*, MacMillan, 26-41.

Rozeff, M. and W. Kinney (1976), "Capital Market Seasonality: The Case of Stock Returns," *Journal of Financial Economics*, 3, 379-402.

Rubinstein, A. (1989), "The Electronic Mail Game: Strategic Behavior Under 'Almost Common Knowledge,'" *American Economic Review*, 79, 385-391.

Ruud, J. (1993), "Underwriter Price Support and the IPO Underpricing Puzzle," *Journal of Financial Economics*, 34, 135-151.

Senbet, L. and J. Seward (1995), "Financial Distress, Bankruptcy and Reorganization," in Jarrow, Maksimovic and Ziemba (1995), 921-961.

Sharpe, W. (1964), "Capital Asset Prices: A Theory of Market Equilibrium under Conditions of Risk," *Journal of Finance*, 19, 425-442.

Shiller, R. (1990), "Speculative Prices and Popular Models," *Journal of Economic Perspectives*, 4, 55-65.

Shin, H. (1996), "Comparing the Robustness of Trading Systems to Higher Order Uncertainty," *Review of Economic Studies*, 63, 39-60.

Shleifer, A. and R. Vishny (1986a), "Large Shareholders and Corporate Control," *Journal of Political Economy*, 94, 461-488.

Shleifer, A. and R. Vishny (1986b), "Greenmail, White Knights, and Shareholders' Interest," *Rand Journal of Economics*, 17, 293-309.

Stattman, D. (1980), "Book Values and Expected Stock Returns," unpublished MBA Honors paper, University of Chicago, Chicago, Illinois.

Stein, J. (1992), "Convertible Bonds as Backdoor Equity Financing," *Journal of Financial Economic*, 32, 3-21.

Stiglitz, J. and A. Weiss (1981), "Credit Rationing in Markets with Imperfect Information," *American Economic Review*, 71, 393-410.

Thakor, A. (1991), "Game Theory in Finance," *Financial Management*, Spring, 71-94.

Thakor, A. (1996), "The Design of Financial Systems: An Overview," *Journal of Banking and Finance*, 20, 917-948.

Titman, S. (1984), "The Effect of Capital Structure on the Firm's Liquidation Decision," *Journal of Financial Economics*, 13, 137-152.

Tobin, J. (1958), "Liquidity Preference as Behavior Toward Risk," *Review of Economic Studies*, 25, 65-86.

Varian, H. (1989), "Differences of Opinion in Financial Markets," in C. Stone (Ed.), *Financial Risk: Theory, Evidence and Implications*, Kluwer Academic Publications.

von Neumann, J. and O. Morgenstern (1947), *Theory of Games and Economic Behavior*, second edition, Princeton University Press, Princeton, New Jersey.

von Thadden, E. (1995), "Long-Term Contracts, Short-Term Investment and Monitoring," *Review of Economic Studies*, 62, 557-575.

Webb, D. (1987), "The Importance of Incomplete Information in Explaining the Existence of Costly Bankruptcy," *Economica*, 54, 279-288.

Welch, I. (1989), "Seasoned Offerings, Imitation Costs, and the Underpricing of Initial Public Offerings," *Journal of Finance*, 44, 421-449.

Welch, I. (1992), "Sequential Sales, Learning, and Cascades," *Journal of Finance*, 47, 695-732

3 GAME THEORY MODELS IN ACCOUNTING

Chandra Kanodia

Historically, much of accounting research has been concerned with analyzing the statistical properties of accounting data, and their decision relevance, as if the data were generated by a mechanical technology driven process. In recent years, a new literature has emerged that replaces this Robinson Crusoe view by a more game-theoretic view that accounting data affects, and is affected by, strategic interaction within and across firms. This view holds that the accounting process alters the strategic interaction among agents, by impinging on their incentives and on the contracts they make to bind their behavior. Thus the new literature approaches the design of measurement rules and accounting processes in a fundamentally different way. The emphasis is more on contracting and incentives and the control of human behavior than on the recording and analysis of *ex post* data.

Naturally, game theory, mechanism design and the theory of contracts are extensively used in this new approach. Using these tools, accountants have worked on the design of performance measures and contracts to alleviate moral hazard and collusion among agents. The literature on the *ex post* investigation of accounting variances has been replaced by the *ex ante* control of agents through stochastic monitoring. Issues of participatory budgeting have been addressed from the perspective of inducing truthful sharing of information and fine tuning of incentive contracts. We now have a better understanding of why subordinates should be allowed to self-select their own standards and quotas, and how compensation parameters should be tied to these self-selected standards. There has been considerable work on the design of transfer prices and budgets to coordinate inter-divisional activities within a firm in the presence of goal conflicts and information asymmetries. Going beyond the internal management of a firm, accounting research has explored the effect of adverse selection in the pricing of audit services. Even at a macro level, we now have a better understanding of how accounting measurements and disclosure mediate the interaction between capital markets and real investment decisions by firms.

My purpose, here, is to sample this literature in order to provide the non-specialist reader with an appreciation of why strategic interaction is important to the study of accounting phenomena, the kinds of phenomena that have been studied, the methods and assumptions used in the analysis, and the central results obtained. I have limited the discussion to topics that have a distinct game-theoretic flavor. This constraint has had the unfortunate consequence of biasing the discussion towards "management accounting" topics and excluding the important and growing literature on financial disclosure to capital markets. Even within this limited domain, I have not attempted to be comprehensive. I have deliberately sacrificed breadth in order to provide a more in-depth examination of a few select topics. Nevertheless, the cross section of work examined here illustrates issues that, in my admittedly biased opinion, are central to accounting. Accounting academics disagree on the proper domain of accounting, on the appropriate methodology to be used, and even on what constitutes a resolution of an accounting issue. Hopefully, this essay will whet the appetite of non-specialist readers and stimulate some cross-fertilization of ideas.

I have chosen three themes. In section 1, I discuss the literature that uses the principal-agent paradigm to provide insights into the design of performance measures to alleviate moral hazard.[1] In section 2, I focus on accounting mechanisms, such as budgeting and transfer pricing, which facilitate coordination among responsibility centers in a firm. These two themes constitute the core of game-theoretic approaches to management accounting. In section 3, I have chosen to discuss the recent literature concerning auditor hiring and audit pricing. This work illustrates game-theoretic applications in an area that has traditionally been viewed as "markets" driven. I have omitted the proofs of formal propositions when I felt they could be reconstructed using standard techniques derived in the mechanism design literature. These proofs can be found in the original papers.

1. Design of Performance Measures

The Mirrlees/Holmstrom formulation of the principal-agent model with moral hazard has been a workhorse for accounting researchers. The model develops tradeoffs in the use of noisy signals to reward and motivate an agent to take actions that decrease the agent's utility but increase the principal's utility, given that these actions themselves are unobservable to the principal. Thus, the model provides a framework for studying the design of incentives and performance measures. Baiman and Demski (1980), Lambert (1985), and Dye (1986) have used principal-agent models to study monitoring problems. Related research by Banker and Datar (1989), Bushman and Indjejikian (1993), and Feltham and Xie (1994) has modeled the aggregation of signals in performance measurement. Relative performance evaluation has been studied by Holmstrom (1982), Wolfson (1985), and Antle and Smith (1986) and the use of bonus pools by Baiman and Rajan (1995). In addition, an extensive body of research by Baiman and Evans (1983), Dye (1983), and Penno (1985) has modeled contracting on unverifiable pre-decision and post-decision information communicated by agents. Finally, collusion among agents has been studied by Demski and Sappington (1984), Baiman, Evans and Nagarajan (1991), Rajan (1992), Arya and Glover (1996), and Suh (1987).

The Basic Principal-Agent Model

In the basic version of the principal-agent model, an agent provides effort α that stochastically affects output x. The relationship between effort and output is described by the probability density function $f(x|\alpha)$, with support that is independent of α. The principal observes output, but cannot observe the agent's effort. The agent is strictly risk and effort averse, with utility $U(w) - v(\alpha)$, $U'(.) > 0$, $U''(.) < 0$, $v'(.) > 0$. The principal is risk neutral. Risk sharing considerations would dictate that all of the risk associated with the uncertain output is borne by the principal, with the agent's compensation being a non-contingent certain amount. But in this case, the agent has no incentive to choose anything other than minimal effort. In order to induce greater effort from the agent, the principal sacrifices some risk sharing benefits and offers the agent a contingent wage $s(x)$. The optimal contract between the principal and the agent trades off risk sharing and incentives.

Formally, the optimal contract is found by solving:

$$\underset{s(x),\, \alpha}{\text{Max}} \int [x - s(x)]\, f(x \mid \alpha)\, dx$$

subject to:

$$\int U(s(x))\, f(x \mid \alpha)\, dx - v(\alpha) \geq \theta$$
$$\int U(s(x))\, f_\alpha(x \mid \alpha)\, dx - v'(\alpha) = 0$$

The first constraint represents the agent's participation requirement that his expected utility from the contract exceeds some reservation amount θ. The second constraint is the incentive requirement that the specified effort α is in the agent's best interests. I have used the "first-order approach" (see Jewitt (1988)) to represent the incentive constraint as the first-order condition to the agent's maximization over α. Let λ and μ be the Lagrange multipliers for the participation constraint and the incentive constraint, respectively. Holmstrom (1979) established that the optimal compensation schedule s(x) is characterized by:

$$\frac{1}{U'(s(x))} = \lambda + \mu \frac{f_\alpha(x \mid \alpha)}{f(x \mid \alpha)} \qquad (1.1)$$

Holmstrom also established that $\mu > 0$, i.e. the optimal compensation schedule is such that the principal's expected payoff is strictly increasing in the agent's action. Equation (1.1) indicates that the properties of the compensation schedule s(x) depend on the properties of the likelihood ratio f_α/f, which statistically describes the information contained in x. For, example s(x) is monotone increasing in x if and only if the likelihood ratio is increasing in x. This assumption, called MLRP (the monotone likelihood ratio property) has become standard in the literature. Milgrom (1981) argued that restrictions on the density function $f(x|\alpha)$ are analogous to restrictions on technology, and that MLRP is an intuitively reasonable restriction since it implies that higher values of x convey "good" news regarding the agent's effort. Rogerson (1985) established that MLRP implies that the distribution of x is shifted to the right (in the sense of first-order stochastic dominance) as α is increased. The model is easily extended to the case of n signals, $x_1,...,x_n$, with joint density $f(x_1,...x_n|\alpha)$. The first-order condition characterizing the agent's optimal compensation contract $s(x_1,...x_n)$ is similar to (1.1) except that f_α/f is now a function of n signals.

I focus on two applications of the basic principal-agent model to the design of performance measures. The first application, monitoring, is concerned with the stochastic augmentation of an initial performance measure by additional information collected at a cost. The second application is concerned with aggregation issues in the construction of a performance measure.

Monitoring

Suppose an agent's unobservable action a is stochastically related to two observable signals x and y, with joint probability density function $h(x, y|a)$. Holmstrom (1979) established that if neither variable is a sufficient statistic for both then the optimal contract written on both signals is strictly Pareto superior to the optimal contract written on one signal alone. In the light of this result, Baiman and Demski (1980) studied the following question: Suppose that signal x was freely available to the principal, and suppose that after having observed x the principal had the opportunity to collect the additional signal y at a cost K. Given that y is incrementally informative (i.e. x is not sufficient for (x, y)), but collection of y is costly, which values of x would trigger collection of the additional information? This question is of practical significance. Performance measures often consist of easily collected summary information. Much more information on the agent's performance can usually be obtained if the principal undertakes a costly investigation. Should the investigation be triggered by seemingly poor performance or by seemingly good performance? Should the investigation be triggered stochastically?

Let $s(x)$ be the compensation paid to the agent when the investigation is not conducted and let $m(x, y)$ be the agent's compensation when the investigation is conducted. Let $p(x) \varepsilon [0, 1]$ be the probability of investigation contingent on the observed value of the initial performance measure. Baiman and Demski assume that x and y are conditionally independent, i.e. $h(x, y|a)$ can be factored and put into the form: $h(x, y|a) = f(x|a)g(y|a)$. This implies,

$$\frac{h_\alpha(x, y|\alpha)}{h(x, y|\alpha)} = \frac{f_\alpha(x|\alpha)}{f(x|\alpha)} + \frac{g_\alpha(y|\alpha)}{g(y|\alpha)}$$

Let X and Y be the supports of x and y, respectively and assume these supports do not change with α. Additionally, assume that $f(.)$ and $g(.)$ possess the monotone likelihood ratio property (MLRP), i.e. f_α/f is strictly increasing in x and g_α/g is strictly increasing in y, at every α. The principal chooses $s(x)$, $m(x, y)$ and $p(x)$ to solve:

$$\text{Max} \int_X \left\{ (1-p(x))[x-s(x)] + p(x)\int_Y [x-m(x,y)-K]g(y|\alpha)dy \right\} f(x|\alpha)dx$$

(1.2)

subject to:

$$\int_X \left\{ (1-p(x))U(s(x)) + p(x)\int_Y U(m(x,y))g(y|\alpha)dy \right\} f(x|\alpha)dx - v(\alpha) \geq \theta$$

(1.3)

$$\int_X (1-p(x))U(s(x))\frac{f_\alpha(x|\alpha)}{f(x|\alpha)} f(x|\alpha)dx$$
$$+ \int_X p(x)\int_Y U(m(x,y))\left[\frac{f_\alpha(x|\alpha)}{f(x|\alpha)} + \frac{g_\alpha(y|\alpha)}{g(y|\alpha)}\right]g(y|\alpha)dy\, f(x|\alpha)dx - v'(\alpha) = 0$$

(1.4)

In the above formulation, (1.3) is the agent's participation constraint, and (1.4) is the incentive constraint expressed as the first-order condition to the agent's maximization over α. The action α is exogenously specified, since there is no additional insight to be gained from making α endogenous. It is assumed that the principal can commit *ex ante* to the monitoring policy $p(x)$. Let λ be a Lagrange multiplier for (1.3) and let μ be the multiplier for (1.4).

At each x the point-wise Lagrangian to the above programming problem is linear in p, implying that the optimal monitoring policy is bang-bang in nature. At each x the principal either monitors with probability one or does not monitor. Differentiating the point-wise Lagrangian with respect to p, yields the result that the principal monitors at x if and only if:

$$\varphi(x) \equiv s(x) - \int_Y m(x,y) g(y|\alpha) dy$$

$$-\left[U(s(x))\left(\lambda + \mu \frac{f_\alpha(x|\alpha)}{f(x|\alpha)}\right)\right] - \int_Y U(m(x,y))\left(\lambda + \mu \frac{f_\alpha(x|\alpha)}{f(x|\alpha)} + \mu \frac{g_\alpha(y|\alpha)}{g(y|\alpha)}\right) g(y|\alpha) dy \geq K$$

(1.5)

Additionally, the first-order conditions with respect to s and m yield:

$$\frac{1}{U'(s(x))} = \lambda + \mu \frac{f_\alpha(x|\alpha)}{f(x|\alpha)} \tag{1.6}$$

$$\frac{1}{U'(m(x,y))} = \lambda + \mu \left(\frac{f_\alpha(x|\alpha)}{f(x|\alpha)} + \frac{g_\alpha(y|\alpha)}{g(y|\alpha)} \right) \tag{1.7}$$

The term $\varphi(x)$ can be interpreted as the expected benefit from investigation given that x has been observed. Assume that the optimal monitoring policy is not degenerate, i.e. K is not so large that it never pays to monitor, nor is it so small that monitoring is desirable for all values of x. Of particular interest is the case where $\varphi(x)$ is monotonic since, if this is true, the monitoring region is either convex and lower tailed or convex and upper tailed. This would correspond to the intuition that investigation is triggered only when the observed value of x is extreme. To examine this possibility, differentiate $\varphi(x)$ and use (1.6) and (1.7). This yields:

$$\varphi'(x) = \left[\int_Y U(m(x,y)) g(y|\alpha) dy - U(s(x)) \right] \mu \frac{\partial}{\partial x}\left(\frac{f_\alpha(x|\alpha)}{f(x|\alpha)} \right)$$

Now since $\mu > 0$, and $\dfrac{\partial}{\partial x} \dfrac{f_\alpha(x|\alpha)}{f(x|\alpha)} > 0$ by MLRP, it follows that $\varphi'(x)$ is positive or negative at every value of x according as:

$$\int_Y U(m(x, y))g(y \mid \alpha)dy - U(s(x)) > 0, \text{ or } < 0, \forall x \qquad (1.8)$$

If $\varphi'(x) < 0$, for all x, then the expected benefits from investigation are strictly decreasing with x, implying that the investigation region is lower tailed. Conversely, if $\varphi'(x) > 0$, for all x, then the expected benefits from investigation are strictly increasing with x, implying that the investigation region is upper tailed. Inequalities (1.8) describe a striking result. The investigation region is lower tailed if and only if investigation is bad for the agent in the sense that investigation is used to decrease the agent's expected utility. Conversely, the investigation region is upper tailed if investigation is used to reward the agent. Thus, (1.8) provides insights into whether a "carrot" or a "stick" should be used to motivate the agent. If investigations are used as a stick, only low values of x are investigated; if used as a carrot only high values of x are investigated.

Dye (1986) proved that, given MLRP and conditional independence of signals, whether investigations are used as a carrot or as a stick depends only upon the risk aversion of the agent. To see this, define:

$$\omega(z) \equiv U(U'^{-1}(1/z))$$

From (1.6), it follows that $U(s(x)) = \omega(\lambda + \mu f_\alpha/f)$ and, from (1.7), it follows that $U(m(x, y)) = \omega(\lambda + \mu f_\alpha/f + \mu g_\alpha/g)$. But $\int g_\alpha(y|\alpha)dy = 0$, for all α. Thus, at each value of x,

$$\lambda + \mu \frac{f_\alpha(x \mid \alpha)}{f(x \mid \alpha)} = E_Y\left[\lambda + \mu \frac{f_\alpha(x \mid \alpha)}{f(x \mid \alpha)} + \mu \frac{g_\alpha(y \mid \alpha)}{g(y \mid \alpha)}\right]$$

Since $\omega(z)$ is an increasing function, it follows from Jensen's inequality that if $\omega(z)$ is strictly concave then $U(s(x)) > \int U(m(x, y))g(y|\alpha)dy$, for all x. On the other hand, if $\omega(z)$ is strictly convex then $U(s(x)) < \int U(m(x, y))g(y|\alpha)dy$, for all x. This proves that if $\omega(z)$ is strictly concave the investigation region is lower tailed, and if $\omega(z)$ is strictly convex the investigation region is upper tailed.[2]

To illustrate how the agent's risk aversion affects the concavity or convexity of $\omega(z)$, consider the class of power utility functions: $U(s) = s^{\gamma}$, $0 < \gamma < 1$. The parameter γ is readily interpreted as the (constant) coefficient of relative risk aversion. For this class of utility functions, $\omega(z) = (\gamma z)^{\gamma/1-\gamma}$. Thus $\omega(z)$ is strictly concave if $\gamma < .5$, and strictly convex if $\gamma > .5$.

It is perplexing why the optimality of lower and upper tailed monitoring depends only on the risk aversion of the agent, and not on the relative information content of x and y. It appears that this result is driven by the rather strong assumption of conditional independence of signals. In some sense, the "informativeness" of y becomes independent of x under this assumption, and therefore appears as a constant in the analysis. In many situations, the value of the initial performance measure x would condition the principal's beliefs regarding what he would discover if he were to conduct an investigation, in which case the conditional independence assumption is violated. Lambert (1985) examined this issue by way of parametric examples and found cases where the optimal investigation region is actually two tailed, i.e. all extreme values of x are investigated.

The bang-bang nature of the optimal monitoring policy, found by Baiman and Demski, also seems to be inconsistent with casual empiricism. Empirically, most audits and investigations are stochastic, and the lay wisdom is that "surprise" is an important element of investigation policy. For example, the IRS is known to use a probabilistic policy for auditing tax returns. Townsend (1979), Kanodia (1985), Border and Sobel (1987), and Mookherjee and Png (1989) found that optimal monitoring policies are stochastic rather than deterministic when the incentive problem concerns the revelation of hidden information in addition to moral hazard. Another restrictive assumption of the Baiman and Demski formulation is the assumption that the principal can commit to a monitoring policy. It is often *ex ante* optimal to threaten an investigation contingent on some observation but *ex post* irrational to execute the threat. In such cases the threat loses credibility. Mukherji (1998) investigated sequentially rational monitoring policies and showed that optimally such monitoring policies are always lower tailed. Melumad and Mookherjee (1989) showed that delegation of audits to an independent party could serve as a commitment device.

Aggregation of Signals in the Design of Performance Measures

Performance measures are usually aggregates of many signals. When a salesman is paid a commission on monthly sales revenue, the sales to individual customers are aggregated. When a divisional manager is rewarded on the basis of his division's profit, many signals on various costs and revenues are aggregated into a single performance measure. When a production supervisor is evaluated on the basis of production costs, number of defective goods produced, and customer satisfaction measures, these different signals are aggregated into some overall performance measure. Banker and Datar (1989) used the basic principal-agent model to provide insights into how an optimal aggregation of signals is related to the relative information content of these signals.

Suppose there are n signals, $x_1,....,x_n$ available to the principal, with joint probability density function $f(x_1,....,x_n|\alpha)$. Holmstrom (1979) established that the optimal compensation contract offered to the agent $s(x_1,..,x_n)$ is characterized by:

$$\frac{1}{U'(s(x_1,...,x_n))} = \lambda + \mu \left(\frac{f_\alpha(x_1,...x_n|\alpha)}{f(x_1,...x_n|\alpha)} \right) \qquad (1.9)$$

Hereafter I use the notation $L(x_1,...x_n)$ to denote the log likelihood ratio f_α/f, and x to denote the vector of n signals $(x_1,...x_n)$. Since $L: \Re^n \to \Re$, $L(x)$ can be viewed as an aggregation of the n signals implied by the optimal compensation contract. In this sense, $L(.)$ is an aggregate performance measure, and the agent's contract can be written as $s(L(x))$. Holmstrom (1979) established that in the special case where a sufficient statistic $T(x_1,...x_n): \Re^n \to \Re$ exists, any aggregation other than $T(x)$ is suboptimal. In the absence of a single sufficient statistic any aggregation of the n signals results in a loss of information, in a statistical sense, but (1.9) indicates that, nevertheless, some aggregation is optimal because all of the information contained in the n signals is not used in the optimal contract.[3]

Banker and Datar (1989) posed the following questions: Under what conditions will $L(x)$ be a *linear* aggregate $\gamma_1 x_1 + \gamma_2 x_2 +...+\gamma_n x_n$ of the n signals, where the weights on individual signals are independent of x (but may depend on the agent's action α)? Second, if $L(x)$ is a linear aggregate, what factors determine the relative weights assigned to each signal? Banker and Datar showed that the weight on a signal is directly proportional to the product of its precision and sensitivity with respect to the agent's action, thus confirming the intuition that the relative information content of each signal is the decisive factor.

For the rest of this discussion I assume, without loss of generality, that $n = 2$ and label the two signals x and y. Consider the exponential family of densities:[4]

$$f(x, y|\alpha) = h(x, y)K(\alpha) \cdot exp[q_1(\alpha)\varphi_1(x, y) + q_2(\alpha)\varphi_2(x, y)]$$

In the above expression, $K(\alpha)$ is a scaling function that makes the density integrate to unity at each α. For this class of densities,

$$L(x, y|\alpha) = K'(\alpha)/K(\alpha) + q_1'(\alpha)\varphi_1(x, y) + q_2'(\alpha)\varphi_2(x, y)$$

It follows immediately that $L(.)$ is a linear aggregate of the signals x and y if φ_1 and φ_2 are linear in these signals. This establishes that performance measures in agencies are linear aggregates of the signals available to the principal if the joint density of signals has the form:

$$f(x, y|\alpha) = h(x, y)K(\alpha) \cdot exp[\gamma_1(\alpha)x + \gamma_2(\alpha)y], \quad (1.10)$$

in which case,

$$L(x, y|\alpha) = K'(\alpha)/K(\alpha) + \gamma_1'(\alpha)x + \gamma_2'(\alpha)y \quad (1.11)$$

Banker and Datar established that the family described in (1.10) is not only sufficient but also necessary if *some* linear aggregate is to be optimal for every specification of α and specification of the agent's utility function. Now we will show that $\gamma_1'(\alpha)$ and $\gamma_2'(\alpha)$ are related to the "informativeness" of x and y, in some statistical sense. Banker and Datar established this link for a subfamily of (1.10), described by:

$$f(x, y|\alpha) = exp[\gamma_1(\alpha)x + \gamma_2(\alpha)y + p(x) + t(y) - r(\alpha)] \quad (1.12)$$

For this family of densities, x and y are conditionally independent, implying $cov(x, y) = 0$, and $L(x, y|\alpha) = \gamma_1'(\alpha)x + \gamma_2'(\alpha)y - r'(\alpha)$.

Proposition 1. For the family of densities (1.12), the weights on signals x and y, in the optimal linear aggregate $L(x, y|\alpha)$ are:

$$\gamma_1'(\alpha) = \frac{\partial \{E(x|\alpha)\}/\partial \alpha}{var(x|\alpha)} \qquad (1.13)$$

$$\gamma_2'(\alpha) = \frac{\partial \{E(y|\alpha)\}/\partial \alpha}{var(y|\alpha)} \qquad (1.14)$$

Proof. The density described in (1.12) can be factored into the product of two marginal densities for x and y, i.e.

$$f(x, y|\alpha) = exp[\gamma_1(\alpha)x + p(x) - r_1(\alpha)] \cdot exp[\gamma_2(\alpha)y + t(y) - r_2(\alpha)]$$

where $r_1(\alpha) + r_2(\alpha) = r(\alpha)$ are scaling functions that make each marginal density integrate to unity. Let $f_X(x|\alpha)$ and $f_Y(y|\alpha)$ be the marginal densities of x and y, respectively. Since $\int \partial \{f_X(x|\alpha)\}/\partial \alpha \, dx = 0$,

$$\int \{\gamma_1'(\alpha)x - r_1'(\alpha)\} \cdot exp[\gamma_1(\alpha)x + p(x) - r_1(\alpha)] \, dx = 0,$$

which implies:

$$E(x|\alpha) \equiv \int x \cdot exp[\gamma_1(\alpha)x + p(x) - r_1(\alpha)] \, dx = r_1'(\alpha)/\gamma_1'(\alpha) \qquad (1.15)$$

$$\frac{\partial \{E(x|\alpha)\}}{\partial \alpha} = \frac{\gamma_1' r_1'' - r_1' \gamma_1''}{(\gamma_1')^2} \qquad (1.16)$$

Now, to calculate $var(x|\alpha)$, differentiate (1.15) with respect to α. This yields,

$$\int x \cdot exp[\gamma_1(\alpha)x + p(x) - r_1(\alpha)]\{\gamma_1'(\alpha)x - r_1'(\alpha)\}dx$$

$$= (\gamma_1' r_1'' - r_1' \gamma_1'')/(\gamma_1')^2,$$

which implies,

$$\gamma_1' E(x^2|\alpha) - r_1' E(x|\alpha) = (\gamma_1' r_1'' - r_1' \gamma_1'')/(\gamma_1')^2$$

Solving for $E(x^2|\alpha)$ and using (1.15) yields,

$$var(x \mid \alpha) = E(x^2 \mid \alpha) - [E(x \mid \alpha)]^2 = \frac{\gamma_1' r_1'' - r_1' \gamma_1''}{(\gamma_1')^3} \qquad (1.17)$$

Dividing (1.16) by (1.17) yields (1.13). The proof for (1.14) is similar.

Banker and Datar interpret $\partial\{E(x|\alpha)\}/\partial\alpha$ as the sensitivity of the signal to the agent's action, while $var(x|\alpha)$ measures the noise in the signal. Propostion 1 indicates that the weight assigned to a signal should be directly proportional to the signal's sensitivity and inversely proportional to its noise. Banker and Datar extend these results to a family of exponential densities that allows correlation among signals and show that similar results hold, except that the sensitivity of a signal is adjusted for covariance terms.

Banker and Datar use Proposition 1 to examine the desirability of rewarding managers on the basis of divisional income, and the desirability of aggregating overhead costs into cost pools for allocation purposes. Neither practice is warranted unless the signals that are aggregated into income and cost pools have equal ratios of sensitivity to variance. For example if the revenues in a division have a greater sensitivity to the managers' efforts than divisional costs, then a performance measure that weights revenues more heavily than costs would be more desirable than divisional income as a measure of the manager's performance.

2. Transfer Pricing and Budgeting

From an accounting perspective, it is useful to conceptualize a firm's technology as a network of activities. These activities are linked in the sense that the outputs of some activities form inputs to other activities, and collections of activities often share a common resource. These kinds of linkages among activities create a need for planning and coordination. Accounting practices such as cost allocations, transfer pricing and budgeting serve to facilitate the coordination of activities. Cost allocation schemes attempt to coordinate activities by assigning the costs of one activity to other activities in proportion to some measure of usage. The belief here is that if downstream activities are charged a proportionate share of the upstream costs they generate, then activity managers will make the right tradeoffs among costs and benefits. Transfer pricing practices carry this principle one step further, and use explicit optimization goals to set internal prices at which upstream activities supply downstream activities. Indeed, the main difference between transfer pricing and cost allocation is that cost allocation is based on *ex post* average observed costs, while transfer prices are based on *ex ante* calculations of marginal cost.

Budgets achieve coordination by assigning targets (cost targets, revenue targets, production targets, etc.) to activity managers. The budgeting exercise is an *ex ante* process in which there is extensive communication and negotiation between activity managers and a central manager. The goal of the central manager is to simultaneously coordinate the firm's activities, assign tasks to each activity manager, and motivate managers to execute their assigned tasks efficiently. Performance is evaluated by assessing deviations from budgets. Unfavorable deviations are penalized and favorable deviations are rewarded.

Coordination of activities is particularly difficult when the relevant information is dispersed among several activity managers, each having their own private costs and benefits. In such settings (typical of large firms), there is the additional complication of motivating truthful communication of private information. This kind of problem has been extensively studied in the accounting literature. Groves (1976) and Groves and Loeb (1979) first applied Groves' demand revealing mechanism to intrafirm coordination problems. However, managers' compensations and divisional incomes were assumed to be equivalent. Banker and Datar (1991) explicitly introduced managerial compensation in a Groves' mechanism under the assumption that only *ex ante* participation constraints need to be satisfied, implying that managers could commit to implementing budget plans even after realizing that doing so would make their compensations negative. Essentially, Groves' scheme assigns the entire profit of the firm to each divisional manager and then taxes away part of this profit by an amount that does not depend on that individual manager's message. Applying this idea, Banker and Datar found that each manager's compensation would equal the difference between the *ex ante* expected profit of the entire firm and the *ex post* realized profits of the firm, and that this compensation scheme would allow the firm to implement first best plans. The implication of this result is that rather than decompose its operations into decentralized organizational units with the aid of transfer prices, divisional budgets and divisional income measurements, the firm ought to motivate coordination among activities by making the rewards of every activity manager contingent only on the global profits of the firm.

Coordination and Budgeting

Kanodia (1993) examined coordination problems under the assumption that managers' participation constraints must be satisfied state by state, rather than in an *ex ante* sense. He found that Groves' type mechanisms are generally suboptimal and that indeed the firm is best off decomposing its operations so that the performance measure of one manager is unaffected by the performance of other managers. The optimal coordination mechanism, derived from first principles, was shown to be a budget based mechanism. The formulation below is based on the Kanodia (1993) model.

Consider two divisions (D1 and D2) in an upstream-downstream relationship within a firm, with each division having its own separate manager. The output of D1 (an intermediate good) is an input into D2's operations. The production costs incurred in D1 are stochastic with a distribution that is affected by its production quantity, its local operating environment, and by cost reducing (or efficiency enhancing) actions controlled by its manager. Specifically, suppose that $\tilde{C}$ is the random production cost with mean:

$$S(q, \alpha, \theta) = (\theta - \alpha)s(q), \quad s'(.) > 0, \, s''(.) > 0,$$

where θ is a parameter measuring the operating environment in D1, α measures the effect of cost reducing actions taken in the department, and q is D1's production quantity. The above specification assumes that $S_{\theta q} > 0$ and $S_{\alpha q} > 0$, i.e. the marginal effects of the operating environment as well as cost reducing actions are greater when output is larger. Also, larger values of θ represent unfavorable local conditions.

The intermediate good, when used in D2, produces benefits for the firm which depend stochastically on D2's local environment μ, and on value enhancing actions β controlled by D2's manager. Specifically, the expected benefits are:

$$R(q, \beta, \mu) = (\mu + \beta)r(q), \quad r'(.) > 0, \, r''(.) < 0$$

Cost reducing and value enhancing actions are privately costly to divisional managers. The net utilities of managers D1 and D2, respectively, are:

$$w_1 - \varphi_1(\alpha), \text{ and } w_2 - \varphi_2(\beta)$$

where w_i is manager i's wage and φ_i is increasing and strictly convex, $i = 1,2$. Participation constraints require each manager's net utility to be non-negative in every environment. I assume that the environments θ and μ are observed before the decision variables q, α, and β are chosen.

First-best production plans and allocations are described by the solution to:

$$\text{Max } (\mu + \beta)r(q) - (\theta - \alpha)s(q) - \varphi_1(\alpha) - \varphi_2(\beta), \quad (2.1)$$

which yields the first-order conditions:

$$(\mu + \beta)r'(q) = (\theta - \alpha)s'(q) \quad (2.2)$$

$$\varphi_1'(\alpha) = s(q) \quad (2.3)$$

$$\varphi_2'(\beta) = r(q) \quad (2.4)$$

Examination of these first-order conditions illustrates the nature of the coordination problem faced by the firm. Equation (2.2) indicates that the production quantity must adjust to the environment of both divisions, that is, $q = q(\theta, \mu)$ where q is strictly increasing in μ and strictly decreasing in θ. This fact, together with (2.3) and (2.4) indicates that α and β are also functions of the environment in both divisions. In fact, it can be established that $\alpha(\theta, \mu)$ and $\beta(\theta, \mu)$ are also strictly decreasing in θ and strictly increasing in μ. Thus, there is a need to coordinate not only the production quantity, but also the cost reducing and value enhancing efforts of both divisional managers. If the environment is favorable in either division, both divisions must work harder to reduce marginal costs and increase marginal revenues. However, it is clear from (2.3) and (2.4) that the dependence of α on μ and the dependence of β on θ arises *only through* the production quantity q. Given q, the optimal value of α minimizes the sum of production and personal costs in D1. Thus, if the production quantity is appropriately coordinated, and if D1 is held accountable for the costs realized in its local operations D1 will choose the appropriate value of α regardless of D2's choice of β. Similar observations hold for D2's choice of β. This suggests that appropriately chosen budgets could decentralize the firm and make each manager's rewards independent of performance in the other division.

Below, I show that a mechanism of this type is optimal in the presence of information asymmetry.

Suppose, now, that D1's environment θ is privately observed by D1's manager, and D2's environment μ is privately observed by D2's manager. Other parties view θ as a drawing from the distribution $F(\theta)$ with strictly positive density function $f(\theta)$ on the interval $[\underline{\theta}\ \overline{\theta}]$ and μ as a drawing from the distribution $G(\mu)$ with strictly positive density function $g(\mu)$ on the interval $[\underline{\mu}\ \overline{\mu}]$. The cost reducing and value enhancing actions taken in these divisions are also unobservable. Only the production quantity, total production costs in D1 and total revenues in D2 are publicly observed. To characterize the optimal mechanism for this setting, I use the methodology developed in Laffont and Tirole (1986). First consider a non-stochastic environment, where $C(q, \alpha, \theta)$ is the realized production cost in D1, and $R(q, \beta, \mu)$ is the realized revenue in D2. The optimal mechanism for this non-stochastic environment will then be extended to the setting where production costs and revenues are random. .

For this non-stochastic environment, the "per unit" quantities: $c = (\theta - \alpha)$, and $v = (\mu + \beta)$ can be calculated *ex post* from observation of production quantity, total cost, and total revenue. The Revelation Principle states that, without loss of generality, attention can be restricted to mechanisms that induce truth telling as a Bayesian Nash equilibrium. However, I use the result in Mookherjee and Reichelstein (1992, Proposition 6) that for a class of environments, which includes the setting under study, an optimal allocation can be implemented equivalently in dominant strategies. An optimal revelation mechanism, in dominant strategies, consists of three decision rules $c(\theta, \mu)$, $v(\theta, \mu)$, $q(\theta, \mu)$, and two wage schedules $w_i(\theta, \mu)$, $i = 1, 2$, that solve:

$$\text{Max} \int_{\underline{\theta}}^{\overline{\theta}} \int_{\underline{\mu}}^{\overline{\mu}} \{v(\theta,\mu)r(q(\theta,\mu)) - c(\theta,\mu)s(q(\theta,\mu)) - w_1(\theta,\mu) - w_2(\theta,\mu)\}\, dF(\theta)dG(\mu)$$

subject to: $w_1(\theta,\mu) - \varphi_1(\theta - c(\theta,\mu)) \geq w_1(\hat{\theta},\mu) - \varphi_1(\theta - c(\hat{\theta},\mu))$

$$\forall\, \theta, \hat{\theta}\, \varepsilon\, [\underline{\theta}\ \overline{\theta}],\ \forall \mu\, \varepsilon\, [\underline{\mu}\ \overline{\mu}] \tag{2.5}$$

$$w_2(\theta,\mu) - \varphi_2(v(\theta,\mu) - \mu) \geq w_2(\theta,\hat{\mu}) - \varphi_2(v(\theta,\hat{\mu}) - \mu)$$

$$\forall \mu, \hat{\mu} \in [\underline{\mu}, \overline{\mu}], \forall \theta \in [\underline{\theta}, \overline{\theta}] \quad (2.6)$$

$$w_1(\theta,\mu) - \varphi_1(\theta - c(\theta,\mu)) \geq 0, \quad \forall \theta, \forall \mu \quad (2.7)$$

$$w_2(\theta,\mu) - \varphi_2(v(\theta,\mu) - \mu) \geq 0, \quad \forall \theta, \forall \mu \quad (2.8)$$

Constraints (2.5) and (2.6) require truth-telling to be a dominant strategy for each divisional manager, and (2.7) and (2.8) are participation constraints that are required to be satisfied for every possible environment in *both* divisions. Thus no manager would want to withdraw his participation at *any* stage of the game, and no manager needs to be concerned about the messages communicated by the other manager. A mechanism with these properties seems more consistent with decentralization.

Using techniques that are now standard in the literature, it can be established:

<u>Proposition 2.</u> A mechanism $\{w_1(\theta,\mu), w_2(\theta,\mu), q(\theta,\mu), c(\theta,\mu), v(\theta,\mu)\}$ satisfies the truth telling and participation constraints in the sense of (2.5) through (2.8) if and only if:

$$w_1(\theta,\mu) = \varphi_1(\theta - c(\theta,\mu)) + \int_\theta^{\overline{\theta}} \varphi_1'(t - c(t,\mu))dt + \Gamma_1(\overline{\theta}, \mu) \quad (2.9)$$

$$w_2(\theta,\mu) = \varphi_2(v(\theta,\mu) - \mu) + \int_{\underline{\mu}}^{\mu} \varphi_2'(v(\theta,\tau) - \tau)d\tau + \Gamma_2(\theta, \underline{\mu}) \quad (2.10)$$

$c(\theta,\mu)$ is nondecreasing in θ for each given μ, \quad (2.11)

$v(\theta,\mu)$ is nondecreasing in μ for each given θ, \quad (2.12)

$$\Gamma_1(\overline{\theta}, \mu) \geq 0, \forall \mu \text{ and} \quad (2.13)$$

$$\Gamma_2(\theta,\underline{\mu}) \geq 0, \forall \theta \qquad (2.14)$$

In the above Proposition, Γ_1 and Γ_2 are the indirect utilities of the D1 and D2 managers, respectively. Specifically, $\Gamma_1(\theta,\mu) \equiv w_1(\theta,\mu) - \varphi_1(\theta - c(\theta,\mu))$, and $\Gamma_2(\theta,\mu) \equiv w_2(\theta,\mu) - \varphi_2(v(\theta,\mu) - \mu)$. The integral expressions in (2.9) and (2.10) constitute informational rents that must be paid to divisional managers to induce truth telling. D1 has a natural incentive to claim that he is operating in an unfavorable environment, i.e. report a high value of θ and benefit from shirking. Truth-telling incentives are provided by increasing D1's wage when low values of θ are reported. D2's wage schedule, characterized in (2.10), is similarly motivated. It can be shown that $\Gamma_1(\theta,\mu)$ is decreasing in θ, and $\Gamma_2(\theta,\mu)$ is increasing in μ. Hence, (2.13) and (2.14) guarantee that the participation constraints will be satisfied for all environments. The monotone requirements on $c(\theta,\mu)$ and $v(\theta,\mu)$ guarantee that local incentives for truth telling are sufficient for global incentives.

Notice that once the decision rules $q(\theta,\mu)$, $c(\theta,\mu)$ and $v(\theta,\mu)$ are specified, D1 and D2's wage schedules can be calculated from (2.9) and (2.10). The next proposition provides a characterization of the optimal decision rules.)

<u>Proposition 3</u>. Given that $\varphi_i'''(.) \geq 0$, $i = 1,2$, $F(\theta)/f(\theta)$ is non-decreasing, and $(1 - G(\mu))/g(\mu)$ is non-increasing, the optimal decision rules, $q^*(\theta,\mu)$, $c^*(\theta,\mu)$ and $v^*(\theta,\mu)$ maximize:

$$vr(q) - c\,s(q) - \varphi_1(\theta - c) - \varphi_1'(\theta - c)\frac{F(\theta)}{f(\theta)} - \varphi_2(v - \mu) - \varphi_2'(v - \mu)\frac{1 - G(\mu)}{g(\mu)}$$

Proposition 3 yields:

$$v^* r'(q^*) = c^* s'(q^*) \qquad (2.15)$$

$$\varphi_1'(\theta - c^*) + \varphi_1''(\theta - c^*)\frac{F(\theta)}{f(\theta)} = s(q^*) \qquad (2.16)$$

$$\varphi_2'(v^* - \mu) + \varphi_2''(v^* - \mu)\left[\frac{1 - G(\mu)}{g(\mu)}\right] = r(q^*) \quad (2.17)$$

Comparing these second-best decision rules to the first-best plan described in (2.2) through (2.4), it is clear that in all but the worst environments marginal costs are strictly bigger and marginal revenues are strictly smaller than first best. Thus, the quantity produced and transferred across divisions is strictly smaller than first best. This inefficiency arises due a need to control the informational rents of divisional managers. Kanodia (1993) shows that if first-best decisions are sought to be implemented, without any attempt to squeeze informational rents, the mechanism characterized above becomes identical to a Groves' mechanism. Groves-like mechanisms do not incorporate any notion of a surplus that accrues to the firm's owners and consequently there is no attempt to control managers' informational rents.

The mechanism characterized above is essentially a budget mechanism even though it doesn't look like one. Production plans $q^*(\theta,\mu)$, cost standards $c^*(\theta,\mu)$, and revenue standards $v^*(\theta,\mu)$ are formulated in consultation with divisional managers. The consultation takes the form of information sharing, but does not go beyond this. The head office commits *ex ante* to how the information that is revealed will and will not be used. In this sense, production quantities and standards are imposed from above rather than negotiated. In this non-stochastic environment, deviations from budget are entirely controllable by divisional managers, so no deviations are permitted by the head office.

I now return to the case of stochastic production costs and revenues, where deviations from budget could arise due to factors that are non-controllable by divisional managers. I will show that essentially the same plans as that derived for the non-stochastic case can be implemented by suitably incorporating *ex post* deviations from budget into managers' compensation schedules. The following cost and revenue budgets are assigned to the managers of D1 and D2, respectively.

$$C^*(\theta,\mu) = c^*(\theta,\mu)s(q^*(\theta,\mu)) \quad (2.18)$$

$$R^*(\theta,\mu) = v^*(\theta,\mu)r(q^*(\theta,\mu)) \quad (2.19)$$

The compensation schedule for D1 is specified as:

$$z_1(\theta,\mu,\tilde{C}) = w_1^*(\theta,\mu) + \gamma_1(\theta,\mu)[C^*(\theta,\mu) - \tilde{C}] \tag{2.20}$$

where,
$$\gamma_1(\theta,\mu) = \frac{\varphi_1'(\theta - c^*(\theta,\mu))}{s(q^*(\theta,\mu))} \tag{2.21}$$

The compensation schedule for D2 is:

$$z_2(\theta,\mu,\tilde{B}) = w_2^*(\theta,\mu) + \gamma_2(\theta,\mu)[\tilde{B} - B^*(\theta,\mu)] \tag{2.22}$$

where,
$$\gamma_2(\theta,\mu) = \frac{\varphi_2'(v^*(\theta,\mu) - \mu)}{r(q^*(\theta,\mu))} \tag{2.23}$$

In the above scheme, the starred schedules, w_1^*, w_2^*, q^*, c^* and v^* are the optimal mechanism for the non-stochastic setting characterized in (2.15) through (2.17). As in the non-stochastic setting, the production schedule $q^*(\theta,\mu)$ is enforced by the head office and no deviations are permitted, but the schedule that is enforced is chosen with divisional participation. A divisional manager will perceive w_i^* as a fixed wage, since it is unaffected by any of his actions (though it is affected by his message). It is clear from (2.16) and (2.17) that $0 < \gamma_i < 1$, so γ_i can be interpreted as a sharing parameter. The manager's compensation scheme is thus a very simple linear scheme, consisting of a fixed salary plus a bonus whose size is proportional to deviations from the assigned budget. However, the parameters of the compensation scheme vary with the messages communicated by divisional managers. It is this feature that induces truthful revelation of information. Deviations from a manager's budget could occur either because the manager has chosen to provide cost reducing or value enhancing actions different from that incorporated in the budget, or because of the uncontrollable randomness in divisional costs and revenues. The head office cannot distinguish between these two causes. This is why incentives for action are provided by penalizing the manager for unfavorable deviations and rewarding him for favorable deviations from budget. Notice that if the manager took the actions that are expected of him, the expected deviation from budget would be zero, so the manager's expected compensation would be the same as in the non-stochastic setting.

<u>Proposition 4.</u> Under the budget-based mechanism described in (2.18) through (2.23), for the stochastic environment, truthful communication is a

dominant strategy for each divisional manager. Conditional on truthful communication, D1's optimal cost reducing actions are $\alpha^*(\theta,\mu)$ and D2's optimal value enhancing actions are $\beta^*(\theta,\mu)$ as in the non-stochastic setting.

Proof. I prove the proposition for manager D1; the proof for D2 is analogous. Given θ, D1 chooses his message $\hat{\theta}$ and his action α to:

$$\text{Max } w_1^*(\hat{\theta}, \mu) + \gamma_1(\hat{\theta}, \mu)[C^*(\hat{\theta}, \mu) - E\{\tilde{C}|\theta,\alpha)\}] - \varphi_1(\alpha)$$

Inserting (2.18) and (2.21) and canceling common terms, the above is equivalent to:

$$\text{Max } w_1^*(\hat{\theta}, \mu) + \varphi_1'(\alpha^*(\hat{\theta}, \mu))[c^*(\hat{\theta}, \mu) - \theta + \alpha] - \varphi_1(\alpha) \qquad (2.24)$$

Conditional on reporting some $\hat{\theta}$, D1's optimal choice of α is given by the first-order condition,

$$\varphi_1'(\alpha) = \varphi_1'(\alpha^*(\hat{\theta}, \mu)),$$

implying that the manager chooses $\alpha = \alpha^*(\hat{\theta}, \mu)$. The sharing parameter γ_1 has been designed so that the manager always chooses an action that is consistent with his report, so truth telling is the only remaining incentive issue. Inserting $c^*(\hat{\theta}, \mu) \equiv \hat{\theta} - \alpha^*(\hat{\theta}, \mu)$ and $\alpha = \alpha^*(\hat{\theta}, \mu)$ into (2.24), the manager's report $\hat{\theta}$ must be the solution to:

$$\text{Max } w_1^*(\hat{\theta}, \mu) + \varphi_1'(\alpha^*(\hat{\theta}, \mu))[\hat{\theta} - \theta] - \varphi_1(\alpha^*(\hat{\theta}, \mu)).$$

When $\hat{\theta} = \theta$ the maximand collapses to $\Gamma_1^*(\theta,\mu)$. Thus, truth telling is a dominant strategy if,

$$\Gamma_1^*(\theta,\mu) \geq \Gamma_1^*(\hat{\theta}, \mu) + \varphi_1'(\alpha^*(\hat{\theta}, \mu))[\theta - \hat{\theta}] \quad \forall \theta, \hat{\theta}, \forall \mu.$$

For $\hat{\theta} > \theta$, the above inequality is equivalent to,

$$-\int_\theta^{\hat{\theta}} \frac{\partial}{\partial t}\Gamma_1^*(t,\mu)\,dt \geq \int_\theta^{\hat{\theta}} \varphi_1'(\alpha^*(\hat{\theta},\mu))\,dt$$

But from (2.9),

$$\frac{\partial}{\partial \theta}\Gamma_i^*(\theta,\mu) = \frac{\partial}{\partial \theta}\int_\theta^{\hat{\theta}} \varphi_1'(t - c^*(t,\mu))\,dt = -\varphi_1'(\alpha^*(\theta,\mu))$$

Therefore, truth telling is a dominant strategy if,

$$\int_\theta^{\hat{\theta}} \varphi_1'(\alpha^*(t,\mu))\,dt \geq \int_\theta^{\hat{\theta}} \varphi_1'(\alpha^*(\hat{\theta},\mu))\,dt, \forall \mu$$

This last inequality holds because $\alpha^*(\theta,\mu)$ is a decreasing function of θ for any fixed μ and $\varphi_1'(\alpha)$ is increasing. The analysis for $\hat{\theta} < \theta$ is similar.

Transfer Pricing

It is important to distinguish between transfer pricing mechanisms and budget mechanisms. Budgets are usually thought of as spending constraints, but, more generally, budgets are targets that may be defined in terms of costs, production quotas, defect rates, or revenue goals. The central authority, in a firm, is closely involved in the setting of budgets and local managers participate by sharing information and negotiating targets with the central authority. A transfer pricing mechanism is a more decentralized form of organization, that relies very strongly on the concept of "divisional income" and its maximization (see Solomons (1968)). Rather than specifying targets and decision rules, the central authority instructs divisional managers to do whatever they think is necessary to maximize their divisional income. Transfers across divisions are priced and units are exchanged as if the transfer was an arms length transaction.

Arrow (1959) and Hirshleifer (1956) first formulated the transfer pricing problem for environments with no uncertainties or information asymmetries. Kanodia (1979) extended the Arrow and Hirshleifer models to uncertain environments incorporating managerial risk aversion and risk sharing. However, except in the case where there is a competitive external market for the intermediate good, these early models arrive at the optimal transfer price in a manner that is inconsistent with decentralization. The central authority first calculates the optimal transfer quantity and then rigs the transfer price to induce this optimal quantity. Decentralization of decisions is a figment here since the central authority could simply mandate the optimal transfer quantity without introducing a transfer price.

In later research, information asymmetries were explicitly introduced to obtain insights into how transfer prices would emerge if the central authority was less informed than divisional managers. However, in the preceding analysis, I have shown that intrafirm coordination mechanisms derived via the Revelation Principle take the form of budget mechanisms. Transfer prices and divisional incomes do not emerge naturally from such formulations. Vaysman (1996) showed how coordination mechanisms can be framed as transfer price mechanisms. Continuing with the preceding setting, Vaysman's mechanism gives D2 the right to choose the production quantity q but D2 is required to compensate D1 by making a transfer payment $T(q,\theta)$ that is calculated as follows:

$$T(q,\theta) = \underset{c}{\text{Min}} \quad cs(q) + \varphi_1(\theta - c) + \varphi_1'(\theta - c)[F(\theta)/f(\theta)]$$

Vaysman interprets the above specification of $T(q,\theta)$ as a "standard cost" imposed on D1, contingent on D1's announcement of θ, and D2 is required to pay D1's standard cost by way of a transfer payment. If all of D1's costs were transferred to D2, D2 would obviously have the right incentives for choosing the production quantity to be transferred, but transfers at realized cost would eliminate all of D1's incentives to provide cost reducing effort. This is the rationale for transferring at standard cost, rather than at realized cost. Given the transfer payment schedule, divisional incomes, Π_1 for D1 and Π_2 for D2, are calculated in the usual way, i.e.

$$\Pi_1 = T(q,\theta) - \tilde{C},$$

$$\Pi_2 = \tilde{B} - T(q,\theta)$$

Managers' compensations are specified as linear functions of their divisional incomes, i.e. manager i is paid:

$$w_i^{**}(\theta,\mu) + \gamma_i^{**}(\theta,\mu)\Pi_i, \quad i = 1, 2$$

Vaysman shows that if the compensation parameters w_i^{**} and γ_i^{**} are chosen appropriately, his transfer price mechanism would achieve the same allocations as the budget mechanism described in the previous section. Unfortunately, in order to calculate these compensation parameters the head office would need to first calculate all of the variables $q^*(\theta,\mu)$, $c^*(\theta,\mu)$, $v^*(\theta,\mu)$, $w_i^*(\theta,\mu)$ of the optimal budget mechanism. This seems to be a backdoor way of constructing a transfer price mechanism with no clear advantages over a budget mechanism.

It seems that the use of revelation mechanisms for coordination purposes will inevitably result in a budget mechanism rather than a transfer pricing mechanism. This is because any revelation mechanism is a highly centralized mechanism with the center specifying detailed decision rules for each agent. Transfer pricing, on the other hand, presupposes a certain degree of autonomy and decentralization. Melumad, Mookherjee and Reichelstein (1992, 1995) showed how limits on communication result in a meaningful theory of decentralized responsibility centers. With limited communication between divisional managers and the central authority, delegation of decisions has the advantage of allowing decisions to be based on the richer information possessed by divisional managers, resulting in a flexibility gain. On the other hand, limited communication limits the ability of the central authority to manipulate the incentives of divisional managers. Vaysman (1996) exploited this idea to show that, with limited communication, there is a flexibility gain associated with cost-based transfer pricing arrangements that could more than offset the control loss. Thus transfer pricing mechanisms could be superior to budget mechanisms when communication is limited. This is a promising approach to a difficult problem. It seems that additional insights into bounded rationality could go a long way in furthering our understanding of transfer pricing practices.

3. Adverse Selection in Audit Pricing

The pricing of audit services is a complex phenomenon that cannot be characterized in terms of the usual demand and supply relationships that apply to generic goods. There is no such thing as a standard unit of audit that could be priced in a competitive market. Though the technology of audits has now become fairly homogeneous across auditors, the audit for a particular client still needs to custom fit the nature of the client's business, the organization of the client's operations, and the client's internal control and accounting systems. The demand side of audit services lacks price sensitivity because audits are mandatory for publicly traded firms, and every audit is required to meet or exceed generally accepted audit standards.

On the supply side, the inputs used to perform a given audit are not readily observed by client firms and audit fees contingent on outcomes are prohibited.

Given these institutional features of the audit industry, it is felt that the key variables that affect audit fees are auditor reputation, business risks of clients and auditor liability, audit operating costs, strategic price competition among auditors, and the relative bargaining power of clients vs. auditors. Adverse selection enters into the pricing of audit services in two ways. First, clients may have superior information about their business risks and therefore the legal liability risks associated with their audits. Second, an auditor who has audited a client repeatedly in the past is likely to be much better informed about the costs and risks of the audit than prospective auditors who compete for the client's business. The models surveyed here examine the effect of such adverse selection on audit pricing, and the related phenomena of auditor turnover, and low balling of initial audit engagements.

Take-It-Or-Leave-It Pricing

Kanodia and Mukherji (1994) formulated a model with the following features. There is a pool of auditors, with identical audit technologies, who compete for the audit business of a client firm. The operating cost of an audit is c per period. Additionally, there is a start up cost of K when an auditor performs a first time audit for the client, and a cost of S when a client switches auditors. The client expects to be in business for two periods, and its financial statements are required to be audited in each of the two periods. The client has all the bargaining power, i.e. he has the power to choose the pricing mechanism, but is limited to signing contracts only one period at a time. The audit cost c depends on the audit technology as well as on client characteristics, neither of which is common knowledge at the start of period one. Therefore, initially all parties view c as a random variable and all parties assess a common prior distribution $F(c)$ with strictly positive density $f(c)$ and support $[a, b]$. An auditor who performs the period-one audit learns the value of c at the end of the audit and is, therefore, informationally advantaged for the period-two audit. The client knows that an incumbent auditor is informationally advantaged and uses the competition in the audit market to limit the informational rents of the incumbent auditor.

The equilibrium is calculated by working backward from period two. Assume that all parties are risk neutral. A revelation mechanism for period two is a triple $\{x(c), p(c), q(c)\}$, where $x(c)$ is the probability of retaining the incumbent auditor, $p(c)$ and $q(c)$ are the audit prices that the client offers to the incumbent and competing auditors, respectively. Let R be the event of replacing the incumbent auditor. Then, if the incumbent is replaced under the rules of the mechanism, all parties update their beliefs in the following way:

$$f(c \mid R) = \frac{[1 - x(c)] f(c)}{\int_a^b [1 - x(t)] f(t) dt} \quad (3.1)$$

An optimal mechanism for period two is a solution to:

$$\text{Min} \int_a^b \{x(c)p(c) + [1 - x(c)][S + q(c)]\}f(c)dc$$

subject to:

$$x(c)[p(c) - c] \geq x(c')[p(c') - c], \ \forall \ c, c' \quad (3.2)$$

$$x(c)[p(c) - c] \geq 0, \ \forall \ c \quad (3.3)$$

$$\int_a^b [q(c) - c - K]f(c \mid R)dc \geq 0 \quad (3.4)$$

Inequalities (3.2) are a continuum of incentive constraints requiring that the mechanism induces truth-telling by the incumbent auditor. Constraints (3.3) and (3.4) are participation constraints for the incumbent and competing auditors conditional on what each auditor knows. Assume that the inverse hazard rate $H(c) \equiv F(c)/f(c)$ is strictly increasing. The optimal pricing mechanism that solves the programming problem is characterized in Proposition 5 below.

Proposition 5. There exists a unique $p^* \in [a, b]$ such that the optimal pricing mechanism for period two is described by:

$$x(c) = \begin{cases} 1 & \text{if } c \leq p^* \\ 0 & \text{otherwise} \end{cases} \quad (3.5)$$

$$p(c) = \begin{cases} p^* & \text{if } c \leq p^* \\ 0 & \text{otherwise} \end{cases} \quad (3.6)$$

$$q(c) = K + E(c|c \geq p^*) \quad (3.7)$$

$$p^* = H^{-1}(K+S) \quad (3.8)$$

If there is no solution to this last equation, then $p^* = b$.

Proposition 5 establishes that, given the informational advantage of the incumbent auditor, the client can do no better than make a take-it-or-leave-it price offer to the incumbent auditor. The latter accepts the offer if his audit cost is less than this price p^* and rejects it otherwise. If the offer is rejected the client goes to the market to hire a new auditor and the new audit price is the start up cost of K plus the expectation of audit operating cost conditional on the information released by rejection of the price offer made to the incumbent auditor.

If $p^* < b$, which will be assumed henceforth, then auditor turnover emerges as a natural consequence of the efficient pricing of audit services. It is the result of exploiting market competition to squeeze the rents of an informationally advantaged incumbent auditor. Since p^* is strictly increasing in $(K + S)$, the model yields the intuitive result that auditor turnover rates would be smaller for clients that have higher audit start up and switching costs.

The incumbent auditor earns a rent in period two whenever $c < p^*$. Thus, incumbency has *ex ante* value. In period one, Bertrand competition among auditors will result in low balling the initial audit engagement in anticipation of such value. Thus,

$$\text{Lowball} = \int_a^{p^*} (p^* - c) f(c) dc = \int_a^{p^*} H(c) f(c) dc \qquad (3.9)$$

Since p^* is strictly increasing in $(K+S)$, it follows that the magnitude of the lowball increases with the transaction costs of replacing incumbent auditors.

Kanodia and Mukherji extend their analysis to a three-period setting in order to obtain insights into the dynamics of audit pricing. Do auditors face take-it-or-leave-it price offers every period? Do auditor turnover rates grow or decay over time? How does the magnitude of the lowball change over time as auditors are replaced? Unfortunately, given the constraint that clients can write contracts only one period at a time, an optimal three-period mechanism cannot be characterized via the Revelation principle. In fact, Kanodia and Mukherji show that there is no sequentially rational pair of audit contracts that would induce an incumbent auditor to fully reveal the true audit cost in period two. The intuition underlying this result is as follows. If the incumbent auditor revealed his true cost in period two, the client would price the period three audit at cost thus eliminating all rents in period three. Given this equilibrium in period three, in order to induce full revelation in period two the client would have to price the period two audit so as make up for the lost rent in period three. This forces the rent for low cost declarations to become so large that high cost auditors do better by pretending that costs are low, making excessive profits in period two, then abandoning the client in period three. This take-the-money-and-run strategy makes it infeasible to satisfy the incentive constraints of a revelation mechanism (as in Laffont and Tirole (1988)).

Kanodia and Mukherji investigate the following mechanism. In period one an audit price p_1 is determined through Bertrand competition among auditors. In period two the client makes a take-it-or-leave-it price offer of p_2 to the incumbent auditor. If this offer is accepted, then all parties (the client and competing auditors) know that $c \leq p_2$. Following this, the client offers a price $p_3 \leq p_2$ in period three. If the incumbent rejects p_2 in period two, the client offers q_2 to competing auditors with knowledge that $c \geq p_2$, and then offers the new incumbent a price of q_3 in period three. If the period-three price offers are rejected, the client seeks yet another auditor conditional on the information released by replacement of the previous auditors. In each period, the client weighs the potential information released by acceptance or rejection of his price offer and the effect of this information on future audit prices and future low balling as well as the costs associated with auditor turnover.

Denote by A1 the auditor chosen in period one, A2 the period-two auditor if there is turnover in period two, and A3 the period three auditor if there is turnover in period three. At the start of period three, p_2 is a given parameter and either A1 or A2 is the incumbent auditor. If A1 is the incumbent auditor, then A1 must have accepted the offer of p_2 in period two thus revealing the information that $c \leq p_2$. The client now chooses a new price of $p_3 \leq p_2$ to minimize the expected audit cost of period three:

$$\text{Min } \frac{F(p_3)}{F(p_2)} p_3 + \left[1 - \frac{F(p_3)}{F(p_2)}\right] [K + S + E(c \mid p_3 < c \leq p_2)] \quad (3.10)$$

In the above, $F(p_3)/F(p_2)$ is the client's assessment of the probability that A1 will accept the lower price of p_3. If p_3 is rejected, then the information $c > p_3$ is revealed and the new audit price becomes $K + E(c \mid p_3 < c \leq p_2)$. Analysis of (3.10) yields:

<u>Lemma 1</u>. If $p_2 > H^{-1}(K+S)$, then $p_3 = H^{-1}(K+S)$. If $p_2 \leq H^{-1}(K+S)$, then $p_3 = p_2$.

The result here is similar to that of the last period in the two-period model, since the tradeoffs are similar, except that the distribution of c is truncated above at p_2.

Now suppose that A2 is the incumbent auditor, implying that A1 rejected the client's offer in period two thus revealing that $c > p_2$. The client now makes a price offer of $q_3 \geq p_2$ for the period-three audit, knowing that if the offer is rejected his audit cost will be $K + S + E(c \mid c \geq q_3)$. The optimal value of q_3 must be a solution to:

$$\text{Min } q_3[F(q_3) - F(p_2)] + \int_{q_3}^{b} cf(c)dc + (K + S)[1 - F(q_3)].$$

Thus, if $q_3 < b$, it must satisfy the first order condition,

$$F(q_3) - f(q_3)(K + S) = F(p_2). \quad (3.11)$$

Analysis of (3.11) yields:

<u>Lemma 2.</u> The optimal value of q_3 satisfies $q_3 > H^{-1}(K+S)$, $\forall\, p_2 > a$. In addition, if $q_3 < b$ then q_3 is a strictly increasing function of p_2.

The different results regarding p_3 and q_3 arise from the fact that p_3 is determined with knowledge that $c \leq p_2$ while q_3 reflects the knowledge that $c > p_2$. Essentially the client's pricing strategy reflects a search for the true audit cost with the cost of search arising from auditor turnover. If it is known that $c \leq p_2$ and p_2 is small enough, then the gains from further search are too small to offset the cost of auditor turnover; so the client optimally sets $p_3 = p_2$. However, when it is known that $c > p_2$ the distribution is truncated below forcing $q_3 > p^* = H^{-1}(K+S)$.

Now, consider the determination of audit prices in period two. If A1 has been dismissed in period two, competition in the audit market will determine a price that incorporates the information that $c > p_2$ and which reflects low balling in anticipation of the period-three rents to incumbency. The magnitude of the lowball is $\text{Prob}(c \leq q_3 \mid c \geq p_2)\, E(q_3 - c \mid p_2 \leq c \leq q_3)$:

$$\text{A2s lowball} = \frac{1}{1 - F(p_2)} \int_{p_2}^{q_3} (q_3 - c) f(c)\, dc \qquad (3.12)$$

In equilibrium, A2 bids his expected cost less his low ball. Thus,

$$q_2 = K + \frac{1}{1 - F(p_2)} \left[\int_{p_2}^{b} c\, f(c)\, dc - \int_{p_2}^{q_3} (q_3 - c) f(c)\, dc \right] \qquad (3.13)$$

The above analysis implies that p_3, q_3 and q_2 are all functions of p_2 and that q_3 does not depend on q_2. This is because A1's response to p_2 has information content while q_2, being the price offered to a non-incumbent, has no information content. The optimal value of p_2 is determined by minimizing the client's expected cost of audits over both periods two and three, taking into account the effect that p_2 has on all subsequent audit prices. Thus p_2 must be a solution to:

$$\operatorname*{Min}_{p_2} F(p_2) [p_2 + p_3 \operatorname{Prob}(c \leq p_3 | c \leq p_2) +$$

$$\{K+S+E(c | p_3 \leq c \leq p_2)\} \operatorname{Prob}(c \geq p_3 | c \leq p_2)]$$

$$+ [1 - F(p_2)] [S + q_2 + q_3 \operatorname{Prob}(c \leq q_3 | c \geq p_2) +$$

$$\{K+S+E(c | c \geq q_3)\} \operatorname{Prob}(c \geq q_3 | c \geq p_2)]$$

In the above expression the term multiplying $F(p_2)$ is the client's expected cost over two periods if auditor A1 accepts the client's period 2 offer, and the term multiplying $[1-F(p_2)]$ is his expected cost if A1 rejects this offer and is replaced by A2. Inserting probability calculations, and inserting the equilibrium value of q_2 derived in (3.13) yields the equivalent program:

$$\operatorname*{Min}_{p_2} p_2 F(p_2) + p_3 F(p_3) + [2 - F(p_3) - F(q_3)] [K+S] + \int_{p_3}^{p_2} cf(c)dc + 2 \int_{p_2}^{b} cf(c)dc$$

The first-order condition characterizing the optimal value of p_2 is,

$$F(p_2) + F(p_3)\frac{dp_3}{dp_2} - [K+S] [f(p_3)\frac{dp_3}{dp_2} + f(q_3)\frac{dq_3}{dp_2}] = 0 \quad (3.14)$$

It is instructive to compare the optimal second-period price in this three-period dynamic model to the optimal second-period price in the two-period static model. The benefit to decreasing p_2 below p^* is that if this lower price is accepted by the incumbent auditor the client obtains the lower price in each of two periods rather than a single period. However, if q_3 is sensitive to p_2 (which is the case if $q_3 < b$), then a decrease in p_2 induces a decrease in q_3, which in turn implies that the probability of auditor turnover is increased in *both* periods two and three. The net effect on p_2 depends on how sensitive the sequentially rational choice of q_3 is with respect to p_2. If q_3 is not very sensitive to p_2 then the benefits to decreasing p_2 more than offset the costs, so $p_2 < H^{-1}(K+S)$. Conversely, when q_3 is sufficiently sensitive to changes in p_2, then optimally $p_2 > H^{-1}(K+S)$. In general, the value of q_3 and the sensitivity of q_3 with respect to p_2 depends in some complex way on the size of $(K+S)$ and the shape and support of the density function. This implies that in a dynamic setting, the evolution of equilibrium audit prices,

Game Theory Models in Accounting 81

low balling and auditor turnover will be distribution specific. The following parametric example provides some insights.

Assume that audit cost is uniformly distributed over the interval $[a, b]$. For this distribution the hazard rate is $H(c) = c-a$, and $H^{-1}(K+S) = a + K + S$. This linear structure permits precise calculations yielding:

<u>Proposition 6</u>. If audit cost, c, is uniformly distributed over the interval $[a,b]$ then equilibrium audit prices are:

(i) $p_3 = p_2 = a+K+S$, and $q_3 = a+2(K+S)$, if $(K+S)/(b-a) < 1/2$

(ii) $p_3 = p_2 = b-(K+S)$, and $q_3 = b$, if $1/2 \leq (K+S)/(b-a) < 2/3$

(iii) $p_3 = p_2 = a+(1/2)(K+S)$, and $q_3 = b$, if $(K+S)/(b-a) \geq 2/3$

Proposition 6 shows that, in the case of the uniform distribution, the interaction between transaction costs and the uncertainty in audit costs is conveniently summarized in the ratio $(K+S)/(b-a)$. It is the *relative* size of transaction costs to the range of possible audit costs that determines the evolution of audit prices, auditor turnover and low balling. The intuition underlying the results in Proposition 6 is as follows. For the uniform distribution, when $q_3 < b$, $dq_3/dp_2 = 1$, implying that the benefit to reducing period 2's price is *exactly* offset by the cost of an increase in auditor turnover in period 3. Consequently, the period-two audit price is unaffected by dynamic considerations, and $p_2 = H^{-1}(K+S)$, as in the static case. This is result (i) in Proposition 6. However, when transaction costs are sufficiently large relative to the range of possible audit costs, $[K+S \geq (2/3)(b-a)]$, it is too costly for the client to risk additional auditor turnover in period 3. The sequentially rational choice in period 3 is to price the audit at the largest possible audit cost, i.e. $q_3 = b$. In this case, variations in p_2 have no effect on q_3 (the price offered the second incumbent auditor in period 3) and the period 2 price is lowered to squeeze the first incumbent auditor over each of two periods, yielding (iii) of Proposition 6. In the intermediate case (item (ii) of Proposition 6), if p_2 is chosen equal to $H^{-1}[(1/2)(K+S)]$, the sequentially rational choice of q_3 is less than b. The client now has an incentive to increase both second and third period prices so as to save on the transaction costs of auditor turnover. Therefore, the second-period price is increased just enough to make $q_3 = b$ sequentially rational.

Proposition 6 shows that, when audit costs are uniformly distributed, $p_3 = p_2$ regardless of the relative size of transaction costs. This suggests that the first incumbent auditor would face a take-it-or-leave-it price offer at most once during his or her tenure. Once this offer is accepted, there is no further auditor turnover arising from pricing considerations unless, of course, there is a shift in the distribution of audit costs (which could arise due to changes in the client's characteristics). If the first auditor is replaced, the succeeding auditor faces a take-it-or-leave-it offer in period 3 only if transaction costs are relatively small [$K+S < (1/2)(b-a)$]. In this case, turnover could occur in both periods 2 and 3, but the probability of turnover in period three is smaller than in period two. In all other cases, the succeeding auditor does not face a take-it-or-leave-it offer and there is no further auditor turnover. This suggests that auditor turnover would decay over time.

Low balling occurs each time a new auditor is hired. In the case of the uniform distribution, the precise magnitude of low balling in periods 1 and 2 can be calculated and compared. Since $p_3 = p_2$, A1's low ball in period one is $2\text{Prob}(c \leq p_2) \text{E}[p_2 - c \mid c \leq p_2]$, while auditor A2's low ball in period two is $\text{Prob}(c \leq q_3 \mid c \leq p_2) \text{E}[q_3 - c \mid p_2 \leq c \leq q_3]$. Using the equilibrium prices in Proposition 6 to compute these values, we find that if $K+S < (1/2)(b-a)$, A1's lowball exceeds A2's lowball. If $K+S \geq (1/2)(b-a)$, A2's lowball exceeds A1's lowball. In this three-period model, A1 has the opportunity to earn rents over two periods while A2 has the opportunity to earn rents over only one period. In spite of this, A2's lowball is larger than A1's lowball when transaction costs are sufficiently large. It would appear that if the model were extended to dispense with an arbitrary last period, A2's lowball would always be larger. Thus the model predicts that when an incumbent auditor is replaced, the magnitude of low balling increases.

Pricing of Audit Risk

Morgan and Stocken (1998) examine the effect of audit risk (i.e. the risk of litigation following an audit report) on audit pricing, auditor turnover and lowballing in a setting where an incumbent auditor acquires information superior to competing auditors about the client's audit risk. The informed incumbent and uninformed competing auditors simultaneously submit sealed bids for the audit. Unlike Kanodia and Mukherji (1994), the client simply accepts the lowest bid submitted without explicitly seeking to design a pricing mechanism that would squeeze the rents of the informed incumbent.

Assume the client can be one of two types, type H (high risk) or type L (low risk). Type H clients have a probability of litigation of $\pi > 0$, which is assumed to be independent of the audit report submitted. Type L clients have zero probability of litigation. The client's financial statements need to be audited over two periods. Initially, all auditors assess the prior probability λ, $0 < \lambda < 1$, that the client is type H. If litigation occurs at the end of the first audit, the appointed auditor incurs an expected liability of B and the game ends. If litigation does not occur, then the game moves into the second period where another audit is required. If litigation occurs at the end of the second audit, again the expected liability of the assigned auditor is B. It is assumed, without loss of generality, that audit operating costs and start up costs are zero and the only cost is legal damages. Thus the net profit of an auditor who wins the audit is his audit fee less legal damages, if any.

In the first period all auditors bid, with symmetric information, for the audit and the lowest bid is accepted. The incumbent auditor has probability α of learning the client's type during the performance of the first-period audit, where α is common knowledge.[5] In the second period, the incumbent and competing auditors bid for the audit knowing that the incumbent auditor has probability α of being informed about the client's true type. A perfect Bayesian equilibrium is derived by working backward from period two.

Clearly, the period-two equilibrium in the bidding game must involve mixed strategies. The non-incumbent auditor is faced with a "lemons" problem since his bid cannot be contingent on the client's type, while the bid of the incumbent auditor could be so contingent. If he uses a pure strategy of bidding at or above the expected audit costs of unknown types but less than the expected audit cost of a type H client, he would be undercut by the incumbent auditor when the client is either of type L or of unknown type, and the only audits he would win are those of type H clients. Thus a pure strategy would guarantee losses for non-incumbent auditors. He must randomize his bid to protect himself from this adverse selection. To characterize the equilibrium in mixed strategies, let $F_L(R)$ and $F_H(R)$ be the distributions from which an informed incumbent draws his bid for type L and type H clients, respectively. Let $F_U(R)$ be the incumbent's bid distribution when he is uninformed and let $F_P(R)$ be the bid distribution of a competing non-incumbent auditor. Now let $\lambda' = [(1 - \pi)\lambda]/[1 - \pi\lambda]$ be the posterior probability that the client is of type H given the absence of litigation in period one. Note that, given the assumption that audit operating costs are zero, the expected cost of auditing a type L client is zero, the expected cost of auditing a type H client is πB, and the expected cost of auditing a client of unknown type is $\lambda'\pi B$. Morgan and Stocken establish that there is a unique Perfect Bayesian equilibrium in the second period auction described by:

Proposition 7. (i) A competing non-incumbent auditor randomizes his bid over the interval $[\lambda'\pi B, \pi B]$, with:

$$F_P(R) = \begin{cases} 1 - \dfrac{\lambda'\pi\beta}{R}, & \lambda'\pi\beta \leq R \leq \dfrac{\lambda'\pi\beta}{1-\alpha+\alpha\lambda'} \\ 1 - \dfrac{\lambda'\pi\beta\alpha(1-\lambda')}{R - \lambda'\pi\beta}, & \dfrac{\lambda'\pi\beta}{1-\alpha+\alpha\lambda'} \leq R < \pi\beta \\ 1, & R \geq \pi\beta \end{cases}$$

(ii) An informed incumbent auditor, who has learned that the client is of type L, randomizes his bid over the interval $[\lambda'\pi B, (\lambda'\pi B)/(1-\alpha+\alpha\lambda')]$, with:

$$F_L(R) = \dfrac{R - \lambda'\pi B}{\alpha(1-\lambda')R}$$

(iii) An informed incumbent auditor, who has learned that the client is of type H bids $R = \pi B$ with probability 1.

(iv) An uninformed incumbent auditor randomizes his bid over the interval $[(\lambda'\pi B)/(1-\alpha+\alpha\lambda'), \pi B]$, with

$$F_U(R) = \dfrac{R(1-\alpha+\alpha\lambda') - \lambda'\pi B}{(1-\alpha)(R - \lambda'\pi B)}$$

Auditors' expected profits can be calculated from the equilibrium bidding strategies in Proposition 7. It is straightforward to verify that the expected profit of a competing non-incumbent auditor is zero, and the expected profit of an uninformed incumbent auditor is $\lambda'\pi B\alpha(1-\lambda')$. The expected profit of an informed incumbent auditor is $\lambda'\pi B$ on type L clients and zero on type H clients. Since incumbency has value it follows immediately that the period one audit contract will be low balled even in the absence of audit start up and auditor switching costs.

Morgan and Stocken derive two additional results that seem consistent with empirical observations. First, it is clear from the equilibrium bidding strategies characterized in Proposition 7 that the non-incumbent auditor has a much higher probability of making the winning bid when the client is of type H than when the client is of type L. Therefore, the model predicts that auditor turnover will be higher for high-risk clients than for low-risk clients. Second, the expected audit price for type H clients is strictly less than the expected audit cost of πB, while the expected audit price of type L clients is strictly greater than $\lambda'\pi B$ even though the expected audit cost is zero. Therefore, the expected litigation costs of high-risk firms are subsidized by low-risk firms. On average, auditors make losses on high-risk audits and make this up from excess profits on low-risk audits.

Auditor Resignations and Audit Pricing

Bockus and Gigler (1998) develop an interesting model that explains why informationally advantaged incumbent auditors resign the audit engagements of high-risk clients rather than price adjust their bids to cover the higher litigation costs associated with such clients, and why a successor auditor would accept the rejected client. They show that any attempt by an incumbent auditor to price out the higher litigation cost would precipitate adverse selection, with less risky clients rejecting the incumbent's bid and only high-risk clients retaining the incumbent.

Suppose there are two types of clients. High-risk clients (type H) represent a potentially significant litigation risk to the auditor, while low-risk clients (type L) have no risk at all. Clients know their type, but auditors have access only to noisy signals that are used to assess probabilities on client type. An auditor can avoid the litigation risk of a type H client by detecting the hidden irregularities in the client's financial statements. If such irregularities are detected, the client suffers a loss of P, while if irregularities go undetected the auditor suffers a litigation cost of L. By deciding how much resources to put into his audit, an auditor chooses the probability of detection $\alpha \in [0, 1]$ at a cost $c(\alpha)$. Assume $c(\alpha)$ is increasing and strictly convex with $c'(0) = 0$ and $c'(1) = \infty$, thus assuring interior solutions for α. The decision on α is made at the time of the actual audit engagement. Thus the choice of α is required to be sequentially rational; no pre-commitments are possible.

Auditors differ in their ability to absorb the liability associated with litigation. Auditors with wealth $W < L$ are "wealth constrained," and have a maximum liability of W. Auditors with wealth $W > L$ are "solvent" and incur the full liability of L. Wealth constrained auditors are viewed as supplying lower quality audits, so that clients hiring wealth constrained auditors incur an exogenous opportunity cost of $B > 0$. Audits are required in each of two periods. Each period, auditors compete for the client's engagement by submitting bids F_i. In the first period all auditors hold the common belief φ^o that the client is of type H. After doing the first-period audit, the incumbent auditor acquires additional private information $\tilde{y}$ about the client's type and holds the posterior belief $\varphi(\tilde{y})$ that the client is of type H. Now, assume that the distribution of $\tilde{y}$ conditional on L and H is such that $0 < \varphi < 1$, $\forall\, y$, so that the only way to know the client's type for certain is to induce clients to self-select. Bids in the second period are assumed to be made sequentially; potential successor auditors bid first and the incumbent auditor bids after observing his rival's bids. This assumption avoids the mixed strategy equilibria found in Morgan and Stocken (1998) and gives the incumbent the ability to retain the client with probability one if he so chooses. Let $F_I(y)$ be the bid of the incumbent auditor and let F_S be the bid of a potential successor. Let $\alpha_I(y)$ and α_S be the irregularity detection probabilities that the incumbent and successor auditors choose.

Bockus and Gigler establish that, in equilibrium, the incumbent auditor (i.e. the auditor chosen in the first period) must be a "solvent" auditor and, if there is auditor turnover, the successor auditor must be "wealth constrained." To simplify the exposition, the model as presented below takes this result as a given fact. Also, for simplicity, I assume that the client knows the value of $\tilde{y}$ observed by the incumbent auditor. The client's strategy is straightforward. A type H client anticipates the probabilities of detection and chooses the minimum of $\{F_I + \alpha_I P,\ F_S + B + \alpha_S P\}$. A type L client chooses the minimum of $\{F_I,\ F_S + B\}$. Note that a type L client is not concerned about the probability of detection since there is nothing to detect. Assume that ties are broken by assigning the audit to the incumbent auditor. Lemma 3 below establishes that, given the client's strategy and given that auditors cannot pre-commit to a value of α, it is infeasible for the incumbent auditor to bid in such a way as to induce self-selection by client types.

Lemma 3. It is infeasible for the incumbent auditor to retain one client type and not the other. Either both client types retain the incumbent auditor or both client types switch auditors.

Proof. Suppose that the incumbent bids to retain only type L clients. Then sequential rationality dictates he must choose $\alpha_I = 0$. Therefore his bid F_I must satisfy:

(i) $F_I \leq F_S + B$, and

(ii) $F_I > F_S + B + \alpha_S P$

The successor auditor must choose $\alpha_S > 0$, since he realizes that only a type H client would hire him. Given $\alpha_I = 0$ and $\alpha_S > 0$, there is no F_I that satisfies (i) and (ii) simultaneously.

Suppose the incumbent bids to retain only type H clients. He must choose $\alpha_I > 0$ and successor auditors must choose $\alpha_S = 0$. Therefore, F_I must satisfy:

(iii) $F_I > F_S + B$, and

(iv) $F_I + \alpha_I P \leq F_S + B$

Once again, both inequalities cannot be satisfied simultaneously. This completes the proof.

Given that the incumbent auditor must either retain or resign both client types, his incentives depend on his assessed posterior probability $\varphi(y)$ that the client is of type H. Let $f_L(y)$ and $f_H(y)$ be the probability density functions of $\tilde{y}$ conditional on a type L and type H client, respectively. Assume that the support of $\tilde{y}$, $[\underline{y}, \overline{y}]$, is independent of the client type and assume the monotone likelihood ratio property: $f_L(y)/f_H(y)$ is strictly decreasing. These assumptions ensure that $\varphi(y)$ is strictly increasing and that $0 < \varphi(y) < 1$, $\forall\, y$. Bockus and Gigler show that there is a critical value y^* such that for $y \leq y^*$ the incumbent auditor bids to retain both client types, and for $y > y^*$ the incumbent auditor resigns the audit engagement. The intuition for this result is the following. Each of the two auditors has some advantage over the other. The incumbent auditor has an informational advantage over the successor, since he privately observes the value of y. Thus, while the incumbent auditor assesses $\varphi_I(y) = \text{Prob}(H \mid \tilde{y} = y)$, the potential successor, anticipating that he wins the audit only when $y > y^*$, must assess $\varphi_S = \text{Prob}(H \mid \tilde{y} > y^*)$. On the other hand, given that the successor auditor is wealth constrained with $W < L$, the successor enjoys the advantage of a smaller litigation cost of W while the incumbent's litigation cost is L. For low values of y (i.e. $y \leq y^*$) the informational advantage dominates the litigation cost advantage since the probability of litigation, as perceived by the incumbent, is low. However, the informational advantage shrinks as y becomes larger and the litigation cost advantage increases in importance because the probability of type H and therefore the probability of litigation becomes larger.

To establish an equilibrium, let:

$$g_I(y) \equiv \min_\alpha\ (1-\alpha)\varphi_I(y)L + c(\alpha),$$

and

$$g_S(y, W) \equiv \min_\alpha\ (1-\alpha)\varphi_S(y)W + c(\alpha),$$

where

$$\varphi_S(y) = Prob(H \mid \tilde{y} \geq y).$$

Let $\alpha_I(y)$ and $\alpha_S(y,W)$ attain g_I and g_S, respectively. These detection probabilities α_I and α_S are the cost-minimizing actions, and g_I and g_S are the minimized audit costs of the incumbent and successor auditors, respectively, given their beliefs. Given that φ_I and φ_S are strictly increasing, and given that $c(\alpha)$ is strictly convex, it is apparent that $\alpha_I(y)$ and $g_I(y)$ are both strictly increasing. Also, $\alpha_S(y, W)$ and $g_S(y, W)$ are strictly increasing in both arguments.

An equilibrium of the type described by Bockus and Gigler exists when there exists some $y^* \ \varepsilon\ (\underline{y}, \overline{y})$ satisfying:

$$g_I(y^*) = g_S(y^*, W) + B, \text{ and} \qquad (3.15)$$

$$\alpha_I(y^*) \leq \alpha_S(y^*, W) \qquad (3.16)$$

To see how (3.15) and (3.16) could be satisfied, notice that $\varphi_I(y) < \varphi_S(y)$, $\forall\ y < \overline{y}$ and therefore $g_I(y) < g_S(y, L)$ and $\alpha_I(y) < \alpha_S(y, L)$, $\forall\ y < \overline{y}$. But, if the litigation cost of the successor auditor is replaced by $W < L$, then g_S and α_S fall, i.e. the auditor puts less resources into the audit and his minimized audit cost falls. Provided that W is not too much smaller than L, it will be true that for sufficiently small values of y, $\alpha_I < \alpha_S$ and $g_I > g_S$. Therefore, the simultaneous satisfaction of (3.15) and (3.16) requires that W is not too small and B is not too big relative to L.

Proposition 8. Let y^* satisfy (3.15) and (3.16). Then, in equilibrium, the wealth constrained successor auditor bids $F_S^* = g_S(y^*, W)$ and takes action $\alpha_S(y^*, W)$ whenever he is awarded the audit engagement. The incumbent auditor bids $F_I^*(y) = F_S^* + B$, retains both client types, and takes action $\alpha_I(y)$, $\forall y \leq y^*$. The incumbent auditor resigns the audit when $y > y^*$.

Proof.[6] Competition among successor auditors reduces their expected profit to zero. Therefore, under the belief that a successor auditor wins the audit engagement only when $y > y^*$, his equilibrium bid must be $F_S^* = g_S(y^*, W)$. Now, for the incumbent auditor to retain both client types when $y \leq y^*$ his bid F_I must satisfy:

$$F_I(y) \leq F_S^* + B, \text{ and} \qquad (3.17)$$

$$F_I(y) + \alpha_I(y)P \leq F_S^* + B + \alpha_S(y^*, W)P \qquad (3.18)$$

The claimed equilibrium bid for the incumbent auditor satisfies (3.17) with equality. Since $\alpha_I(y)$ is strictly smaller for smaller y, (3.16) implies that (3.18) is satisfied as an inequality at each $y \leq y^*$. Therefore, given auditors' bids, both client types retain the incumbent auditor at each $y \leq y^*$. At y^* the expected profit of the incumbent auditor is: $F_S^* + B - g_I(y^*) = 0$, given that y^* satisfies (3.15). Now, consider $y > y^*$. Since $g_I(y)$ is strictly increasing, the incumbent auditor must raise his bid above $F_S^* + B$ if the audit is to be profitable. But when he does so, (3.17) is violated implying that a type L client will reject the incumbent's bid. Since at y^*, (3.18) could be a strict inequality, the only clients that may accept the higher bid are type H clients. In the absence of the subsidy provided by type L clients, the incumbent auditor would make a loss on the audit if his bid were accepted. This completes the proof.

Bockus and Gigler point out that their model also predicts some of the anecdotal claims made by practicing auditors. An increase in auditor liability (i.e. an increase in L) would result in more frequent auditor resignations. The probability of fraud detection would increase for clients retained by an incumbent auditor but would decrease for the clients for which the auditor resigns. Thus, contrary to popular belief, the overall probability of fraud detection might actually decline as auditor liability increases.

4. Concluding Remarks

It would be vain to pretend that the game-theoretic approach to accounting has yielded definitive answers that can be implemented in a practical sense. Though important insights have been obtained, much additional work needs to be done before the literature can be translated into policy prescriptions. For example, consider the implications of the principal-agent framework for the design of contracts and performance measures. The theory indicates that contracts should be agent specific and fine tuned to the agent's risk and work aversion. There are at least two problems with implementing this prescription. To illustrate the first, suppose a retailing firm hires 20 salespersons, with apparently similar job qualifications, to sell similar products in similar environments. Theoretically, the firm should offer different contracts to each salesperson even though they are all engaged in essentially the same job. Political and administrative realities make this recommendation difficult to implement. A second, and perhaps more serious, limitation of the theory is the assumption that the principal knows each agent's preferences perfectly and that the contract can be changed every time the agent's preferences change. Realistically, an individual's preferences are so private and so volatile that even long association with that individual does not fully reveal those preferences. These are serious limitations; yet the theory does provide valuable insights into practical issues that no firm can escape. The theory establishes that the cost of relying on noisy and imprecise performance measures arises from the imposition of risk on the agent, not from a lack of control. The theory provides the insight that alternative performance measures should be evaluated in terms of statistical measures of informativeness (such as sensitivity and precision). But the theory would be enriched by incorporating notions of "robustness" and contract simplicity. Contracts that are easy to calculate, whose implications are transparent to agents, and which work reasonably well over a wide range of agent preferences are more easy to implement.

On the other hand, the literature concerning incentive problems that arise from hidden information (or hidden types) is more immediately implementable. The insights underlying the construction of a "menu" of contracts that achieve control through self-selection is a major new development with exciting possibilities. The enormous success of such menu arrangements has been amply demonstrated in insurance markets, in airline pricing, and in home mortgage markets. There is no reason why similar success cannot be achieved in the internal arrangements of a firm.

For example, costly industrial engineering studies and negotiations for setting standards can be replaced by self-selected standards, targets and budgets. The theory indicates that the efficient way to control the manager's choice of standards is to vary the manager's compensation parameters with the standard he self-selects. There is some anecdotal evidence reported by management consultants that firms are moving in this direction by attaching a "difficulty factor" to goals that are self-selected by subordinates.

The accounting literature seems over-committed to the assumptions of complete contracting, unlimited and costless communication, and costless decision making. Relaxing these assumptions would go a long way to providing implementable prescriptions. Much of the important information in a firm is too soft to be verifiable and contractible, making complete contracting a dubious assumption. Without complete contracts, real world issues like the delegation of decision rights, contract renegotiation, reputation and corporate culture become important. The costs associated with decision making have been completely ignored in the literature. A policy that specifies a decision for every hypothetical situation has beneficial incentive effects, but is much more costly to calculate than an optimal decision for a specific situation that has materialized. Real world managers will refuse to formalize an infinite list of every contingency that could conceivably arise and every message that could be received and *ex ante* calculate and commit to how they would respond in each situation. These bounded rationality considerations are difficult to formalize but are essential to understanding the internal management of an organization.

References

Amershi, A. and J. Hughes (1989), "Multiple Signals, Statistical Sufficiency, and Pareto Orderings of Best Agency Contracts," *The Rand Journal of Economics*, 20, 102-112.

Antle, R. and A. Smith (1986), "An Empirical Investigation of the Relative Performance Evaluation of Corporate Executives," *Journal of Accounting Research*, 24, 1-39.

Arrow, K. (1959), "Optimization, Decentralization and Internal Pricing in Business Firms," in *Contributions to Scientific Research in Management*, University of California Press, Berkeley, California.

Arya, A., and J. Glover (1996), "The Role of Budgeting in Eliminating Tacit Collusion" *Review of Accounting Studies*, 1, 191-206.

Baiman, S. (1982), "Agency Research in Managerial Accounting," *Journal of Accounting Literature: A Survey* (Spring), 154-210.

Baiman, S. (1990), "Agency Research in Managerial Accounting: A Second Look," *Accounting Organizations and Society*, 15, 341-371.

Baiman, S. and J. Demski. (1980), "Economically Optimal Performance Evaluation and Control Systems," *Journal of Accounting Research* (supplement), 184-220.

Baiman, S. and J.H. Evans (1983), "Pre-Decision Information and Participative Management Control Systems," *Journal of Accounting Research*, 21, 371-395.

Baiman, S. and M. Rajan (1995), "The Informational Advantages of Discretionary Bonus Schemes," *The Accounting Review*, 70, 557-580.

Baiman, S., J.H. Evans, and N.J. Nagarajan (1991), "Collusion in Auditing,," *Journal of Accounting Research* 29, 1-18.

Banker, R. and S. Datar. (1989), "Sensitivity, Precision, and Linear Aggregation of Signals for Performance Evaluation," *Journal of Accounting Research*, 27, 21-39.

Banker, R. and S. Datar (1992), "Optimal Transfer Pricing under Post-contract Information," *Contemporary Accounting Research* (Spring), 329-52.

Bockus, K. and F. Gigler. (1998). "A Theory of Auditor Resignation," *Journal of Accounting Research*, 36, 191-208.

Border, K. and J. Sobel. (1987), "Samurai Accountant: A Theory of Auditing and Plunder," *The Review of Economic Studies*, 54, 525-540.

Bushman, R. and R. Indjejikian. (1993), "Accounting Income, Stock Price and Managerial Compensation," *Journal of Accounting and Economics*, 16, 3-23.

Demski, J. and D. Sappington (1984), "Optimal Incentive Contracts with Multiple Agents," *Journal of Economic Theory*, 33, 152-171.

Dye, R. (1986), "Optimal Monitoring Policies in Agencies, *Rand Journal of Economics*, 17(3), 339-350.

Dye, R. (1983), "Communication and Post-decision Information," *Journal of Accounting Research*, 21, 514-533.

Feltham, G. and J. Xie (1994), "Performance Measure Congruity and Diversity in Multi-Task Principal/Agent Relations," *The Accounting Review*, 69, 429-454.

Groves, T. (1976), "Incentive Compatible Control of Decentralized Organizations," in Y. Ho and S. Mitters (Eds.) *Directions in Large Scale Systems: Many Person Optimization and Decentralized Control*, Plenum, New York.

Groves, T. and M. Loeb. (1979), "Incentives in a Divisionalized Firm," *Management Science*, 25, 221-230.

Hirshleifer, J. (19??), "On the Economics of Transfer Pricing," *Journal of Business*, 29, 172-184.

Holmstrom, B. (1979), "Moral Hazard and Observability," *Bell Journal of Economics*, 10, 74-91.

Holmstrom, B. (1982), "Moral Hazard in Teams," *Bell Journal of Economics*, 13, 324-340.

Jewitt, I. (1988), "Justifying the First-Order Approach to Principal-Agent Problems," *Econometrica*, 56(5), 1177-1190.

Kanodia, C. (1985), "Stochastic Monitoring and Moral Hazard," *Journal of Accounting Research*, 23(1), 175-194.

Kanodia, C. (1993), "Participative Budgets as Coordination and Motivational Devices," *Journal of Accounting Research*, 31, 172-189.

Kanodia, C. (1979), "Risk Sharing and Transfer Price Systems Under Uncertainty," *Journal of Accounting Research*, 17, 74-98.

Kanodia, C. and A. Mukherji (1994), "Audit Pricing, Lowballing and Auditor Turnover: A Dynamic Analysis," *The Accounting Review*, 69, 593-615.

Laffont, J.J. and J. Tirole (1988), "The Dynamics of Incentive Contracts," *Econometrica*, 56, 1153-1175.

Laffont, J.J. and J. Tirole (1986), "Using Cost Observation to Regulate Firms," *Journal of Political Economy*, 94, 614-641.

Lambert, R. (1985), "Variance Investigation in Agency Settings," *Journal of Accounting Research*, (23)2, 633-647.

Melumad, N. and D. Mookherjee (1989), "Delegation as Commitment: The Case of Income Tax Audits," *The Rand Journal of Economics*, 20, 139-163.

Melumad, N., D. Mookherjee, and S. Reichelstein (1992), "A Theory of Responsibility Centers," *Journal of Accounting and Economics*, 15, 445-484.

Melumad, N., D. Mookherjee, and S. Reichelstein (1995), "Hierarchical Decentralization of Incentive Contracts," *Rand Journal of Economics*, 26, 654-672.

Milgrom, P. (1981), "Good News and Bad News: Representation Theorems and Applications," *Bell Journal of Economics*, 12, 380-391.

Mookherjee, D.,and I. Png (1989), "Optimal Auditing, Insurance, and Redistribution," *The Quarterly Journal of Economics*, 104, 399-415.

Mookerjee, D. and S. Reichelstein (1992), "Dominant Strategy Implementation of Bayesian Incentive Compatible Allocation Rules," *Journal of Economic Theory* (April), 378-399.

Morgan, J. and P. Stocken (1998), "The Effects of Business Risk on Audit Pricing," *Review of Accounting Studies*, 3, 365-385.

Mukherji, A. (1988), "Optimal Monitoring with Incomplete Contracts," Working paper, University of Minnesota.

Penno, M. (1985), "Informational Issues in the Financial Reporting Process," *Journal of Accounting Research*, 23, 240-255.

Rajan, M. (1992), "Cost Allocation in Multiagent Settings," *The Accounting Review*, 67, 527-545.

Rogerson, W. (1985), "The First Order Approach to Principal-Agent Problems," *Econometrica*, 53, 1357-1368.

Solomons, D. (1968). *Divisional Performance: Measurement and Control,* Richard D. Irwin, Illinois.

Suh, Y. (1987), "Collusion and Noncontrollable Cost Allocation," *Journal of Accounting Research*, 25 (Supplement), 22-46.

Townsend, R. (1979), "Optimal Contracts and Competitive Markets with Costly State Verification," *Journal of Economic Theory*, 22, 265-293.

Vaysman, I. (1996), "A Model of Cost-based Transfer Pricing," *Review of Accounting Studies*, 1, 73-108.

Wolfson, M. (1985), "Empirical Evidence of Incentive Problems and their Mitigation in Oil and Tax Shelter Programs," in J. Pratt and R. Zeckhauser (Eds.), *Principals and Agents: The Structure of Business*, Harvard Business School Press, Boston, MA.

Notes

[1] For other applications of agency models to accounting, readers can refer to two excellent surveys by Baiman (1982, 1990).

[2] This result should be interpreted with caution since nothing is known about the validity of the first-order approach to principal agent models when $\omega(z)$ is strictly convex. On the other hand, Jewitt (1988) has shown that concavity of $\omega(z)$ together with MLRP and some other conditions are sufficient to justify the first-order approach.

[3] See Amershi and Hughes (1989) for an elaboration of this issue.

[4] Amershi and Hughes (1989) contains an excellent discussion of the special importance of the exponential family of distributions to the study of principal-agent problems.

[5] In Morgan and Stocken (1998), the choice of α by the incumbent auditor is endogenous. For simplicity, I have made α an exogenous parameter.

[6] The proof here is slightly different from that in Bockus and Gigler.

4 GAME THEORY MODELS IN OPERATIONS MANAGEMENT AND INFORMATION SYSTEMS

Lode Li and Seungjin Whang

1. INTRODUCTION

The Operations Management and Information Systems (OM/IS) field has been slow to apply game theory. This is partly because of the complexity involved in coordinating large-scale activities for creation of goods and services in a fast- and ever-changing and increasingly competitive environment. Not surprisingly, the field has focused primarily on analyzing and improving the performance of physical systems (e.g. queueing or inventory systems) from the decision-theoretic perspective. This approach assumes that there exists a single body possessing the information set and decision making authority on behalf of the system as a whole. Recently, however, the field has expanded to address various issues of inter-person and inter-firm dynamics. Examples include the design of performance systems for managers who may have conflicting incentives, the design of contracts between supply chain members in the presence of incomplete information, and market competition with positive or negative externalities.

This chapter gives an overview of the existing OM/IS literature using game theory. We have chosen to focus on five topics: (1) time-based competition, (2) priority pricing for a queueing system, (3) manufacturing/marketing incentives, (4) incentives for information sharing within oligopolistic competition, and (5) competition in the software market highlighting network externalities. On each topic, we review one or two works at some length and list other related works as references. Admittedly, the list or the coverage is biased around the authors' taste and research interest.

Time-based competition (in Section 2) is a subject of great interest to the OM/IS field where the response time (as well as the price) is an important dimension of market competition. In the operation of a manufacturing or service facility, e.g., a computer/communication system, queueing delays arise in a nonlinear fashion as the utilization of the facility increases. The focus is on how queueing delays change the outcome of competition in markets where customers are sensitive to delay and what role a firm's operations strategy plays in such competition.

Priority pricing (in Section 3) has a similar setting as time-based competition, but the concern is the mechanism design under asymmetric information. Queueing delays as a form of negative externalities create an incentive for each individual to overcrowd the system in the absence of any control. Moreover, each individual user is better informed about her own usage. The question is how to use the pricing scheme and induce selfish and better-informed users to achieve the overall efficiency of the system under the informational asymmetry.

Manufacturing/Marketing Incentives (in Section 4) deals with goal congruency within a manufacturing firm which makes to stock. There exist three types of goal conflicts. First, in setting the inventory level, a potential conflict exists between the manufacturing manager who wants to minimize inventory cost and the marketing managers who want to minimize stockouts. Second, different marketing managers in charge of different products compete over the fixed capacity of the manufacturing facility. Lastly, there exists the traditional principal-agent problem in which managers (in the absence of appropriate incentives) would exert lower efforts than the owner would like. The objective is to design an internal compensation scheme that mitigates the conflicting incentives.

Information Sharing (in Section 5) addresses whether competitors in oligopolistic competition would sincerely disclose their demand and/or cost information to competitors. The tradeoff facing a player is between the efficiency gain by having more information and the strategic gain or loss caused by the reaction of other players to the changed information allocation.

Software competition (in Section 6) highlights positive network externalities associated with a software product. As the installed base of a software product grows, it becomes more attractive to other users, thus growing the installed base even more. This bandwagon effect can distort the efficiency of market competition and result in excess inertia (i.e., inefficient nonadoption of a new technology) and/or excess momentum (i.e., inefficient adoption of a new technology). In analyzing such network competition, each user's net present value is a function of the present and future size of the installed base of the technology. Accordingly, one user's decision depends on the decisions by other users past and future - a natural setting for a game theoretic model.

The remainder of the chapter discusses the above five topics in detail, and concluding remarks are provided in the last section.

2. COMPETITION IN TIME-SENSITIVE MARKETS

Production becomes strategic when consumers are concerned not only about the price to pay but also about possible delays in delivery, namely, when the market is time-sensitive. For example, a firm that charges a low price may capture a large market share, but it may have difficulty in delivering the demanded quantity to consumers in a reasonable amount of time because of limited capacity. Delays and shortages might also be consequences of variability in demand, supply, and production, even when average capacity is higher than the average demand rate - the queueing phenomenon. When firms compete to supply time-sensitive customers in the presence of such variability, all design and operational decisions such as inventory, scheduling, and capacity become strategic.

Beckmann (1965), Levitan and Shubik (1972), and Kreps and Scheinkman (1983) study the Edgeworth (1897) "constrained-capacity" variation on Bertrand competition. In the Edgeworth-Bertrand price competition, the surrogate for market sensitivity to delivery time is the assumption that unsatisfied demand goes to the firm naming the second lowest price. Kreps and Scheinkman show that capacity choice is indeed strategic - the firm with a larger capacity will charge a higher price and enjoy a higher profit. De Vany and Savings (1983), Reitman (1991), Loch (1991), Kalai, Kamien and Rubinovitch (1992), Li (1992), Li and Lee (1994), and Lederer and Li (1997) employ queueing models to study the strategic interaction among time-sensitive consumers and competing firms. In this line of research, the merger of queueing theory and game theory is a natural one: queueing systems depict the relationships among consumers' choices, firms' decisions and delivery performance, whereas game theory addresses the incentives and behavior of consumers and firms.

We shall present a general model for time-based competition and two special cases to illustrate the important role that a firm's delivery capability plays in such competition. The examples, a duopoly model and a competitive equilibrium model, are based on Loch (1991) and Lederer and Li (1997) respectively. Section 2.1 describes the modeling framework, and the results for the two special cases, a duopoly with homogeneous customers and a competitive equilibrium model with heterogeneous customers, are developed in Sections 2.2 and 2.3 respectively.

2.1 A GENERAL MODEL

Firms compete to sell goods or services to customers. Denote the set of firms by $N = \{1, 2, ..., n\}$ and the set of customer-types by $M = \{1, 2, ..., m\}$.

Each firm can set and/or improve the design parameters of its *production technology*. Let α_i be a vector of firm i's design decisions on technology that affect its processing capability as well as production costs. The choice of processing capability may involve specifying parameters for a probability distribution of the firm's processing time. The firms' products or services, though substitutable, may have different levels of *quality* perceived by customers, denoted by $v_i, i \in N$. Firms also specify *prices* for each type of customer. Let $p_i(k)$ be the price firm i specifies for customers of type $k, k \in M$, and $\vec{p}_i = (p_i(k))_{k \in M}$. After firms choose production technologies, quality levels and prices, each firm then chooses a *scheduling/inventory policy* to specify how jobs are sequenced and/or how processing rate should be controlled at any time. Denote by f_i the scheduling/inventory policy employed by firm i.

The demand from type-k customers arises over time according to a Poisson process. Customers are differentiated by the goods or services desired, and by sensitivity to the price paid and delay from the time an order is made until the time of delivery. In particular, a customer of type k perceives a good or service of quality v worth $u_k(v)$ and has a cost of $c(k)$ per unit delay time. Thus, if p is the price charged and T is the delivery lead time, the net value to customer-type k is

$$U_k(v, p, T) = u_k(v) - p - c(k)T.$$

The interactions between customers and firms go as follows. First, firms make their strategic choices such as price, quality, capacity and other operations strategies. When the demand arises, each customer chooses a firm to contract its job so as to maximize its net value based on its information about the price, quality and delivery speed of each firm. That is, type-k customer places an order with firm i if

$$E[U_k(v_i, p_i, T_i)|\mathcal{F}_t(k)] = \max_{j \in N} E\{U_k(v_j, p_j, T_j]|\mathcal{F}_t(k)],$$

where $\mathcal{F}_t(k)$ represents the information available to a customer of type k at time t, and t is the time prior to the completion of its order, at which the customer has a choice of which firm to contract with. Note that the actual time T_i a type-k customer spends waiting for a product or service from firm i, is a function of firm i's design parameters, α_i, workload and scheduling/inventory policy, f_i, and the processing requirement of the job. Queueing theory predicts that delays arise in a nonlinear fashion as the utilization of a firm increases. That is, as more customers purchase from a firm charging the lowest price, the delivery performance of the firm deteriorates at an even faster rate, and customers sensitive to delay will change their choice at some point. This "negative externality" in time-based competition is often referred to as "endogenous quality."

The customer behavior described above leads to a mapping from firms' decisions on technology, $\alpha = (\alpha_i)_{i \in N}$, quality, $v = (v_i)_{i \in N}$, price, $\vec{p} = (\vec{p}_i)_{i \in N}$, and scheduling/inventory policy, $f = (f_i)_{i \in N}$, into the expected sales (demand) rate for each firm and for each customer type, $\lambda_i(k, \alpha, v, \vec{p}, f)$ for $i \in N$ and $k \in M$. Let $\lambda_i(\alpha, v, \vec{p}, f) = (\lambda_i(k, \alpha, v, \vec{p}, f))_{k \in M}$ be the sales rates of firm i for all customer types. Thus, the expected profit rate for firm i is

$$\pi_i(\alpha, v, \vec{p}, f) = \sum_{k \in M} p_i(k) \lambda_i(k, \alpha, v, \vec{p}, f) - C_i(\lambda_i, \alpha_i, v_i, f_i) - \bar{C}_i(\alpha_i, v_i),$$

where $C_i(\lambda_i, \alpha_i, v_i, f_i)$ is the production cost rate for given technology, quality, and scheduling policy, and $\bar{C}_i(\alpha_i, v_i)$ is the amortized cost of setting design parameters and quality level.

The firm i's strategy set can then be defined as a vector of decisions, $s_i \subseteq \{\alpha_i, v_i, \vec{p}_i, f_i\}$. A Nash equilibrium is a vector of strategies for the firms, $s^* = (s_i^*)_{i \in N}$, such that for each firm i, $\pi_i(s^*) = \max_{s_i} \pi_i(s_i, s_{-i}^*)$, where $s_{-i} = (s_j)_{j \in N \setminus i}$.

Derivation of the sales rates for each firm crucially depends on the level of information available to customers when their choices are made. The general model can be specialized into two cases: 1) customers can observe the congestion levels of the firms, and their choices are dynamic; 2) customers cannot observe the congestion levels of the firms, and their choices are only based on long-term aggregate information. The first case includes Kalai, Kamien and Rubinovitch (1992) for processing rate competition, Li (1992) for inventory competition, and Li and Lee (1994) for price competition.

Both examples discussed in the next two subsections assume that customer choices are based on long-term aggregate information. The key assumption for this class of model is that customers' expected net values are given by

$$E[U_k(v,p,T)|\mathcal{F}_t(k)] = u_k(v) - p - c(k)W,$$

where W is the expected waiting time. Following De Vany and Savings (1983), customer type k is said to have an expected *full price* equal to $p + c(k)W$ when the price paid is p and the expected wait is W. Each customer buys from the firm offering the least full price for its type. Consequently, customers of the same type must face the same full price in equilibrium. That is, for each type-k customer ($k \in M$), the following holds

$$p_i + c(k)W_i = p_j + c(k)W_j, \text{ for all } i,j \in N.$$

2.2 A BERTRAND TIME-BASED COMPETITION

Let us consider a Bertrand duopoly competing for homogeneous customers in a make-to-order fashion. Since all customers are of the same type, the index k for customer-types is dropped from the notation. Demand arises in a Poisson fashion with an average rate $d(p + cW)$ which is a function of the full price, $p + cW$. Each customer demands one unit of work at a time. There are two firms in the market, i.e. $N = \{1,2\}$. Firm i's processing time for a unit of work is independently distributed with mean $1/\mu_i$ and variance σ_i^2. For simplicity, we assume each firm's production cost is zero, $C_i = 0$.

Firms compete by specifying *prices*, $p_i, i \in N$. Once the prices are posted, the expected demand rates λ_i for firm i for all $i \in N$, are determined by the following two equations:

$$P(\lambda_i + \lambda_2) = p_1 + cW_1(\lambda_1),$$
$$P(\lambda_1 + \lambda_2) = p_2 + cW_2(\lambda_2),$$

where $P(\lambda) = d^{-1}(\lambda)$ is the full price as a function of the total demand rate, and $W_i(\lambda_i)$ is the expected waiting time for customers served by firm i. We assume $P(\lambda)$ is strictly positive on some bounded interval $(0, \Lambda)$, on which it is twice-continuously differentiable, strictly decreasing, and $P(\lambda) = 0$ for $\lambda \geq \Lambda$. According to queueing theory (Heyman and Sobel (1982), p. 251),

$$W_i(\lambda_i) = \frac{1}{\mu_i} + \frac{\lambda_i(1 + \mu_i^2 \sigma_i^2)}{2\mu_i(\mu_i - \lambda_i)}.$$

Note that the demand rate for each firm is a function of the prices, the processing rate μ_i, and the processing time variance $\sigma_i^2, i \in N$. We denote firm i's demand rate by $\lambda_i(p_i, p_j)$ for $i, j \in N$ and $i \neq j$. Then, the expected profit rate for firm i is

$$\pi_i(p_i, p_j) = p_i \lambda_i(p_i, p_j).$$

Loch (1991) shows that the game has a unique Nash equilibrium in pure strategies when firms are symmetric, i.e., $\mu_1 = \mu_2$ and $\sigma_1 = \sigma_2$. However, the pure-strategy equilibrium may not exist when firms are asymmetric. Dasgupta and Maskin (1985) establish the existence of equilibria for asymmetric firms (possibly in mixed strategies). We denote the equilibrium strategy for firm i by $\psi_i(p)$, where $\psi_i(p)$ is the probability distribution function for the strategy of firm i. Also let λ_i^* and π_i^* be the expected equilibrium sales rate and profit for firm i respectively.

Proposition 1. *If $P(\lambda)$ is concave, then a Nash equilibrium exists. Furthermore, suppose $\mu_i \geq \mu_j$ and $\mu_i \sigma_i \leq \mu_j \sigma_j$ for $i \neq j$. Then, ψ_i stochastically dominates ψ_j, $\lambda_i^* \geq \lambda_j^*$ and $\pi_j^* \geq \pi_j^*$.*

We call firm i a *faster producer* than firm j if $\mu_i \geq \mu_j$, and a *lower variability producer* if $\mu_i \sigma_i \leq \mu_j \sigma_j$ (i.e., a producer whose processing time distribution, with mean $1/\mu_i$ and standard deviation σ_i, has a lower coefficient of variation, $\mu_i \sigma_i$). Then the above result says that a faster and lower variability firm always enjoys a price premium, a larger market share and a higher profit. Firms realize strictly positive profits in a Bertrand duopoly as long as customers value delivery performance (i.e., customers have a strictly positive waiting cost, $c > 0$). The traditional Bertrand equilibrium is a special case of our model when customers place zero weight on delivery time ($c = 0$).

2.3 A COMPETITIVE EQUILIBRIUM MODEL

We now consider a case where both customers and firms are heterogeneous, and firms supply customers in a make-to-order fashion.

Customers demand identical goods or services, each of which requires one unit of work, and they are differentiated by sensitivity to price and delay. A customer of type k experiences cost of $c(k)$ dollars per unit delay time, and demand of customer-type k arises according to a Poisson process with an average rate $d_k(p + c(k)W)$. Assume $d_k(\cdot)$ is differentiable and strictly decreasing. Firms are differentiated by processing time distribution and production cost. Firm i's processing time for a unit of work is independent and exponentially distributed with mean $1/\mu_i$.

Firms compete by specifying *prices and production rates* for each type of customer, and *scheduling policies*. That is, each firm i chooses a price $p_i(k)$ and a production rate $\lambda_i(k)$ for each customer-type k, and a scheduling policy, f_i. Given firm i's production and scheduling decisions, the expected waiting time for a type-k customer served by firm i is denoted by $W_i(k, \lambda_i, f_i)$. Also, we denote firm i's production cost by $C_i(\sum_{k \in M} \lambda_i(k))$ where $C_i(\cdot)$ is continuously differentiable, increasing and convex.

We assume that there are enough firms in the market so that each firm's influence on full prices is negligible. Therefore, each firm takes the market's full prices as given. Because each firm is a *full price taker* and not a *price taker*, it can adjust its prices, but firm i's price and production rates obey

$$P(k) = p_i(k) + c(k)W_i(k, \lambda_i, f_i),$$

where $P(k)$ is the full price in the market for type-k customers. Given market full prices, $P = (P(k))_{k \in M}$, firm i's choice of $(\vec{p}_i, \lambda_i, f_i)$ is equivalent to the choice of (λ_i, f_i). Hence, its actual prices $\vec{p}_i$ are determined by $p_i(k) = P(k) - c(k)W_i(k, \lambda_i, f_i)$ for $k \in M$.

The expected profit rate for firm i is a function of full prices, production rates and scheduling decisions as follows:

$$\pi_i(P, \lambda_i, f_i) = \sum_{k \in M} [P(k) - c(k)W_i(k, \lambda_i, f_i)]\lambda_i(k) - C_i(\sum_{k \in M} \lambda_i(k)).$$

Each firm maximizes its profit by making production rate and scheduling decisions. A competitive equilibrium is a vector of full prices, production rates and scheduling policies such that each firm maximizes its profit, and each customer-type's demand rate is equal to the aggregate production rate for that customer type, i.e., $\sum_{i \in N} \lambda_i(k) = d_k(P(k))$ for all k. The market clears in long-run average because we assume that firms produce only to orders, and customers will not withdraw their orders after they decide to get goods or services from the firms with the right full price. This is a "short term" competitive equilibrium, assuming that firms' cost functions are fixed over time and there is no entry.

Note that given P and λ_i, the profit equation implies that firm i's profit maximizing scheduling policy solves the problem:

$$\min_{f_i} \sum_{k \in M} c(k)\lambda_i(k)W_i(k, \lambda_i, f_i).$$

From queueing theory (Federgruen and Groenevelt (1988)), the static preemptive priority rule (often referred to as the $c\mu$ rule) is the optimal solution to the problem for any λ_i. The static preemptive priority rule for firm i assigns customers of type k a higher priority than customers of type k' if $c(k) > c(k')$, and allows preemption when higher priority work arrives. Then, we can conclude that all firms employ the static preemptive priority rule. For simplicity of exposition we order the customer-types by decreasing priority so that

$$c(1) \geq c(2) \geq \ldots \geq c(m).$$

Since the optimal scheduling policy is known, f_i is dropped from the notation, and from queueing theory (Jaiswal (1968), p. 96):

$$W_i(k, \lambda_i) = \frac{\mu_i}{(\mu_i - \sum_{l=1}^{k-1} \lambda_i(l))(\mu_i - \sum_{l=1}^{k} \lambda_i(l))}.$$

Thus, firm i's problem becomes choosing production rates λ_i to maximize its profits for given full prices P:

$$\max_{\lambda_i \in \Re_+^m} \sum_{k \in M} P(k)\lambda_i(k) - \left[\sum_{k \in M} c(k)\lambda_i(k) W_i(k, \lambda_i) + C_i(\sum_{k \in M} \lambda_i(k)) \right].$$

Define the *full cost* for firm i as the sum of the total delay cost for customers served by firm i and production cost incurred by firm i (the term in the brackets in the above expression). Also define the *marginal full cost* of firm i with respect to customer-type k as the marginal increase in the full cost when firm i increases its production rate for type k.

Proposition 2.

1. *There exists a unique competitive equilibrium. Denote the competitive equilibrium by $(P^*, (\lambda_i^*)_{i \in N})$. In the competitive equilibrium, each customer-type's full price is equal to the marginal full cost of each firm serving the customer-type, and further, firms' production rates minimize the total cost of firms' production and customers' delay subject to meeting demand.*

2. *For each $i \in N$, the equilibrium price $p_i^*(k)$ decreases in k, and the average delivery time $W_i(k, \lambda_i^*)$ increases in k, while the equilibrium full price $P^*(k)$ decreases in k.*

3. *Suppose two firms, i and j, $i \neq j$, have constant marginal production costs $C_i'(\cdot) = \kappa_i$ and $C_j'(\cdot) = \kappa_j$ respectively, and $\mu_i \geq \mu_j$. Also, suppose that $\lambda_i^* > 0$ and $\lambda_j^* > 0$. Then,*

   (a) $\lambda_i^*(k) \geq \lambda_j^*(k)$ *for* $k = 1, \ldots, m-1$; $p_i^*(1) \geq p_j^*(1)$, *and* $p_i^*(k) = p_j^*(k)$ *for* $k = 2, \ldots, m-1$; $W_i(1, \lambda_i^*) \leq W_j(1, \lambda_j^*)$, *and* $W_i(k, \lambda_i^*) = W_j(k, \lambda_j^*)$ *for* $k = 2, \ldots, m-1$.

(b) If $\kappa_i \leq \kappa_j$, then $\lambda_i^*(m) \geq \lambda_j^*(m)$; $p_i^*(m) \leq p_j^*(m)$; $W_i(m, \lambda_i^*) \geq W_j(m, \lambda_j^*)$; $p_i^*(k) - \kappa_i \geq p_j^*(k) - \kappa_j$ for all $k \in M$, $\rho_i^* \geq \rho_j^*$ where $\rho_i^* \equiv (\sum_{k=1}^M \lambda_i^*(k))/\mu_i$ for $i \in N$.

(c) If $\kappa_i \geq \kappa_j$, then $p_i^*(m) \geq p_j^*(m)$; $W_i(m, \lambda_i^*) \leq W_j(m, \lambda_j^*)$.

The existence of a competitive equilibrium follows from the Brouwer fixed point theorem, and the uniqueness of an equilibrium relies on a monotonicity property of the supply function: if full prices for some customer-types rise, total production for those types also rise. The rest of the results come from the first order conditions for the firms' problems of maximizing their profits subject to the market clearing conditions.

2.4 DISCUSSION

Models in time-based competition produce many interesting and important implications for competitive strategy analysis, operations strategy in particular. For example, Proposition 2 shows how firms differentiate their competitive strategies (e.g., price and/or delivery service) according to their own competencies (e.g., cost and/or processing capability) and characteristics of customers. Part 3(a) of Proposition 2 shows that a *faster producer* always has a larger market share, delivers at least as quickly and prices no lower than its competitor for all customer types, except possibly for the customers least sensitive to delay. Part 3(b) shows that a faster producer with *lower costs* differentiates its competitive strategy according to the market segmentation: it competes for time-sensitive customers using *faster delivery* while it competes for time-insensitive customers using *lower price*. Not surprisingly, the firm captures a larger market share and earns a larger contribution margin for every customer-type, and its capacity is better utilized. Part 3(c) shows that a faster producer without a cost advantage shifts its competitive priority to *delivery* completely, providing faster (or equal) delivery times and charging higher (or equal) prices to all customer-types, while a lower cost producer without a capacity advantage charges a lower price in all market segments. One remarkable part of the proposition is price and delivery *matching* displayed in 3(a): competing firms that jointly serve many customer-types match price and service for all but two extreme types, regardless of the differences in firms' delivery capability and cost function. This behavior is markedly different from that seen in the one segment analysis, where asymmetric firms offer different prices and delivery services.

In the competitive equilibrium analysis, the derivation of the equilibrium assumes that firms possess aggregate information about *customer-types* such as the delay costs and demand functions, and set prices and schedules accordingly. However, enforcing the equilibrium prices and schedules requires the information about *individual customers*. Suppose firms do not know the individual customer's delay cost $(c(k))$. An equilibrium is enforceable only if the customer has the incentive to reveal its true type, i.e., the equilibrium is *self-enforcing*. Lederer and Li (1997) show that the competitive equilibrium is *incentive compatible*, i.e., each customer has incentive to truthfully report its type given all other customers truthfully report their types. However, the result holds under one of the two conditions: 1) all customer-types have an identical processing time distribution with each firm; 2) individual customers' processing times are known to the firms who can differentiate their types with the information. If there is an additional uncertainty about customer service requirements in a single firm, Mendelson and Whang (1990) propose an optimal incentive-compatible priority pricing scheme based on actual service time, which will be discussed in the next section.

3. PRIORITY PRICING FOR A QUEUEING SYSTEM

Consider a single firm that owns a manufacturing or service system. As in the previous section, the system behaves as a queueing system where a job waits for the server if the server is busy serving another job. In a queueing system, jobs impose negative externalities in the form of queueing delays even if the average demand is strictly less than the capacity. As Pigou (1920) pointed out many years ago, the existence of negative externalities can potentially result in a market failure. The concern of this section is how to control queues to maximize the net value of the system as a whole. Various control or market mechanisms were investigated by different schools of thought like Pigou (1920), Knight (1924), Coase (1960), and Groves (1976). In the specific context of queueing system, Kleinrock, (1967), Naor (1969), Knudsen (1972) and Mendelson (1985) contribute to our understanding of various economic issues.

The queueing system offers a natural setting for game theory for two reasons. First, queueing phenomena arise due to interactions among multiple decision makers. Next, negative externalities create discrepancies between individual optimization and social optimization, since an individual decision maker will not consider the negative effect she imposes on other users.

We offer a game theoretic model showing how pricing can induce privately-informed selfish users to make decisions congruent with the social objective. The work is based on Mendelson and Whang (1990), but we closely follow the summary version of Wilson (1993). For further reading, interested readers are referred to Balachandran (1972), Marchand (1974), Dolan (1978), Dewan and Mendelson (1990), and Masuda and Whang (1997).

3.1 THE MODEL

Consider an M/M/1 queueing system like a mainframe computer, a production facility, or a network system serving many different customers. Each customer's service requirement is sufficiently small relative to the capacity of the system that she can ignore the effect of her jobs on the operating performance of the system. But in the aggregate, customers' jobs impose non-negligible externalities in the form of congestion. Each job is characterized by a triplet (v, t, c) comprising its service value v, its service time t, and the cost c per unit time of delay. Thus, if completion of the job is delayed by d in addition to the actual processing time t, and the price p is charged for service, then the net value to the customer v is $v - c(d+t) - p$.

For simplicity, the set of possible types is assumed to be finite and we use $k = 1, 2, \cdots, m$ to index the m possible types. Jobs of type k have four characteristics:

1. Jobs of type k arrive in a stationary Poisson process at an average rate of λ_k per hour. λ_k is endogenously determined.

2. The aggregate benefit associated with the rate λ_k of type-k jobs is captured by $V_k(\lambda_k)$, which is concave, non-decreasing. Its derivative $V'_k(\lambda_k)$ is the marginal value function, which captures the distribution of different service values v of type-i jobs.

3. Each type-k job requires a service time that is random and distributed according to the exponential distribution with the average service time t_k. (The server's capacity μ is normalized at 1.) Service times are independent and identically distributed within types. For simplicity, the marginal cost of completing each job is zero.

4. For a type-k job, the cost per hour of delay in completing service is c_k.

We consider the system manager who has the objective of maximizing the net value of the system as a whole. His control variables are the (non-preemptive) processing priority $r_k \in \{1, 2, \cdots, m\}$ and the arrival rate λ_k, for each type k. If all the relevant information is available to the system manager, the problem can be formulated as

$$\max_{\lambda, r} \sum_{k=1}^{m} \{V_k(\lambda_k) - c_k \lambda_k W_k(\lambda)\},$$

where $W_k(\lambda)$ is the mean waiting time of type-k jobs in the system (either waiting in queue or in service) when $\lambda = (\lambda_1, \lambda_2, \cdots, \lambda_m)$. According to queueing theory (Heyman and Sobel (1982), p. 435),

$$W_k(\lambda) = W_k^q(\lambda) + t_k = \frac{\Lambda_m}{\bar{S}_{k-1} \cdot \bar{S}_k} + t_k,$$

where $W_k^q(\lambda)$ is the mean waiting time in the queue for a type-k job at λ, $\Lambda_k = \sum_{j=1}^{k} t_j^2 \lambda_j$, $S_k = \sum_{j=1}^{k} t_j \lambda_j$, $S_0 = 0$, $\bar{S}_k = 1 - S_k$, and summation from m to $(m-1)$ is interpreted as zero.

Also for any arrival rates, average delay costs are minimized by serving jobs according to a priority order in which type k is served before type k' if $c_k/t_k > c_{k'}/t_{k'}$ - i.e., the $c\mu$ (or c/t in our notation) rule. By relabeling the types such that $c_1/t_1 > c_2/t_2 > \cdots > c_m/t_m$, the optimal priority rule grants priority k to each type-k job. Once the priority rule is set, the optimal arrival rate λ_k^* for type k jobs can be obtained from the first-order condition.

To implement this 'first-best solution,' however, the system manager needs to know (v, c_k, t_k) for each arriving job. Obviously, this assumption is unrealistic, and the next subsection studies the pricing scheme that induces the first-best solution under relaxed assumptions.

3.2 AN OPTIMAL PRIORITY PRICING SCHEME

The system manager is now assumed to know the aggregate statistics (i.e., the values c_k, t_k, V_k), but not specific values for each job. Only the individual customer knows her type and the value of her job.

Consider a priority scheme that operates as follows. The system manager first posts the pricing scheme for each class of priority. Then, each customer decides whether to submit the job or not, and if so, she chooses any priority and pays according to the announced pricing rule. The pricing scheme must only depend on observable variables. We here consider a pricing scheme in which each job is charged according to the priority selected by the customer and the actual processing time of the job. Let

$P_k(t)$ be the charge imposed for a k-th priority job that actually takes t to process.

The equilibrium concept is the Stackelberg equilibrium (or the subgame perfect equilibrium) with the system manager playing the leader and the customers the followers. The difference is, however, that customers simultaneously make the second stage decisions in the Nash way, since each customer's strategies interact with other customers' strategies through the waiting time W.

A type-k customer having a job worth v prefers to submit it if and only if its net value is positive when it is assigned the priority (denoted by r_k) that yields the least total cost of charges and delays. That is, the job is submitted if $v > v_k$, where (assuming a pure strategy within each type)

$$v_k = \min_{r_k}(E[P_{r_k}(t)|k] + c_k[W_{r_k}^q(\lambda) + t_k]),$$

and the expected charge $E[P_{r_k}(t)|k]$ is calculated conditional on the actual type k of the job. If we denote the optimal submission rates by λ^*, optimality requires that the resulting submission rates are $\lambda = \lambda^*$, then, for each type k,

$$v_k = V_k'(\lambda_k^*) \text{ and } r_k^* = k.$$

Proposition 3. *A pricing scheme that produces optimal submission rates and priority selections has the quadratic form*

$$P_k(t) = A_k t + \frac{1}{2} B t^2, \quad k = 1, 2, \cdots, m,$$

where, letting $a_k = c_k \lambda_k^* \Lambda_m^*$,

$$B = \sum_{k=1}^{m} \frac{c_k \cdot \lambda_k^*}{\bar{S}_{k-1} \cdot \bar{S}_k}$$

$$A_k = \frac{a_k}{\bar{S}_{k-1} \bar{S}_k^2} + \sum_{j=k+1}^{m} a_j \cdot \left(\frac{1}{\bar{S}_{j-1}^2 \bar{S}_j} + \frac{1}{\bar{S}_{j-1} \bar{S}_j^2} \right)$$

with S_k evaluated at the optimum.

3.3 DISCUSSION

Note that under this pricing scheme, the price for each priority level can be decomposed into two parts: a *basic charge* and a *priority surcharge*. The basic charge, equal for all priority classes, corresponds to the price of the lowest priority-class, $A_m t + \frac{1}{2} B t^2$. Note that it is *quadratic* in the processing time. The priority surcharge $(A_k - A_m)t$, is *proportional* to the processing time, with a coefficient that increases strictly as the priority level goes up. This demonstrates how the factors of job length and priority each contribute to the overall price.

Qualitatively speaking, the pricing formula represents the expected delay costs imposed on other jobs before and after this job starts its service. To be exact, $P_k(t)$ corresponds to the (conditional) expected externality of a job joining priority class i conditioned on its service requirement t. A salient feature of the pricing formula is the role of the quadratic term. To give a rough idea of the source of the quadratic term, consider a job of high type with processing requirement t. During its processing time, it delays all the jobs of the same or inferior priorities that arrive during its service period. The expected number of type-k jobs arriving during the processing time t is $\lambda_k t$, and on average each of these is delayed by $t/2$; hence, the expected delay costs are proportional to t^2.

The quadratic term indicates that long jobs should be charged (or penalized) more than in proportion to their resource requirements. It is interesting to note that Nielson (1968, p.230) reports an actual computer system (Stanford University Computing Center) which uses a convex, piecewise-linear pricing schedule: "the base rate increased by 50% for all time in excess of five minutes and by 100% for all time in excess of ten minutes."

A major weakness of the model is its limited applicability to the functional area. Most manufacturing or service systems in reality have multiple servers which jobs visit or skip in some sequence. Hence, a network of queue would be more appropriate than an M/M/1 queue. Unfortunately, however, analysis of service protocols in a network of queue is extremely complicated even without any consideration of multi-person interactions. While it is deemed difficult to derive a specific functional form of optimal pricing in a network model, it would be feasible to expect some structural results in future research.

4. MARKETING/MANUFACTURING INCENTIVES

The conflict between manufacturing and marketing has been part of management folklore for some time (e.g., Ackoff, 1969). Manufacturing complains that marketing wants too much produced, and marketing complains that manufacturing produces products too little, too late. Marketing seems mainly concerned about satisfying customers whereas manufacturing seems mainly interested in factory efficiency. In the same vein conflicts also arise among different product divisions within marketing since they compete over the limited capacity of the manufacturing facility. These differences are due to the incentive structures established by the firm's management. If manufacturing is rewarded for efficiency and marketing is rewarded for satisfying customers, it is not surprising that conflicts arise. In this section we review work by Porteus and Whang (1991), who propose a framework that can be used to understand this classic conflict.

One of the earliest reports on the incentive problems in the manufacturing/marketing environment is provided by Ackoff (1967). He describes (without formal analysis) the conflicts of interest between the sales manager who wants to secure a sufficient amount of goods on hand to prevent stockout and the production manager whose incentive it is to minimize the inventory costs. Eliashberg and Steinberg's (1987) work also addresses the joint decision problem for marketing and manufacturing. Other works analyzing incentive schemes in a multi-product environment include Farley (1964), Srinivasan (1981), Lal and Staelin (1987), and Lal and Srinivasan (1988), and Harris, Kriebel and Raviv (1982).

4.1 THE MODEL

A firm produces and markets N types of goods indexed by $i \in \{1, 2, \cdots, n\}$. Product manager (PM) i is in charge of marketing product i, for $i = 1, 2, \cdots, n$. The manufacturing manager (MM), or manager 0, manufactures all products using an existing facility with finite capacity. The model is a single-period model and resembles the classical multi-product newsvendor problem with resource constraints.

Production of each unit of product i costs $c_i(>0)$ and requires $a_i(>0)$ units of capacity. All products are made to stock, unfilled sales are lost, and leftover stocks are disposed of at price r_i per unit. The quantity of product i produced is denoted by y_i. A unit of product i is sold at price p_i. Assume $p_i > c_i > r_i \geq 0$. The realized contribution from product i is given by

$$f_i^{RC}(y_i, D_i) \equiv p_i \min(y_i, D_i) - c_i y_i + r_i \max(y_i - D_i, 0),$$

where D_i is the realized demand for product i.

Demand D_i is stochastically affected by effort e_i chosen by PM i. Let $D_i(e_i)$ denote the demand for product i under effort e_i, and $\Phi_i(\cdot|e_i)$ its distribution function. The expected contribution function is given by

$$f_i(y_i, e_i) \equiv \int_{\xi=0}^{\infty} f_i^{RC}(y_i, \xi) d\Phi_i(\xi|e_i).$$

We assume that the net return function $f_i(y_i, e_i) - g_i(e_i)$ is jointly concave in (y_i, e_i) for every product i.

The capacity b of the facility is also stochastically determined by the manufacturing manager (MM)'s effort. Each unit of effort by the MM creates another unit of capacity, and available capacity is subject to an additive shock: $b = b_0 + e_0 + \epsilon$, where ϵ is a random variable with a distribution function $H(\cdot)$.

All managers as well as the owner are risk-neutral. Manager i's utility u_i depends both on the expected pecuniary compensation $E(s_i)$ and on the effort e_i in a separable way. That is, for $i = 0, 1, 2, \cdots, n$,

$$u_i(s_i, e_i) = E(s_i) - g_i(e_i).$$

The disutility of making efforts, $g_i(e_i)$, is an increasing, convex, differentiable real-valued function defined on R_+. The compensation function $s_i(\cdot)$ is defined later as a function of some jointly observable outcomes (e.g., realized contribution from product i, realized demand, realized capacity, and produced inventory). For managers to stay with the firm, each should be guaranteed a non-negative (expected) utility level.

4.2 THE FIRST-BEST SOLUTION

For the moment, we assume that the owner of the firm can observe the effort level of each manager. She would determine the effort levels e (or the reward structure to enforce them) and the inventory policy $y = (y_i, y_2, \cdots, y_n)$. This problem can be solved in two stages - first the effort levels, and then the stocking decisions.

To start with the stocking decisions, suppose that the capacity is realized at b. The stocking decision can be formulated as follows:

$$\max_{y} \sum_{i=1}^{n} f_i(y_i, e_i)$$

subject to the capacity constraint

$$\sum_{i=1}^{n} a_i y_i \leq b.$$

Let $y^*(b, e)$ be the optimal stock level, $\lambda(b, e)$ the Lagrange multiplier to the capacity constraint, and $v(b, e)$ the optimal value of the objective function.

Then, moving back to the first stage, the optimal effort levels are decided according to

$$\Pi(e) \equiv \sum_{i=1}^{n} E_\epsilon f_i(y^*(b,e), e_i) - \sum_{i=1}^{n} g_i(e_i).$$

Note that since efforts e are directly observable by the principle, an optimal compensation scheme will always extract all the profit from managers by letting $s_i = g_i(e_i)$.

Then, the solution should satisfy, for each $i = 1, 2, \cdots, n$,

$$y_i^*(b,e) = \Phi_i^{-1}(\frac{p_i - c_i - a_i \lambda(b,e)}{p_i - r_i}|e_i);$$

$$\frac{\partial v(b,e)}{\partial b} = \lambda(b,e); \quad \text{and}$$

$$\frac{\partial \Pi(e)}{\partial e_i} = g_i'(e_i).$$

4.3 MORAL HAZARD AND INCENTIVE PLAN

We now consider a case in which each manager's effort level is hidden from the owner and all other managers. The capacity shock is revealed only to the MM. We offer a specific incentive plan to induce the same expected returns as the first best solution. The timing is as follows. First, the owner specifies an incentive plan s^* for each manager. Second, managerial effort is exerted. Third, the realization of available capacity is observed by all. Fourth, each PM i selects the stock level y_i. Fifth, demands, sales and contributions are realized. Finally, the owner pays the managers according to the incentive plans. In the given sequence of events, the first-best outcome can be attained through a subgame-perfect equilibrium (s, e, y) as follows.

Proposition 4. *The first-best outcome is achieved if*
(a) *PM* $i(i = 1, 2, \cdots, n)$ *is offered the following incentive plan:*

$$s_i^*(y_i, D_i, b) \equiv [f_i^{RC}(y_i, D_i) - \lambda(b, e^*) a_i y_i] - \beta_i,$$

where $\beta_i \equiv E_\epsilon[f_i(y_i^*(b, e^*), e_i^*) - \lambda(b, e^*) a_i y_i^*(b, e^*)] - g_i(e_i^*);$ *and*
(b) *the MM is offered*

$$s_0^*(b) \equiv E_\epsilon \lambda(b^*, e^*) - \beta_0,$$

where e^* is the first-best (scheduled) effort levels by managers, b is the realized capacity, and $b^* \equiv b_0 + e_0^* + \epsilon$ is the theoretical capacity to be realized when the MM exerts the scheduled effort level, and $\beta_0 \equiv b_0 + e_0^*)E_\epsilon \lambda(b^*, e^*) - g_0(e_0^*)$.

1.4 DISCUSSION

The existence of a decentralized mechanism achieving the first-best solution is not surprising, particularly in light of the risk-neutrality assumption. In fact, there are other ways of implementing the same outcome. But the structure of the mechanism provides some insights. The mechanism consists of three components. First, each PM pays a fixed lump sum β_i to the owner for the right to operate the business and keeps the revenue f_i^{RC} - a franchise contract. Second, each PM pays the owner at the rate $\lambda(b, e^*)$ per unit of capacity a_i used for producing q_i units of product i. The rate equals the *realized* shadow price of the capacity constraint. Third, the MM is paid for producing capacity b at the rate $E_\epsilon \lambda(b^*, e^*)$ - the *expected* shadow price. Combining the last two components, the owner is operating a "futures market": she buys all the realized capacity from the MM at the *expected* rate and resells it to PMs at the *realized* rate. To see why a futures market is needed, suppose instead that the MM directly sells the capacity to PMs. Then, the transfer price should be set equal to the *realized* shadow price since it optimally allocates the resource to PMs. But this would give the MM *disincentives* to work, since the realized shadow price decreases in the realized capacity. In the extreme case, if the MM works hard and creates slack capacity, the shadow price will be zero, and the MM will be paid zero. Thus, the futures market is required to achieve the dual objectives of optimally allocating the resource and encouraging the MM to work hard. The owner loses money on average by operating the futures market, but she can extract all the managers' surplus profit through the fixed lump sum.

Therefore, the model derives an incentive scheme that achieves the first best outcome. That is, under the proposed incentive schemes, all managers exert the right levels of effort and choose the right stock levels, while the owner of the firm keeps all the economic rent. This powerful result is, however, mostly due to the strong assumption that every manager is risk neutral and that there is no informational asymmetry (except the effort level). In a more realistic model, each manager may hold some private information about his productivity factor, so that the owner does not derive the correct level of efforts for each manager. As future work, one can design a 'signaling' mechanism to induce the managers to share the private information and make the right capacity and effort allocation.

5. INFORMATION SHARING IN OLIGOPOLY

The presence of unpredictable variability imparts value to information. Forecasting techniques and information technology that convert the unpredictable into the predictable are an essential and integral part of a production system. In a decision theoretic framework, the value of information is the added expected utility an individual can realize by possessing it, without considering how the actions of others will affect it or what information others possess. In a general conflict situation, namely in a game, the information and actions available to others require careful consideration. Suppose that one decision-maker acquires some additional information about certain aspects of the game. This, on the one hand, enables the player to make more informed decisions. On the other hand, other decision-makers, knowing the fact that the player is informed, may change their strategies accordingly. The net effect may or may not be beneficial to the informed decision-maker. Thus, incentives for information acquisition need to be reexamined on a case-by-case basis. Similarly, decision-makers may or may not have an incentive to disclose their private information to others in the presence of strategic conflict.

The issue of information sharing in a game is prevalent in operations and information systems management. For example, can supply chain partners derive gains by sharing information through Electronic Data Interchange (EDI)? A better understanding of the issue can also generate important implications for voluntary disclosure of accounting information and anti-trust policy on information collusion.

We offer a model to investigate whether competitors in an oligopolistic industry have incentives to share their information concerning the common market condition and firm-specific production technologies. The work is based on Li (1998). We also discuss the results in the literature of information sharing (Novshek and Sonnenschein (1982), Clark (1983), Vives (1984), Gal-Or (1985, 1986), Li (1985), and Shapiro (1986)) and information acquisition (Li, Mckelvey and Page (1987)) in oligopoly.

5.1 THE MODEL

Consider a Cournot oligopoly with n firms producing a homogeneous product. Let $N = \{1, \ldots, n\}$ be the set of the firms. The demand curve is linear (price as a function of quantity), $p = A - q$, where A

is a random variable. Before making its quantity decision, each firm i observes a signal Y_i about A. Firm i produces at a constant marginal cost C_i per unit, and C_i is the private information of the firm. Assume the joint probability distributions, $G(a, y_i, \ldots, y_n)$ for $(A, Y_1, \ldots, Y_n)$ and $F(c_1, \ldots, c_n)$ for $(C_1, \ldots, C_n)$, are common knowledge. Also assume that the demand and firm-specific costs are independent, namely, G and F are independent. In the absence of information sharing activity, each firm will determine its output quantity based on its private information (Y_i, C_i).

To investigate whether firms have incentive to share their private information, we consider a two-stage noncooperative game. In the first stage of the game before learning its private signals, each firm decides whether to disclose its demand information or cost information or both to other firms and whether to acquire other firms' private information. For example, consider a duopoly case. If firm j decides to disclose its cost information (C_j) to firm i and firm i decides to acquire such information, we say an information disclosure agreement (for C_j) is reached, and firm i will later make its quantity decision based on (Y_i, C_i, C_j). That is, when an information disclosure agreement is reached, then information transmission will be truthful (maybe conducted by an "outside agency"). If either firm decides to do otherwise, the agreement becomes null and void, and each firm will make its quantity decision only with its private information. In the second stage of the game, each firm makes the output decision based on its private signals and the additional information acquired in the first stage of the game. We denote the demand and cost signals observed by firm i by two vectors of random variables X_i^d and X_i^c respectively (with $Y_i \in X_i^d$ and $C_i \in X_i^c$), and $X_i \equiv (X_i^d, X_i^c)$. In the above duopoly example, $X_i^d = (Y_i), X_i^c = (C_i, C_j)$ and $X_i = (Y_i, C_i, C_j)$. Assume that information disclosure and acquisition are costless.

The sequence of events and decisions is as follows:

1. Each firm decides (simultaneously and independently) whether to disclose its private information to and acquire information from other firms.

2. Nature selects $(Y_1, \ldots, Y_n)$ and $(C_1, \ldots, C_n)$, and signals are observed by firms according to the information disclosure agreements made earlier.

3. Based on the available information, firms choose their production levels. To facilitate analysis, some assumptions on the information structure are necessary.

(A1) $E[Y_i|A] = A$ and $E[A|X_i^d] = \alpha_0 + \alpha_i X_i^d$ for all i, where α_i are vectors of constants. $Y_i, i \in N$, are independent, conditional on A.

(A2) $E[C_i|C_{-i}] = \beta_i^i + \sum_{j \neq i} \beta_j^i C_j$ where $C_{-i} = (C_1, \ldots, C_{i-1}, C_{i+1}, \ldots C_n)$ and $\beta_j^i \geq 0$ for all i, j.

The first assumption is general enough to include a variety of interesting prior-posterior distribution pairs such as normal-normal, gamma-Poisson, beta-binomial. In the second assumption, $\beta_j^i \geq 0$ imply that C_i's are positively (nonnegatively) correlated - a natural case to consider.

5.2 QUANTITY COMPETITION

We first investigate the firms' quantity competition in the second stage of the game, for each of the possible information disclosure arrangements made in the first stage. Recall that $X_i = (X_i^d, X_i^c)$ is the information firm i has when making its quantity decision. The expected profit for firm i, given its information, is

$$E[\pi_i|X_i] = q_i(E[A|X_i] - q_i - \sum_{j \in N \setminus i} E[g_j|X_i] - C_i).$$

Note that this is a game of private information, or a Bayesian game. Each firm's equilibrium strategy is a function of its private information X_i. We denote firm i's equilibrium strategy by $q_i^*(X_i)$.

The results for a Bayesian-Cournot duopoly game with normally distributed demand uncertainty is first established by Basar and Ho (1973). Similar results are true for our more general model (see, e.g., Li (1985)):

Proposition 5. *Given any information disclosure agreements reached in the first stage of the game, there is a unique Bayesian-Cournot equilibrium to the second-stage subgame. The equilibrium strategy for each firm is linear (affine) in its private information as well as the information disclosed by other firms.*

More specifically, firm i's equilibrium quantity is of the form:

$$q_i^*(X_i) = e_i + a_i X_i^d + b_i X_i^c$$

where a_i and b_i are vectors of constants and are functions of the parameters of the game and the information structure, namely, the joint probability distributions G and F. Note that a Nash equilibrium must satisfy two conditions: 1) each firm's strategy maximizes the firm's expected profit given its conjecture (expectation) of other firms' strategies;

2) the conjectures are correct (fulfilled). To solve for the constants in the above equilibrium strategies, we first assume that each firm's conjecture of the other firms' strategies is of the above linear form with the coefficients e_i, a_i, and b_i to be determined. Then, we substitute such conjectures into the first-order condition to the preceding objective of each firm (the conditional expected profit), and then compute the conditional expectations and collect terms for each component of X_i. Since each firm i's first-order condition must hold for any value of each random variable (a component of X_i), the coefficients of each random variable in the first-order condition equation must be zero. The resulting system of linear equations determines the values of e_i, a_i and b_i for all i.

By applying the first-order condition, we can write the expected profit of firm i in the second-stage subgame as a function of its equilibrium strategy:

$$E[\pi_i^* | X_i] = [q_i^*(X_i)]^2.$$

The assumptions of linear conditional expectations ((A1) and (A2)) are critical, which, together with the assumptions of linear demand and cost functions, ensure a unique, close-form solution to the second-stage quantity competition.

5.3 INFORMATION SHARING

We proceed to solve the first-stage game assuming the firms will follow their equilibrium strategies in the second-stage subgame given their choices in the first stage. Note that the information firms possess before making their quantity decisions in the second-stage subgame, $(X_i)_{i \in N}$, corresponds to a particular n-tuple of all firms' information disclosure and acquisition decisions made in the first stage of the game. For each such set of decisions, i.e., each $(X_i)_{i \in N}$, we can derive the expected payoffs for the first stage of the game for all $i \in N$:

$$\Pi_i \equiv E[E[\pi_i^* | X_i]] = E[(q_i^*(X_i))^2].$$

The game is the one with certainty in which each firm chooses whether to disclose or to acquire each piece of information, and the outcome of the firms' choices is determined by the above expression.

To obtain the results in Proposition 6, we need one more assumption:

(A3) $Y_1, \ldots, Y_n$ are identically distributed and so are $C_1, \ldots, C_n$.

That is, firms' private signals are symmetric in probability distribution. Let ρ be the correlation coefficient between C_i and $C_j, i \neq j$.

Proposition 6. *There is a Nash equilibrium in the first stage of the game, in which no firm discloses its demand information to any other firm while each firm exchanges its cost information with all other firms. Furthermore, this equilibrium Pareto-dominates other possible equilibria, and there exists a $\rho^* > 0$ such that for $\rho \in [0, \rho^*)$ the equilibrium is unique.*

5.4 DISCUSSION

The method we use in this study is an extension of the decision theoretic approach for quantifying the value of information. The expected value of information disclosure to a disclosing party is the difference in *her* expected profit between disclosing the information and not doing so, whereas that to a receiving party is the difference in *his* expected profit between having the information and without it. A disclosure arrangement is reached only if both parties receive positive values. For example, disclosure of demand information never takes place because each firm is worse off by disclosing its demand information, although every firm has an incentive to acquire more information about demand. On the other hand, disclosure of cost information is an equilibrium outcome because *both* the disclosing party and the receiving party are better off with the arrangement.

One shortcoming of the early literature on information sharing in oligopoly is that all firms are assumed to always make use of the disclosed information without regard to whether such an action is beneficial to them. When information disclosure is considered as a bilateral agreement, complete sharing of cost information might not always be the unique equilibrium. This weakens the results from the early literature.

Consider an example of a duopoly only with cost uncertainty. There are two pieces of private information, C_1 and C_2, with $E[C_i] = c$ and $Var[C_i] = \sigma^2$ for $i = 1, 2$. Assumptions (A2) and (A3) imply that $E[C_i|C_j] = c + \rho(C_j - c)$ for $i = 1, 2$ and $j \neq i$ where ρ is the correlation coefficient between C_1 and C_2. In the first stage of the game, each firm i has two decisions to make, whether to disclose its own private information C_i and whether to acquire its opponent's private information $C_j, j \neq i$. As consequences of the first-stage decisions, there are four possible information arrangements. In the first case, both C_1 and C_2 remain private. (Note that C_i remains private either when firm i decides not to disclose it or when firm j declines to acquire it). Then, the equilibrium quantity decision in the second stage and the expected profit (denoted by Π_0) are:

$$q_i^*(C_i) = \frac{a-c}{3} - \frac{1}{2+\rho}(C_i - c),$$

$$\Pi_0 = E[(q_i^*(C_i))^2] = \left(\frac{a-c}{3}\right)^2 + \sigma^2\left(\frac{1}{2+\rho}\right)^2.$$

In the next two cases, one firm's cost information (say C_i) is disclosed (by firm i) and used (by firm j) and the other firm's information ($C_j, j \neq i$) remains private. Then, for $i = 1, 2$,

$$q_i^*(C_i) = \frac{a-c}{3} - \frac{2-\rho}{3}(C_i - c),$$

$$q_j^*(C_j, C_i) = \frac{a-c}{3} - \frac{1}{2}(C_j - c) + \frac{2-\rho}{6}(C_i - c), \quad j \neq i.$$

Because of the symmetric nature of the game, we denote the expected profit for the firm whose information is disclosed (firm i) by Π_D and that for the firm whose information remains private (firm j) by Π_P, and

$$\Pi_D = E[(q_i^*(C_i))^2] = \left(\frac{a-c}{3}\right)^2 + \sigma^2\left(\frac{2-\rho}{3}\right)^2,$$

$$\Pi_P = E[(q_j^*(C_j, C_i))^2]$$
$$= \left(\frac{a-c}{3}\right)^2 + \sigma^2\left(\left(\frac{2-\rho}{6}\right)^2 - \rho\left(\frac{2-\rho}{6}\right) + \left(\frac{1}{2}\right)^2\right).$$

In the last case, both C_1 and C_2 are disclosed and used, and for $i = 1, 2$,

$$q_i^*(C_i, C_j) = \frac{a-c}{3} - \frac{2}{3}(C_i - c) + \frac{1}{3}(C_j - c),$$

$$\Pi_C = E[(q_i^*(C_i, C_j))^2] = \left(\frac{a-c}{3}\right)^2 + \sigma^2\left(\frac{5-4\rho}{9}\right)$$

where Π_C denotes the expected profit in the case when information is completely shared.

It can be shown that for $0 \leq \rho \leq 1$,
$$\Pi_C \geq \Pi_D \geq \Pi_P, \Pi_0.$$

This implies that disclosing the cost information is a (weakly) dominant strategy. First, note that a disclosing decision becomes null and has no effect on the firm's profitability if the other firm declines to acquire such information. It suffices to consider the case in which the firm always acquires the information whenever it is disclosed. If the other firm's information is disclosed, the firm gets Π_C by disclosing its information and Π_P by not doing so, and if the other firm's information is not disclosed, the firm gets Π_D by disclosing its information and Π_0 by not doing so. In either case, the firm is better off by disclosing its information since $\Pi_C \geq \Pi_P$ and $\Pi_D \geq \Pi_0$.

Therefore, we may assume that each firm always discloses its information, and focus our analysis on whether firms have incentive to acquire the disclosed information. It is easy to see that complete information sharing is an equilibrium, i.e., if firm i acquires C_j, then firm j is better off by acquiring C_i since $\Pi_C \geq \Pi_D$. However, this might not be the only equilibrium outcome. Note that

$$\Pi_P - \Pi_0 = \frac{\sigma^2(1-\rho^2)(16-12\rho-7\rho^2)}{36(2+\rho)^2},$$

and $\Pi_P > \Pi_0$ if $0 \leq \rho < \rho^*$, $\Pi_P = \Pi_0$ if $\rho = \rho^*$, and $\Pi_P < \Pi_0$ if $\rho^* < \rho < 1$ where $\rho^* = 0.8808$ is the positive solution to the quadratic equation $16-12\rho-7\rho^2$. Hence, there is another Nash equilibrium in which neither firm acquires other firm's information if $\rho \in [\rho^*, 1]$. Because, given that firm i does not acquire C_j, firm j gets Π_P by acquiring C_i and gets Π_0 otherwise, but $\Pi_P \leq \Pi_0$ for $\rho \in [\rho^*, 1]$. This represents a classic example in which additional information may have negative value to the informed decision-maker in a game. By acquiring C_i, firm j could make a more informed quantity decision. On the other hand, firm i, knowing that firm j is informed, would change its quantity decision by increasing the weight on $(C_i - c)$ from $1/(2+\rho)$ to $(2-\rho)/3$. This would hurt firm j's profitability. Therefore, when information is not too valuable in a statistical sense (i.e., ρ is close to one) and when C_j remains private, firm j will be worse off to learn C_i. The example demonstrates that it is imperative to examine both the incentives for information acquisition and those for information disclosure in an information-sharing game.

There are also definite answers to the question whether firms have an incentive to share information in a Bertrand oligopoly. That is, all firms have an incentive to exchange their demand information but no firm has any incentive to disclose its cost information in a Bertrand oligopoly. The result (a reversal of that for Cournot oligopoly) should not be surprising if we notice that firms' decisions are strategic substitutes in a Cournot game and strategic complements in a Bertrand game.

Lack of information pooling, in the case of demand uncertainty in Cournot competition, is socially undesirable. However, Li (1985) and Palfrey (1985) show that when the number of firms becomes large, the equilibrium price with privately held information converges to the price in the pooled-information situation. That is, the firms behave as if the information is shared when competition intensifies.

The work presented here is only concerned about information disclosure "horizontally" between firms competing in a single market. Many information disclosure activities take place in "vertical linkages," for example, EDI between suppliers and retailers, disclosure of a company's accounting information to investors, etc. However, in many cases, vertical information disclosures may lead to horizontal information "leakage". Thus, to study incentives for vertical information disclosure, one must incorporate the effects of the disclosed information on horizontal competition. The game theoretical approach discussed above should apply.

6. COMPETITION IN THE SOFTWARE MARKET

The software market has some aspects distinct from many industrial goods, and compatibility is one of them. Compatibility is a product's capability of working together with other products, and plays a critical role in customers' acquisition decision. Compatibility creates demand-side economics of scale (i.e., positive network externalities), since a popular software product will attract many compatible products to be available in the market. As a result, the competition of software products shows certain non-traditional phenomena, such as excess inertia and high market concentration. A rich literature has developed around network externalities.[2] Examples include: Rohlfs (1974), Littlechild (1975), Oren and Smith (1981), Katz and Shapiro (1985, 1986), and Farrell and Saloner (1985, 1986). We choose to review Farrell and Saloner (1986) to see how game theory is used to analyze the market competition of software products.

6.1 THE NETWORK EXTERNALITY MODEL

Suppose there exists only one technology U in the market until time T^* when another technology V becomes available (Farrell and Saloner 1986). Users are infinitesimal and arrive continuously over time with arrival rate $n(t) \geq 0$. The strategy for each user arriving after T^* is whether to choose U or V. If everyone from T^* adopts V, we call this the *adoption* outcome. If nobody adopts it, we call it *nonadoption*.

Denote by $u(x)$ a user's flow of benefits from technology U when the installed base of U is x. The presence of network externalities implies $u'(\cdot) > 0$. For simplicity, we assume that $n(t) = 1$ and $u(x) = a + bx$. Here a represents the intrinsic benefits of U, arising independent of the network. At time t, the network grows to size t, giving rise to network-generated benefits of bt. We define $\bar{u}(T)$ as the present value of U to the user who adopts U at time T, assuming V will never be adopted. Also, let $\tilde{u}(T)$ be the net present value of benefits to a user who adopts U at time T and is the last user to adopt it. Then, for the discount rate r,

$$\bar{u}(T) = \int_T^\infty (a+bt)e^{-r(t-T)}dt = (a+bT)/r + (b/r^2)$$

$$\tilde{u}(T) = (a+bT)\int_T^\infty e^{-r(t-T)}dt = (a+bT)/r.$$

Similarly, let $v(x)$ be the utility function corresponding to technology V. Assume that $v(x) = c + dx$. Also, let $\bar{v}(T)$ be the present-value benefit of V to a user who adopted V at time T, assuming that all new adopters after time T^* adopt V. Let $\tilde{v}(T)$ be the net present value of benefits to a user who adopts V at time T and is the last user to adopt it. Then, for $T \geq T^*$, it can be shown that

$$\bar{v}(T) = \int_T^\infty (c + d(t-T^*))e^{-r(t-T)}dt = \frac{c+d(T-T^*)}{r} + \frac{d}{r^2}$$

$$\tilde{v}(T) = (c+d(T-T^*))\int_T^\infty e^{-r(t-T)}dt = \frac{c+d(T-T^*)}{r}.$$

Note that each user's adoption decision not only depends on the size of the installed base of each technology as of the arrival time, but also on the (anticipated) decisions of future users. This offers a natural setting for a game theoretic model.

6.2 THE EQUILIBRIUM AND ITS EFFICIENCY

In the above adoption game, the subgame perfect equilibrium is characterized as follows.

Proposition 7. *Adoption of V is a subgame perfect equilibrium if and only if $r(c-a) + d - brT^* \geq 0$. Nonadoption of V is a subgame perfect equilibrium if and only if $r(c-a) - b(rT^* + 1) \leq 0$. One or both of these conditions will hold, so there will exist either a unique equilibrium or multiple equilibria.*

Note that adoption of V is likely only when V's intrinsic value is far superior to U, and its market introduction is not too late. In addition, the network-generated benefit factor b of the existing technology plays a critical role in ensuring that nonadoption prevails as a unique equilibrium.

Farrell and Saloner (1986) define the net present value G of the net gain in welfare from adoption of V as the benchmark of efficiency. Specifically,

$$G \equiv \int_{T^*}^{\infty} [\bar{v}(t) - \bar{u}(t)] e^{-r(t-T^*)} dt - bT^*/r^2,$$

where the first term is the gain (loss) to users who arrive after T^*, and the second term is the loss to the installed base due to the adoption of V. After some manipulation, they show that

$$G = [2(d-b) - 2brT^* + r(c-a)]/r^3.$$

Note that the efficiency condition (i.e., $G \geq 0$ and V *should* be adopted) does not coincide with the equilibrium condition of Proposition 7 (i.e., V *will* be adopted). We call an outcome "excess inertia" when adoption of V can take place in equilibrium while it is inefficient (i.e., inefficient adoption). Likewise, "excess momentum" is defined as inefficient nonadoption. Then, the following result reports the environment in which excess inertia arises.

Proposition 8.

(1) *If $d > 2b$, then there is a region in which adoption would be efficient but it is not in equilibrium. There is excess inertia.*

(2) *If $b/2 < d < 2b$, then adoption is an equilibrium wherever it is efficient. However, it need not be the unique equilibrium. There may be excess inertia.*

(3) *If $d < b/2$, then adoption is the unique equilibrium whenever it is efficient. There cannot be excess inertia.*

6.3 DISCUSSION

The model shows that competition under network externalities can lead to excess inertia in switching to a new superior product, so a software product can enjoy monopoly power for a longer time than is efficient for the economy. In particular, note from Proposition 7 that nonadoption is an equilibrium regardless of d. To see this intuitively, consider time T^* when V is newly introduced to the market. If $\bar{u}(T^*) \geq \tilde{v}(T^*)$, new users near T^* may expect later users to adopt U, so they find U more attractive and adopt it. This in turn makes U even more attractive to late users, thereby fulfilling the expectation. For efficiency, however, a large d (with, say, $a = c$ for simplicity) will improve the welfare of the economy in the long run. Hence arises excess inertia, and late users may lose due to early adopters' decisions. Casual observations confirm the conclusion. A software product (like the Windows operating system) that has a large network-generated benefit factor and has been adopted for a long time may be hard to displace in the market unless the new product delivers a far superior value over the existing one.

The paper also reports that excess momentum, the opposite phenomenon of excess inertia, is also an equilibrium, if the installed base and the intrinsic value of V (relative to U) are both large. In conclusion, market competition will not necessary achieve social efficiency in the software product market due to network externalities.

One driver of network externalities in the software market is the high cost of learning how to use it. Once a user (e.g., programmer) invests in learning a software product (e.g., a mainframe's job control language), the labor market will create demand-side scale economies; more learning leads to more adoption, and vice versa. This can create monopoly power and a high market share for a dominant supplier (e.g., Microsoft and Intuit in their respective market). By contrast, Whang (1995) attempts to explain the high market concentration of custom-made software market through the *suppliers'* learning effect. When one supplier gets the first project (e.g., an income tax processing system for Arizona State), she acquires the expertise to do next projects (e.g., an income tax processing system for New York State) of similar type in a more cost effective way. Thus, suppliers have an incentive to low ball on the first contract. Once the first winner is selected by low balling and some luck, the winner may continue to maintain the cost advantage and forever dominate the market segment.

Jones and Mendelson (1998) offer an alternative game-theoretic explanation of the high market concentration of the software market. Even network externalities aside, they show, high market concentration of software can be explained by the cost structure of zero production cost and high development cost. Jones and Mendelson (1998) show (using a Hotelling-type model) that duopoly competition in price and quality result in an outcome more like that of monopoly.

These three research themes offer alternative models to the peculiarity of the software market. In practice, however, all the three forces - network externalities, cost structures and learning effects - may be at work in a hybrid form. These models all contribute to our understanding of the software market through the applications of game theory.

7. CONCLUSION

This paper has reviewed five applications of game theory in the OM/IS field. While this field has been relatively slow to exploit game theory, it is now rapidly changing as the field is expanding its interest to multi-party coordination (for example, supply chain management). Some new subjects that we have not discussed in this chapter due to space constraints are: software contracting (Richmond, Seidmann and Whinston 1992, Whang 1992), channel coordination (Jeuland and Shugan 1983, Monahan 1984, Lee and Rosenblatt 1986, Pasternack 1985, Weng 1995), stock allocation under potential shortage (Cachon and Lariviere 1996, Lee, Padmanabhan and Whang 1997), and performance measurement in a decentralized multi-echelon inventory control setting (Lee and Whang 1992, Chen 1997). OM/IS offers natural settings for game theoretic applications, since multiple players interact through market competition, contracts and agreed-upon protocols. A key challenge is how to deal with the mathematical complexity one faces when more sophisticated concepts of game theory are applied to more realistic settings of OM/IS. But the efforts will be well justified since it will enrich the field with insights into multi-person dynamics.

References

Ackoff, "Management Misinformation Systems," *Management Science,* **14**(4), 1967, pp. 147-156.

Balachandran, K.R., "Purchasing Priorities in Queues," *Management Science,* **18**(5), Part I, 1972, pp. 316-326.

Basar, T. and Ho, Y.C., "Information Properties of the Nash Solutions of Two Stochastic Nonzero-Sum Games," *Journal of Economic Theory,* **7**, 1973, pp. 370-387.

Bechmann, M., "Edgeworth-Bertrand Duopoly Revisited" in R. Henn, ed., *Operations Research-Verfahren, III,* Meisenheim: Verlag Anton Hein, 1965, pp. 55-68.

Cachon, G. and M. Lariviere, "Capacity Choice and Allocation: Strategic Behavior and Supply Chain Performance," Working Paper, Duke University, 1996.

Chen, F., "Decentralized Supply Chains Subject to Information Delays," Graduate School of Business, Columbia University, 1997.

Clarke, R., "Collusion and Incentives for Information Sharing," *Bell Journal of Economics,* **14**, 1983, pp. 383-394.

Coase, R., "The Problem of Social Cost," *Journal of Law and Economics,* **3**, October 1950, pp. 1-44.

DasGupta, P., and E. Maskin, "Existence of Equilibrium in Discontinuous Games, 1: Theory," *Review of Economic Studies,* **53**, 1986, pp. 1-27.

De Vany, A.S. and T.R. Saving, "The Economics of Quality," *Journal of Political Economy,* **91**, 1983, pp. 979-1000.

Dewan, R. and H. Mendelson, "User Delay Costs and Internal Pricing for a Service Facility," *Management Science,* **36**, pp. 1502-1517.

Dolan, R.J., "Incentive Mechanisms for Priority Queueing Problems," *Bell Journal of Economics,* **9**(2), 1978, pp. 421-436.

Edgeworth, F., "La Teoria Pura del Monopolio," *Giornale degli Economisti,* **40**, 1887, pp. 13-31. Reprinted in English as "The Pure Theory of Monopoly," in F. Edgeworth, *Papers Relating to Political Economy,* **1**, London: MacMillan & Co, Ltd., 1925, pp. 111-142.

Eliashberg and Steinberg, "Marketing-Production Decisions in an Industrial Channel of Distribution," *Management Science* **33**, 1987, pp. 981-1000.

Groves, T., "Information, Incentives, and the Internalization of Production Externalities," *Theory and Measurment of Economic Externalities,* S. Lin (editor), Academic Press, New York, 1976.

Farley, J.U., "An Optimal Plan for Salesmen's Compensation," *Journal of Marketing Research,* **16(1)**, 1979, pp. 133-140.

Farrell, J. and G. Saloner, "Standardization, Compatability, and Innovation," *Rand Journal of Economics,* **16**, Spring 1985, pp. 70-83.

Farrell, J. and G. Saloner, "Installed Base and Compatibility: Innovation, Product Preannouncements, and Predation," *American Economic Review,* **76(5)**, 1986, pp. 940-955.

Federgruen, A. and H. Groenevelt, "Characterization and Optimization of Achievable Performance in General Queueing Systems," *Operations Research,* **36**, 1988, pp.733-741.

Fruhan, W.E., Jr., *The Flight for Competitive Advantage: A Study of the United States Domestic Trunk Carrier,* Division of Research, Harvard Business School, Boston, 1972.

Gal-Or, E., "Information Sharing in Oligopoly," *Econometrica,* **53**, 1985, 329-343.

Gal-Or, E., "Information Transmission: Cournot and Bertrand Equilibria," *Review of Economic Studies,* **53**, 1986, 85-92.

Harris, M., C. Kriebel and A. Raviv, "Asymmetric Information, Incentives and Intrafirm Resource Allocation," *Management Science,* **28(6)**, 1982, pp. 604-620.

Heyman, D.P. and M.J. Sobel, *Stochastic Models in Operations Research, Volume I: Stochastic Processes and Operating Characteristics,* McGraw-Hill, New York.

Jaiswal, N.K., *Priority Queues,* Academic Press, New York and London, 1968.

Jeuland, A. and S. Shugan, "Managing Channel Profits," *Marketing Science* **2**, 1983, pp. 239-272.

Jones, R. and H. Mendelson, "Product and Price Competition for Information Goods," Graduate School of Business, Working Paper, January 1998.

Kalai, E., M.I. Kamien and M. Rubinovitch, "Optimal Service Speeds in a Competitive Environment," *Management Science*, **38**, 1992, pp. 1154-1163.

Katz, M. and C. Shapiro, "Network Externalities, Competition, and Compatibility," *American Economic Review* **75**, 1985, pp. 424-440.

Katz, M. and C. Shapiro, "Technology Adoption in the Presence of Network Externalities," *Journal of Political Economy*, **94**(4), 1986, pp. 822-841.

Kleinrock, L., "Optimal Bribing for Queue Position," *Operations Research*, **15**, 1967, pp. 304-318.

Knudsen, N.C., "Individual and Social Optimization in a multi-server Queue with a General Cost-Benefit Structure," *Econometrica*, **40**(3), 1972, pp. 515-528.

Kreps, D.M. and J.A. Scheinkman, "Quantity Precommitment and Bertrand Competition Yield Cournot Outcomes," *Bell Journal of Economics*, **14**, 1983, pp. 326-337.

Lal, R. and V. Srinivasan, "Salesforce Compensation Plans: A Dynamic Perspective," Technical Report, Graduate School of Business, Stanford University, June 1988.

Lederer, P.J. and L. Li, "Pricing, Production, Scheduling and Delivery-Time Competition," *Operations Research*, **45**(3), 1997, pp. 407-420.

Lee, H. and M. Rosenblatt, "A Generalized Quantity Discount Pricing Model to Increase Supplier Profits," *Management Science*, **32**, 1986, pp. 1177-1185.

Lee, H., V. Padmanabhan and S. Whang, "Information Distortion in a Supply Chain: The Bullwhip Effect," *Management Science*, **43**, 1997, pp. 546-558.

Lee, H. and S. Whang, "Decentralized Multi-echelon Inventory Control Systems: Incentives and Information," Working Paper, Stanford University, 1992.

Levitan, R. and M. Shubik, "Price Duopoly and Capacity Constraints," *International Economics Review*, **13**, 1972, pp. 111-122.

Li, L., "Cournot Oligopoly with Information Sharing," *Rand Journal of Economics*, **16**, 1985, pp. 521-536.

Li, L., "The Role of Inventory in Delivery-Time Competition," *Management Science*, **38**, 1992, pp. 182-197.

Li, L., "Information Disclosure and Acquisition in Oligopoly," Hong Kong University of Science and Technology and Yale University, 1998.

Li, L. and Y.S. Lee, "Pricing and Delivery-Time Performance in a Competitive Environment," *Management Science*, **40**, 1994, pp. 633-646.

Li, L., Mckelvey, R.D. and Page, T., "Optimal Research for Cournot Oligopolists," *Journal of Economic Theory*, **42**, 1987, pp. 140-166.

Littlechild, S.C., "Two-part Tariffs and Consumption Externalities," *Bell Journal of Economics*, **6**(2), 1975, pp. 661-670.

Loch, C., "Pricing in Markets Sensitive to Delay," Doctoral Dissertation, Stanford University, 1991.

Marchand, M., "Priority Pricing," *Management Science*, **20**(3), 1974, pp. 1131-1140.

Masuda, Y. and S. Whang, "Dynamic Pricing for Network Service: Equilibrium and Stability," Graduate School of Business, Stanford University, 1997.

Mendelson, H., "Pricing Computer Services: Queueing Effects," *Communications of the ACM*, **28**(3), 1985, pp. 312-321.

Monahan, J.P., "A Quantity Discount Pricing Model to Increase Vendor Profits," *Management Science*, **4**, 1984, 166-176.

Myerson, R.B., *Game Theory: Analysis of Conflict*, Harvard University Press, Cambridge, MA, 1991.

Naor, P., "On the Regulation of Queue Size by Levying Tolls," *Econometrica*, **37**(1), 1969, pp. 15-24.

Nielson, N.R., "Flexible Pricing: an Approach to the Allocation of Computer Resources," AFIPS, 1968 Fall, pp. 521-531.

Novshek, W. and Sonnenschein, H., "Fulfilled Expectations Cournot Duopoly with Information Acquisition and Release," *Bell Journal of Economics*, **13**, 1982, 214-218.

Oren, S. and S. Smith, "Critical Mass and Tariff Structures in Electronic Communications Markets," *Bell Journal of Economics*, **12**, 1981, pp. 467-486.

Palfrey, T., "Uncertainty Resolution, Private Information Aggregation, and Cournot Competitive Limit," *Review of Economic Studies*, **52**, 1985, pp. 69-84.

Pasternack, B.A., 1985, "Optimal Pricing and Return Policies for Perishable Commodities," *Marketing Science*, **17**, pp. 166-176.

Pigou, A.C., *The Economics of Welfare*, first edition, Macmillan, London, 1920.

Porteus, E. and S. Whang, "On the Manufacturing/Marketing Incentives," *Management Science*, **37**(9), 1991, pp. 1166-1181.

Reitman, D., "Endogenous Quality Differentiation in Congested Markets," *Journal of Industrial Economcis*, **39**, 1991, pp. 621-647.

Richmond, W., A. Seidmann and A.B. Whinston, "Incomplete Contracting Issues in Information Systems Development Outsourcing," *Decision Support Systems*, **8**(5), 1992, pp. 459-477.

Rohlfs, J., "A Theory of Interdependent Demand for a Communication Service," *Bell Journal of Economics and Management Science*, **5**(1), 1974, pp. 16-37.

Shapiro, C., "Exchange of Cost Information in Oligopoly," *Review of Economic Studies,* **53**, 1986, pp. 433-446.

Srinivasan, V., "An Investigation of the Equal Commission Rate Policy for a Multi-Product Salesforce, " *Management Science,* **27**(7), 1981, pp. 731-756.

Stalk, G., Jr., "Time - the Next Source of Competitive Advantage," *Harvard Business Review,* July-August 1988, pp. 41-51.

Stalk, G., Jr. and T.M. Hout, *Competing Against Time,* The Free Press, 1990.

Vives, X., "Duopoly Information Equilibrium: Cournot and Bertrand," *Journal of Economic Theory,* **34**, 1984, pp. 71-94.

Weng, Z.K., "Channel Coordination and Quantity Discounts," *Management Science,* **41**, 1995, pp. 1509-1522.

Whang, S., "Contracting for Software Development," *Management Science,* **38**(3), 1992, pp. 307-324.

Whang, S., "Market Provision of Custom Software: Learning Effects and Low Balling," *Management Science,* **41**(8), 1995, pp. 1343-1352.

Wilson, Robert, *Nonlinear Pricing,* Oxford University Press, 1993.

Notes

1. The authors would like to thank anonymous referees and the editor for many valuable comments and suggestions.

2. Software is not the only product experiencing network externalities. Hardware products requiring an interface standard or long-term training (e.g., broadcasting equipment, QWERTY keyboards, telephone, and modems) are other examples.

5 INCENTIVE CONTRACTING AND THE FRANCHISE DECISION

Francine Lafontaine and Margaret E. Slade

The modern theory of the internal organization of firms — the ownership, management, and structure of production — has its roots in the writings of Knight (1921) and Coase (1937). Knight emphasized the role of risk and uncertainty and the need to insure workers and consolidate managerial-decision making, whereas Coase focused on the costs of transacting in different organizational environments, particularly the costs of writing contracts. Over time, these notions have been expanded and formalized. Moreover, in the process, two distinct but related branches of literature have emerged. The first concentrates on the tradeoff that a principal must make between providing an agent with insurance against risk and giving that agent incentives to work efficiently, e.g., Williamson (1971), Alchian and Demsetz (1972), Mirlees (1976), and Holmstrom (1982). The second emphasizes the market failures that accompany relationship-specific assets and the associated need to assign property and residual-decision rights correctly, e.g., Klein, Crawford, and Alchian (1978), Williamson (1979, 1983), Grossman and Hart (1986), and Hart and Moore (1990).

On the empirical side, efforts to test these theories have been channeled into areas that satisfy two criteria. First, the institutional regularities must correspond to the assumptions that underlie the theories, and second, sufficient data must be available. Three areas that satisfy these constraints have received a large fraction of the attention of applied contract theorists:

executive compensation, sales-force and franchise contracting, and industrial procurement.

Executive-compensation packages provide a rich laboratory in which to test the insurance/incentive aspects of contract theory.[1] Incentive pay is a nontrivial fraction of top-management compensation, where it takes the form of, for example, performance-based bonuses, stock ownership, and options to purchase shares in the firm. Furthermore, the details of executive-compensation packages are often publicly available.

Incentive pay is less prevalent, however, for low-level managers and production workers inside the firm. Nevertheless, it surfaces at this level of the hierarchy in at least one area where it takes a somewhat different form.[2] Franchise contracting is an increasingly popular method of organization for retail markets. Rather than employ an agent to sell a product and give that agent high-powered incentives within the firm, companies often choose a less integrated form of organization that allows them to share their risks and profits with their local managers or agents in a flexible way. In particular, principals can control the incentive/insurance tradeoff and minimize transaction costs by proper choice of sales-force compensation and franchise contract terms. The principal's problem is thus whether to use internal or external salespeople and, in the latter case, how to structure the external contract.

Finally, the theory of relationship-specific investment and the associated need to assign property rights has been most extensively tested in the area of input procurement.[3] When firms require specialized inputs that have higher value inside the buyer/seller relationship than in a more general market, they must decide if they will produce those inputs themselves or purchase them from an independent supplier. In the latter case, they must also decide whether to interact in a spot market or enter into a long-term contract. Moreover, the tradeoff between productive efficiency and the severity of the holdup problem can be dealt with through the choice of the terms of the procurement contract, specifically its length and flexibility.

In this chapter, we look at the second of the above areas of empirical research, franchise contracting and sales-force compensation, and we examine different aspects of the incentive/insurance tradeoff in that context.[4] We do this in two ways. First, we construct the simplest theoretical model that is capable of capturing the effect of our focus, and second, we examine the empirical evidence from published studies that have assessed this aspect of the problem.

The models that we construct are based on the standard principal/agent paradigm. We make no effort to be theoretically sophisticated. Instead, we choose convenient functional forms that lead to definite solutions to the contracting problem. Furthermore, we construct models that involve only a few parameters, and we examine the models' comparative statics with respect to those parameters. Finally, we use the comparative statics from the theoretical exercise to organize our discussion of the empirical evidence.

The object of our exercise is to determine how well the simple theories perform in predicting the empirical regularities. It turns out that the empirical evidence is very consistent. In other words, coefficients from different studies that focus on a particular aspect of the contracting problem are usually of the same sign. This means that there is a set of stylized facts that should be explained. Unfortunately, the agreement between theoretical predictions and empirical regularities is less satisfactory than the robustness of the empirical findings. For this reason, when we discover that theory and evidence do not agree, we attempt to modify the simple model by introducing neglected aspects of the problem that move the theory in the direction of the data.

The organization of the chapter is as follows. In the next section, we develop some background material on the environment in which franchising operates and the constraints that franchising data impose on the analysis.

In section 2, which is the heart of the chapter, we decompose the contract-choice problem into components that are amenable to econometric investigation. We make use of a standard agency model to organize our discussion of nine aspects of the contracting problem and how each affects the choice of organizational form. These aspects are: local-market risk, the importance of the agent's effort, the size of the outlet, the difficulty of monitoring the agent, the importance of the principal's effort, the nature of product substitutability, spillovers among units of the chain, strategic delegation of the pricing decision, and the division of the agent's effort among tasks. We model each of these factors with a different specification of the effort/sales relationship in an otherwise standard model, and then examine the relevant evidence. We conclude this section with a short overview of studies that assess the effects of these same factors but have focused on contract terms rather than contract choice.

In section 3, we turn to some loose ends that need tying. In particular, we touch upon the consequences of contract choice for the level of product prices, its effect on firm performance, the lack of contract fine tuning in most real-world markets, why royalties are based on sales rather than profits, and the relevance of asset specificity for retail-contract choice. Finally, section 4 summarizes and concludes.

1. Background

Manufacturers of retail products must decide whether to sell their products to consumers themselves (vertical integration) or to sell via independent retailers (vertical separation). When manufacturers do not perform the sales function internally, but want exclusive retailers, they choose some form of franchising or employ an independent sales force.

Within the realm of franchising, there are two commonly used modes. Traditional franchising, which involves an upstream producer and a downstream reseller (e.g., gasoline), accounts for the larger fraction of sales revenues. Business-format franchising, however, is the faster growing of the two. With business-format franchising, the franchisor provides a trademark, a marketing strategy, and quality control to the franchisee in exchange for royalty payments and up-front fees. Production, however, usually takes place at the retail outlet (e.g., fast-food).[5]

Not all sales agents that are separated from the parent firm are franchisees. Some industrial companies choose between an internal sales force, which is known as "direct" sales, and an external sales representative. A manufacturer's external sales representative is an independent business entity that offers selling services and receives commissions on realized sales. This agency often serves a number of non-competing manufacturers whose products form a package or product line. Moreover, the agency is normally each principal's exclusive representative for a designated set of customers.

Both the use of franchising and independent sales forces normally involve profit and risk sharing. As a consequence, much of the agency-theoretic literature in the retail-contracting area focuses on explaining the size of the share parameter in a franchise or sales contract, where the share parameter determines the partition of residual-claimancy rights between principal and agent. In particular, the literature shows how this parameter should vary as a function of the specific characteristics of the agent, the principal, the outlet, and the market.[6]

Incentive Contracting and the Franchise Decision 137

In real-world markets, in contrast, instead of offering contracts tailored to the characteristics of each unit, location, and agent, most firms employ a limited set of contracts, often just two — a separated and an integrated contract. In doing this, they reduce the problem of choosing the contract terms for any particular unit from a continuum of options to a simpler dichotomous choice.[7] Consequently, much of the empirical literature has analyzed the dichotomous choice between company operation or in-house sales force (vertical integration, which is associated with lower-powered incentives) and franchising or sales representatives (vertical separation, which is associated with higher-powered incentives) using arguments that were developed to explain how firms should choose the terms of their contracts. In what follows, we focus mostly on the findings from the literature that examines this dichotomy. However, we discuss the more limited literature on the determinants of the terms of franchise contracts at the end of Section 2. We also return later, in Section 3, to the reasons why firms employ a set of standard contracts, and discuss in some detail how the dichotomous choice between franchising and integration then relates to the issue of high and low powered incentives within contracts.

Our analysis of the empirical evidence concerning retail contracting makes use of two sorts of studies. Data for the first sort are at the level of the upstream firm (or sector) and describe the extent to which managers choose to contract out (i.e., their proportion of franchised units). These data are typically cross sections of either a large number of firms from a broad range of industries or from a number of narrowly defined retail sectors.[8] Data for the second type are either at the level of the downstream unit or the sales force in a district and refer to whether this unit is integrated with the upstream firm. These data are typically cross sections from a few upstream firms in a single industry.[9] In other words, with the first type of study, an observation is an upstream firm, whereas with the second, it is a contract. The two sets of studies also differ in that the first involves mostly business-format franchising, whereas the second includes many industries in which the principal is a manufacturer.

Tables 1 to 6 summarize the findings of studies that assess the choice between integration and separation. In all these tables, the signs in the final columns show the observed effect of a variable of interest on the tendency towards vertical separation. A minus sign thus indicates a negative correlation with the extent of franchising in a chain or with the use of "separated" sales representatives in the sales-force-integration problem.

Moreover, in all tables, an asterisk next to a plus or minus sign indicates that the finding is statistically significant at the 0.05 level based on a two-tailed test.

In what follows, each table is discussed in the subsection that presents the corresponding theory. One should be aware that the authors of the empirical studies do not always interpret their results in the way that we do. However, since we try to organize the empirical evidence using the framework of our model, we make no attempt to reconcile their interpretations and ours.

2. Factors that Influence Contract Choice

The Basic Model

We have identified nine factors that frequently surface in empirical investigations of the determinants of retail contracting. These factors are not necessarily the most important, since the list is constrained by considerations of measurability and data availability. To illustrate, the agent's degree of risk aversion plays an important role in the theoretical incentive-contracting literature. Unfortunately, from an empirical point of view, it is virtually impossible to measure this factor directly. For this reason, we do not include it on our list.

In performing our analysis of the factors, we use the following standard principal/agent model. An agent exerts an effort, a, that results in an outcome, q, according to the relationship

$$q = f(a, \varepsilon, \Theta), \quad \varepsilon \sim N(0, \sigma^2). \tag{1}$$

In equation (1), ε is a random variable that determines risk, and Θ is a vector of parameters. We identify the outcome, q, with sales, which is indistinguishable from sales revenue since we normalize product price to one (with some exceptions, clearly noted). The functional form of $f(.)$ will vary, depending on the aspect of the incentive-contracting problem that we examine. Indeed, it is our principal method of distinguishing the various factors whose effects we analyze below. [10]

The agent bears a private cost of effort, $C(a) = a^2/2$, and receives utility from his income y, $u(y) = -\exp(-ry)$, where r is his coefficient of absolute risk aversion. It is well known that in this setup, the agent behaves as if he were maximizing his certainty-equivalent income, CE, which is $E(y) - (r/2)\text{Var}(y)$, where E is the expectation operator, and Var is the variance function.

The risk-neutral principal offers the agent a linear contract, $s(q) = \alpha q + W$, where α is a commission rate, and W is a fixed wage.[11] In other words, αq is the agent's incentive pay, whereas W is his guaranteed income. One can write the contract in an alternative but equivalent form that corresponds more closely to a business-format franchise contract. Without restricting the signs of α and W, we can express the agent's payment as $s(q) = (1-\rho)q - F$, where F is the franchise fee, and ρ is the royalty rate. As we want our model to describe both types of franchising as well as industrial selling, we choose to use the former notation. Then, the agent's income is $y = \alpha q + W - a^2/2$.

The parameter α plays a key role in the analysis as it determines the agent's share of residual claims. Two limit cases are of special interest. When $\alpha = 0$, the agent is a salaried worker who is perfectly insured, whereas when $\alpha = 1$, the agent is the residual claimant who bears all of the risk. One expects that, in general, $0 \leq \alpha \leq 1$. We identify α with the power of the agent's incentives. Moreover, we assume that inside the firm these incentives are low, whereas the contracts that are written with non-employees are higher powered. In theory, this need not be the case.[12] In practice, however, it is a strong empirical regularity.[13]

We also restrict attention to linear contracts. Clearly, linearity is associated with mathematical tractability, which is desirable from our point of view. Unfortunately, however, optimal contracts are rarely linear. Nevertheless, linearity is the rule, not the exception, when one examines the contracts that are written in real-world situations.[14]

We do not attempt to explain these two observed phenomena — low-powered incentives inside firms and linear contracts. Instead, we take them as empirical regularities that can be used to simplify the model. Furthermore, as a way to focus the chapter more specifically on the theories and factors of interest, we relegate most of the mathematical derivations to the appendix.

We now turn to the factors of interest, the first of which is risk.

Risk

One can use the simplest possible form of the effort/sales relationship to capture the effect that risk has on the form of the agent's contract. Specifically, let

$$q = a + \varepsilon. \tag{2}$$

The random variable, ε, is a proxy for either demand or supply uncertainty. In other words, one can interpret (2) as a demand equation (with price suppressed) where the role of effort is to increase sales. On the other hand, one can view (2) as an effort/output production function.[15]

With this form of the effort/sales function, the agent's certainty-equivalent income is given by

$$CE = \alpha a + W - \frac{a^2}{2} - \left(\frac{r}{2}\right)\alpha^2\sigma^2, \tag{3}$$

where the last term, $-(r/2)\text{Var}(y)$, is the agent's risk premium. Given a contract (α, W), the agent will choose effort to maximize equation (3), which leads to the first-order condition

$$a = \alpha. \tag{4}$$

The principal is assumed to maximize the total surplus, which she can extract from the agent with the fixed payment, W. Alternatively, W can be used to divide the surplus between principal and agent when some rent is left downstream.[16] We do not model the choice of W, which we leave intentionally vague. Then, the principal's problem is to

$$\max_{\alpha, a}\left[a - \frac{a^2}{2} - \left(\frac{r}{2}\right)\alpha^2\sigma^2\right] \tag{5}$$

subject to the agent's incentive constraint (4), and a participation constraint that we also do not model.[17]

After equation (4) is substituted into equation (5), the first-order condition for the maximization of (5) with respect to α shows that, in the optimal contract,

$$\alpha^* = \frac{1}{1+r\sigma^2} \tag{6}$$

Equation (6) implies that when either risk or the agent's degree of risk aversion increases, α^* falls.

The standard agency model of retail contracting therefore suggests that, as the level of uncertainty increases, so does the cost of agent insurance and thus the desirability of vertical integration. In other words, the firm will choose to integrate its retail activities more when facing more uncertainty because the higher-powered incentives used outside the firm expose the agent to the vagaries of the market, and the risk premium that the firm must pay consequently rises.

The notion of uncertainty or risk that is relevant in this context is the risk that is borne by the agent, not by the manufacturer. In other words, it is risk at the outlet or downstream level. Unfortunately, data that measure outlet risk are virtually nonexistent. For this reason, imperfect proxies are employed. The two most common are some measure of variation in detrended sales per outlet, and some measure of the fraction of outlets that were discontinued in a particular period of time.[18] Furthermore, data are more often available at the level of the sector rather than at the level of the franchisor or upstream firm. While this is an advantage from the point of view of resolving endogeneity issues, it can be a disadvantage if firm and sector risks are likely to be very different.

Table 1 gives the details of five studies that assess the role of risk in determining the tendency towards franchising (i.e., vertical separation). In all but one of these studies, contrary to prediction, increased risk leads to more franchising (increased separation). Moreover, this positive association does not depend on the measure of risk that is used. These results suggest a robust pattern that is unsupportive of the standard agency model.[19]

The finding that risk is positively associated with vertical separation in the data is indeed a puzzle. Moreover, allowing effort to interact with risk in the model only makes matters worse: with such specifications, increased

Table 1: The Effect of Risk on the Propensity to Contract Out

Author	Year	Data	Measure	% Contracted
Anderson & Schmittlein	1984	Electronics Components by Product Line and Territory	% Forecast Error of Product-Line Sales by Territory	+
John & Weitz	1988	Industrial Firms with Sales above $50 million	Index Capturing Environmental Uncertainty	−
Martin	1988	Sectoral Panel — All US Franchising	Coefficient of Variation of Detrended Sectoral Sales	+*
Norton	1988	Restaurants and Motels by State & Sector	Variance of Detrended % Change in Sectoral Sales by State	+*
Lafontaine	1992	Bus. Format Franchising Firms from All Sectors	Fraction of Outlets Discontinued in Sector	+*

Note: * indicates a result that is significant at the 0.05 level based on a two-tailed test.

incentives can cause effort to fall, making high-powered incentives particularly costly to the principal, and thus especially undesirable.

Some authors, e.g. Martin (1988), have concluded from this that franchisors shed risk onto franchisees. This could be optimal if franchisors were more risk averse than franchisees. However, if franchisors were indeed more risk averse, there would be less need to balance franchisee incentive and insurance needs, and hence less need to use a share contract to start with. At the extreme, franchising would involve franchisees paying only lump-sum fees to franchisors, a situation that is rarely observed in practice.

An alternative, and we believe more satisfactory, explanation for the observed risk/franchising phenomenon surfaces when one considers that market uncertainty can be endogenous and that the power of incentives can influence sales variability. Indeed, franchisees often have superior information concerning local-market conditions (separate from ε). Moreover, since franchising gives retailers greater incentives to react to these conditions, one is likely to find more sales variability across franchised than across company-owned units. In that sense, the positive relationship between risk and franchising can be understood as support for incentive-based arguments for franchising.[20]

Agent Effort

Not all agents are equally important in determining the success or failure of a retail outlet. For example, consider the case of gasoline retailing. Some station operators are merely cashiers who sit in kiosks and collect payment from customers. Others, in contrast, offer a range of services that can include pumping gas, washing windows, checking oil, selling tires, batteries, and other automobile-related items, and repairing cars. Still others manage affiliated convenience stores.

To capture the notion that there are varying degrees of agent importance, we amend the effort/sales function as follows,

$$q = \eta a + \varepsilon, \tag{7}$$

while keeping the rest of the model intact. In equation (7), the parameter η, which is positive by assumption, is a proxy for the importance of the agent's effort.

After performing the same set of calculations as in the previous subsection, one finds that, with the new effort/sales function,

$$\alpha^* = \frac{\eta^2}{\eta^2 + r\sigma^2}. \tag{8}$$

Moreover, differentiating (8) with respect to η shows that $d\alpha^*/d\eta > 0$. The theory thus predicts that increases in the importance of the retailer's input should be associated with more separation and higher-powered contracts. In other words, when the agent's job is more entrepreneurial in nature, his payment should reflect this fact.

From a practical point of view, the measures that have been used to capture this effect have been determined both by data availability and by the industry being studied. Proxies for the importance of the agent's effort (or its inverse) have included various measures of labor intensity (either employees/sales or capital/labor ratios) as the agent is the one who must oversee the provision of labor. Researchers have also used a measure of the agent's value added, or discretion over input choices, and a variable that captures whether previous experience in the business is required. Finally, two studies of gasoline retailing rely on a dummy variable that distinguishes full from self service.

Table 2: The Effect of the Importance of the
Agent's Effort on the Propensity to Contract Out

Author	Year	Data	Measure	% Contracted
Caves and Murphy	1976	Sectoral Data	"Personalized Service" Dummy	+*
Norton	1988	Restaurants and Motels by State & Sector	Employees/Sales	+*
Lafontaine	1992	Bus. Format Franchising Firms from All Sectors	1- (Sales - Franchisor Inputs) / Sales for Sector 2- Previous Experience Required	+ −
Shepard	1993	Gasoline Service Stations in Massachusetts	Full Service	+
Scott	1995	Bus. Format Franchising Firms from All Sectors	Capital/labor Ratio	(−*)
Slade	1996	Gasoline Service Stations in Vancouver	Full Service	+*
Bercovitz	1998b	Individual outlets from 20 Fast-Food and Retailing Chains	Discretionary Inputs = (Costs - $ Value of Franchisor Inputs) / Costs	+*

Notes: Parentheses in the last column indicate that the variable is an inverse measure of agent effort and is therefore expected to have a sign opposite to the others. * indicates a result that is significant in the original study at the 0.05 level based on a two-tailed test.

Table 2 summarizes the results from seven studies that assess the effect of the importance of the agent's effort. In every case where the coefficient of the agent-importance variable is statistically significant, its relationship with separation from the parent company is positive, as predicted by standard agency considerations and other incentive-based arguments. In other words, when the agent's effort plays a more significant role in determining sales, franchising is more likely.

Outlet Size

Modeling the effect of outlet size is less straightforward than for the previous two factors, and model predictions are more sensitive to specification as a consequence. We confess that the particular specification that we adopt was chosen so that results are consistent with the empirical regularity that we present below. Indeed, it is necessary that we model size as interacting with risk in order to obtain our prediction.[21] With this caveat, we specify the effort/sales relationship as a production function whose arguments are franchisee effort, a, and outlet size or capital, k,

$$q = \eta a + (\gamma + \varepsilon)k. \qquad (9)$$

All other assumptions are as before.

There are two things to note about equation (9). The parameter γ measures the direct effect of capital in the production function, whereas k is a proxy for the amount of capital invested. Furthermore, our specification assumes that a larger outlet is associated with increased agent risk. This does not mean that the market is riskier *per se*; it simply means that more capital is subject to the same degree of risk.

After the standard set of manipulations, we obtain

$$\alpha^* = \frac{\eta^2}{\eta^2 + r\sigma^2 k^2}. \qquad (10)$$

Note that γ does not appear in this solution. Thus outlet size, if it enters the production function in an additive way, has no effect on optimal contract terms. However, when interacted with risk, k does matter. In other words, the amount of capital invested in the outlet rather than its importance in determining sales directly is what matters here.

Furthermore, differentiating α^* with respect to k yields a negative relationship, which implies that the agent should be given lower-powered incentives when the size of the capital outlay increases. This presumes that it is the agent's capital, not the principal's, that is at risk. In other words, the larger the outlet, the more capital the franchisee has at stake and the more insurance he requires. Thus the solution implies a lower share for the agent, or more vertical integration. Furthermore, vertical integration in this context has the added advantage that it substitutes the principal's capital for the agent's.[22]

Unlike the factors discussed above, the empirical measurement of size is fairly straightforward. Common measures are average sales per outlet and the initial investment required. Table 3 shows that, with one exception, greater size leads to less separation or increased company ownership. In other words,

Table 3: The Effect of Outlet Size on the Propensity to Contract Out

Author	Year	Data	Measure	% Contracted
Brickley & Dark	1987	Selected Franchising Firms	Initial Investment	_*
Norton	1988	State Level Sectoral Data for Restaurants and Motels	Sales/Outlet	+*
Martin	1988	Sectoral Panel	Sales/Outlet	_*
Brickley, Dark and Weisbach	1991	1- State Level Sectoral Data 2- Outlet Data from 36 Chains	Initial Investment Initial Investment	_* _*
Lafontaine	1992	Bus. Format Franchising Firms from All Sectors	1- Initial Investment 2- Sales/Outlet for Sector	_* _*
Thompson	1994	Bus. Format Franchising Firms from All Sectors	Initial Investment	_*
Lafontaine	1995	Individual Fast-Food Restaurants in the Pittsburgh and Detroit Metropolitan Areas	Number of Seats in an Outlet	_*
Scott	1995	Bus. Format Franchising Firms from All Sectors	Initial Investment	_
Kehoe	1996	Individual Hotels from 11 Major Chains	Number of Rooms	_*

Note: * indicates a result that is significant in the original study at the 0.05 level based on a two-tailed test.

as the model above predicts, people responsible for large outlets tend to be company employees who receive low-powered incentives.[23]

It is reasuring to see that theory and evidence agree. Nevertheless, as noted above, it is possible to argue for the opposite relationship in an equally convincing manner. Indeed, when an outlet is large, the agent has more responsibility. For this reason, outlet size has been used in the empirical literature to measure the importance of the agent's input. Not surprisingly then, it is often claimed that an agency model should predict that an increase in size will be associated with more separation and higher-powered incentives (see note 21). Furthermore, as Gal-Or (1995) shows, in a model with spillovers across units of the same chain, smaller outlets have a greater tendency to free ride since outlets with larger market shares can internalize more of the externality. In this model, small units would be more likely to be vertically integrated.[24] The data, however, contradict this prediction.

Costly Monitoring[25]

The idea that monitoring the agent's effort can be costly or difficult for the principal is central to the incentive-based contracting literature. In fact, if monitoring were costless and effort contractible, there would be no need for incentive pay. The agent's effort level would be known to the principal with certainty, and a contract of the following form could be offered: If the agent worked at least as hard as the first-best effort level, he would receive a salary that compensated him for his effort, whereas if his effort fell short of this level, he would receive nothing.[26] In equilibrium, the agent would be fully insured, and the first-best outcome would be achieved.

Given the centrality of the notion of costly monitoring, it is somewhat surprising that there exists confusion in the literature concerning the effect of an increase in monitoring cost on the tendency towards company operation. For example, consider the following statements from the empirical literature:

> *The likelihood of integration should increase with the difficulty of monitoring performance.*
> Anderson and Schmittlein (1984, p. 388).

> *Franchised units (as opposed to vertical integration) will be observed where the cost of monitoring is high.*
> Brickley and Dark (1987, p. 408), text in parentheses added.

These contradictory statements imply that monitoring difficulties should both encourage and discourage vertical integration.

To reconcile these predictions, we modify the standard agency model to include the possibility that the principal can use not only outcome (i.e., sales) information to infer something about the agent's effort, but also a direct signal of the agent's effort.[27] Furthermore, the principal can base the agent's compensation on both signals.

We consider two types of signals because, in most real-world manufacturer-retailer relationships, it is possible to supervise the actions of a retailer directly by, for example, testing food quality, assessing the cleanliness of the unit, and determining work hours.[28] This direct supervision provides the manufacturer with information on retailer effort that supplements the information contained in sales data. In general, the informativeness principle (Holmstrom (1979), Milgrom and Roberts (1992, p. 219)) suggests that compensation should be based on both sales data and signals of effort obtained via direct monitoring.

To model this situation, we replace the effort/sales relationship (1) with two functions to denote the fact that the principal receives two noisy signals of the agent's effort.[29] First, the principal observes retail sales of the product, q, and second, the principal receives a direct signal of effort, e,

$$q = a + \varepsilon_1$$
$$e = a + \varepsilon_2 \qquad \varepsilon \sim N(0, \Sigma) \qquad (11)$$

where $\varepsilon = (\varepsilon_1, \varepsilon_2)^T$, $\Sigma = (\sigma_{ij})$, $\sigma_{ij} = \sigma_{ji}$, and $\sigma_{ii} > \sigma_{ij}$, $i = 1, 2$, $j \neq i$.

The principal offers the agent a contract that includes, in addition to the fixed wage W, an outcome-based or sales commission rate, α_1, and a behavior-based commission rate, α_2, related to the direct signal of effort. The agent's certainty-equivalent income becomes $(\alpha_1 + \alpha_2)a + W - a^2/2 - (r/2)\alpha^T \Sigma \alpha$, where α is the vector of commission rates, $\alpha = (\alpha_1, \alpha_2)^T$. The agent's incentive constraint for this problem is $a = \alpha_1 + \alpha_2$.

As before, the risk-neutral principal chooses the agent's effort and the commission vector to maximize the total surplus subject to the agent's incentive constraint. When the two first-order conditions for this problem are solved, they yield

$$\alpha_1^* = \frac{\sigma_{22} - \sigma_{12}}{\sigma_{11} + \sigma_{22} - 2\sigma_{12} + r(\sigma_{11}\sigma_{22} - \sigma_{12}^2)}, \qquad (12)(a)$$

and

$$\alpha_2^* = \frac{\sigma_{11} - \sigma_{12}}{\sigma_{11} + \sigma_{22} - 2\sigma_{12} + r(\sigma_{11}\sigma_{22} - \sigma_{12}^2)}. \qquad (12)(b)$$

When the noisy signals are uncorrelated, so that $\sigma_{ij} = 0$, equation (12) takes the simpler form

$$\alpha_1^* = \frac{1}{1 + r\sigma_{11} + \sigma_{11}/\sigma_{22}}, \qquad (13)(a)$$

and

$$\alpha_2^* = \frac{1}{1 + r\sigma_{22} + \sigma_{22}/\sigma_{11}}. \qquad (13)(b)$$

In this form, the solution shows that the optimal contract described in equation (6) must now be amended to account for the relative precision of the two signals. In other words, the new optimal compensation package places relatively more weight on the signal with the smaller variance. Thus equation (6) is a special case of (13) in which σ_{22} is infinite (direct monitoring contains no information).

We are interested in the effect of increases in the two sorts of uncertainty on the size of α_1^* since this is the incentive-based pay that appears in the data. Differentiating equation (12) with respect to the two variances shows that $d\alpha_1^*/d\sigma_{11} < 0$ and $d\alpha_1^*/d\sigma_{22} > 0$. Increases in the precision of sales data $(1/\sigma_{11})$ thus lead to a higher reliance on outcome-based compensation (higher α_1^*) which corresponds to less vertical integration. However, increases in the precision of the direct signal of effort $(1/\sigma_{22})$ lead to less outcome-based compensation (lower α_1^*) or more vertical integration.

While the above model does not explicitly include monitoring costs, it should be clear that if the upstream firm can choose some action that reduces σ_{11} (increases the precision of sales as a signal of effort) at some cost, it will do so to a greater extent the lower this cost is. The resulting decrease in σ_{11} will in turn lead to a greater reliance on sales data in the compensation scheme. In other words, when the cost of increasing the precision of sales data as an indicator of effort is low, we should observe more reliance on sales data in the compensation scheme, which means less vertical integration. On the other hand, when the cost of behavior monitoring, or of reducing σ_{22}, is low, the firm will perform more of this type of monitoring. A low σ_{22} will then lead the firm to choose a lower α_1, which amounts to more vertical integration.[30]

To summarize, our comparative statics show that the effect of monitoring on the degree of vertical integration depends on the type of information garnered by the firm in the process. If this information gives a better direct signal of effort, it reduces the need to use sales-based incentive contracting. If, on the other hand, monitoring increases the value of sales data as a signal of agent effort by increasing its precision, then incentive contracting becomes more attractive.

Turning to the empirical evidence, we separate the studies in two groups in Table 4 based on their interpretation of monitoring costs. The first part of the table shows results obtained in the sales-force compensation literature, where the focus has been on the usefulness of observed sales data as an indicator of agent effort. The second part of Table 4 contains empirical results from the franchising literature, where authors have focused on the cost of behavior monitoring.

In the first part of the table, in the first two studies, researchers asked managers to respond to various statements: In Anderson and Schmittlein (1984), they responded to "it is very difficult to measure equitably the results of individual salespeople" while in Anderson (1985), the measure was tabulated from responses to "(1) team sales are common, (2) sales and cost records tend to be inaccurate at the individual level, and (3) mere sales volumes and cost figures are not enough to make a fair evaluation." In John and Weitz (1988), the length of the selling cycle was used on the basis that a long lag between actions and market responses makes it difficult to attribute output to effort. In addition, these authors included a measure of environental uncertainty, which captures the extent to which agents "control" sales

Table 4, Part I: The Effect of Monitoring Difficulty
on the Propensity to Contract Out

Author	Year	Data	Measure	% Contracted
Anderson & Schmittlein	1984	Electronics Components by Product Line and Territory	Index indicating that it is difficult to measure results of individuals	_*
Anderson	1985	Electronics Components by Product Line and Territory	Index indicating that 1) team sales are common, 2) records are inaccurate and 3) sales and cost figures are insufficient for a fair evaluation	_*
			Importance of non-selling activities	_*
John & Weitz	1988	Industrial Firms with Sales above $50 million	Length of Selling Cycle	_*

Note: * indicates a result that is significant in the original study at the 0.05 level based on a two-tailed test.

Table 4, Part II: The Effect of Monitoring Difficulty on the Propensity to Contract Out

Author	Year	Data	Measure	% Contracted
Brickley & Dark	1987	Selected Franchising Firms	Distance From Monitoring Headquarters	+*
Norton	1988	Restaurants and Motels by State & Sector	Fraction of State Population Rural	+*
Minkler	1990	Taco Bell Restaurants in Northern California and Western Nevada	1- Distance From Monitoring Headquarters 2- Outlet Density = Number of Outlets within a 5 Mile Radius	+* (+)
Brickley, Dark and Weisbach	1991	1- State Level Sectoral Data 2- Outlet Data from 36 Chains	Density: Units per Square Mile Density: Company's Units in County	(−*) (−*)
Carney and Gedajlovic	1991	Canadian Bus. Format Franchising Firms from all Sectors	Density: Proportion of Outlets in Quebec	(−*)
Lafontaine	1992	Bus. Format Franchising Firms from All Sectors	Number of States in which the Chain has Established Outlets	+*
Lafontaine	1995	Fast Food in Pittsburgh and Detroit Metropolitan Areas	Outlet Density = Number of Outlets from the Same Chain in same Zip Code	(−*)
Scott	1995	Bus. Format Franchising Firms from All Sectors	Number of States in which the chain has established outlets	+*
Kehoe	1996	Individual Hotels from 11 Major Chains	Density: Number of Hotels from the Same Chain in Same City	(−*)
Bercovitz	1998b	Individual Outlets from 20 Fast-Food and Retailing Chains	1- Miles to Monitoring HQ 2- Density: Inverse of the Average Distance of the Four Closest Units from the Same Chain	+* (−)

Notes: Parentheses in the last column indicate that the relevant variable is an inverse measure of monitoring cost and is therefore expected to have a sign opposite to the others.
* indicates a result that is significant in the original study at the 0.05 level based on a two-tailed test.

outcomes. Using scores thus obtained as measures of the cost of monitoring sales and inferring effort from it, researchers found that higher monitoring costs lead to more vertical integration, as predicted by our model.

The second part of Table 4 includes studies in which authors have used a variety of measures of behavior-monitoring costs, including some notion of geographical dispersion (captured in one case by whether the unit is more likely to be in a mostly urban or rural area) or of distance from monitoring headquarters. These measures are proxies for the cost of sending a company representative to visit the unit to obtain data on cleanliness, product quality, etc. Outlet density has also been used as an inverse measure of behavior-monitoring cost. One can see that when behavior-monitoring costs are measured either directly by dispersion or distance, or inversely by density, in all cases where coefficients are significant, higher monitoring costs lead to more vertical separation. This reflects the fact that when behavior monitoring is costly, firms rely on it less, and rely more on residual claims to compensate their agents. Again the evidence is consistent with the model.

It should be clear then that the two types of measures used in the empirical literature have captured different types of monitoring costs: the fit of sales data to individual effort versus direct monitoring that is a substitute for sales data. Taking this difference into account, the seemingly contradictory results obtained and claims made by these researchers are in fact consistent with each other as well as with standard downstream-incentives arguments for retail contracting.

Franchisor Effort

The standard agency model assumes, as we have, that only one party, the agent, provides effort in the production (or sales-generation) process. In reality, success at the retail level often depends importantly on the behavior of the upstream firm or principal. For example, franchisees expect their franchisors to exert effort towards maintaining the value of the trade name under which they operate, via advertising and promotions, as well as screening and policing other franchisees in the chain. If this behavior is not easily assessed by the franchisee, there is moral hazard on both sides — the franchisee's and the franchisor's — and the franchisor, like the franchisee, must be given incentives to perform.[31]

To capture the effect of franchisor effort on the optimal contract, we amend the effort/sales relationship to include not only franchisee effort, a, but also franchisor effort, b,

$$q = \eta a + \theta b + \varepsilon, \tag{14}$$

where the parameter $\theta > 0$ is a proxy for the importance of the franchisor's effort. Assume that the franchisor's private cost of effort is $C(b) = b^2/2$, the same as for the franchisee. The franchisor still chooses the share parameter, α, in the first stage, but now the contract must satisfy incentive compatibility for both parties. As before, the first-order condition for the franchisee's effort gives $a = \alpha \eta$. In turn, the first-order condition for the franchisor's choice of effort is $b = (1-\alpha)\theta$. Substituting these into the total-surplus function, one obtains the optimal-share parameter

$$\alpha^* = \frac{\eta^2}{\eta^2 + \theta^2 + r\sigma^2}. \tag{15}$$

Differentiating α^* with respect to η shows that the optimal share, or the extent of vertical separation, goes up as the franchisee's input becomes more important. However, differentiating α^* with respect to θ yields the opposite effect. When the input of the franchisor becomes more important, her share of output, $(1-\alpha^*)$, or the extent of vertical integration, must rise.

Table 5 shows the results of six studies that consider how the importance of the franchisor's inputs affects the optimal contract choice. The importance of these inputs is measured by the value of the trade name (proxied by the number of outlets in the chain or the difference between the market and the book value of equity), the amount of training or advertising provided by the franchisor, or the number of years spent developing the business format prior to franchising. The table shows that, in all cases where franchisor inputs are more important, less vertical separation is observed, as predicted.

One proxy for the importance of the franchisor's input that has been used in the literature but is not included in Table 5 is the chain's number of years of franchising (or business experience). The idea is that more years in franchising (or business) lead to a better known, and thus more valuable, trade name. However, this variable is also a proxy for the extent to which franchisors have access to capital as well as for learning and reputation effects. Furthermore, the empirical results that pertain to this variable are mixed. Using panel data at the franchisor level, Lafontaine and Shaw (1999b) find that, after the first few years in franchising, the proportion of corporate units within chains levels off and remains quite stable. They conclude that a firm's years in franchising is not a major determinant of the extent of vertical integration in franchised chains.[32]

Table 5: The Effect of the Importance of the
Franchisor's Effort on the Propensity to Contract Out

Author	Year	Data	Measure	% Contracted
Lafontaine	1992	Bus. Format Franchising Firms from All Sectors	1- Weeks of Training 2- Lagged No. of Outlets 3- % Time Not Franchising	_* _* _*
Minkler and Park	1994	Panel of Publicly Traded Bus. Format Franchising Firms from All Sectors	Market Minus Book Value of Equity	_*
Thompson	1994	Bus. Format Franchising Firms from All Sectors	Number of Years in Business Prior to Franchising	_*
Scott	1995	Bus. Format Franchising Firms from All Sectors	Days of Training	_
Bercovitz	1998b	Individual Outlets from 20 Fast-Food and Retailing Chains	Franchisor Media Advertising	_*
Lafontaine and Shaw	1999b	18 Year Panel of Bus. Format Franchising Firms from All Sectors	1- Franchisor Media Advertising 2- Number of Years in Business Prior to Franchising	_* _*

Note: * indicates a result that is significant in the original study at the 0.05 level based on a two-tailed test.

Spillovers Within Chains

The standard incentive-cum-insurance model of retail contracting does not usually consider the competitive environment in which the principal/agent relationship operates. Instead, this relationship is modeled as if the market were perfectly competitive and price were exogenous to the firm. Alternatively, the franchisor is modeled as a monopolist, an assumption that also eliminates the importance of rivals. Most markets in which franchising is prevalent, however, are better characterized as monopolistically competitive. Usually, there are several firms that produce similar but not identical products, and firms as well as units within firms face downward-sloping demand. In this and the next two subsections, we consider the consequences of endogenous prices.

One reason for the prevalence of chains rather than independent sales outlets is that there are externalities that are associated with the brand or chain name. Although spillovers can be beneficial, they can also create problems for both franchisees and franchisors. For example, one form that a spillover can take is a demand externality. With this sort of spillover, a low price at one outlet in a chain increases demand, not only at that outlet but also for other franchisees in the same chain. Conversely, a high price can cause customers to switch their business to another chain rather than merely seek a different unit of the same chain.

In order to investigate the effect of demand spillovers, we amend the effort/sales relationship to include own price, p, and $\bar{p}$, the price charged by another outlet in the same chain,

$$q = 1 - p - \mu\bar{p} + a + \varepsilon. \qquad (16)$$

Equation (16) is a standard linear demand equation, with a parameter, μ, that represents the extent of demand spillovers. Thus we assume that $\mu > 0$, which means that a high price at a given unit causes an erosion of the sales of all members of the chain. We also assume that the franchisor chooses downstream prices as well as the share parameter in this version of the model.[33] All other model assumptions are as before.

None of the modifications of the model affects the agent's incentive constraint, which still gives $a = \alpha$. Using this to eliminate a, one finds that, in a symmetric equilibrium,

$$\alpha^* = \frac{1}{(1+r\sigma^2)2(1+\mu)-1} \qquad (17)$$

and $d\alpha^*/d\mu < 0$. Thus, when there are demand externalities of the type one normally associates with branding, integration becomes more desirable. This is because the chain internalizes spillovers external to the individual unit.

There are other sorts of spillovers, such as franchisee free riding. Indeed, as noted by Klein (1980), Brickley and Dark (1987), and Blair and Kaserman (1994), once an agent is given high-powered incentives via a franchise contract, the franchisee can shirk and free ride on the trade name. The problem is that the cost of the agent's effort to maintain the quality of the trademark is private, whereas the benefits of these activities accrue, at least partially, to all members of the chain. Here, the spillover works through effort, not price.

Whether the externalities work through price and/or effort, spillover problems are exacerbated in situations where consumers do not impose sufficient discipline on retailers, namely in cases of non-repeat businesses. The franchisor may therefore decide to operate directly those units in transient-customer locations, such as those around freeway exits, or to operate more outlets directly if involved in markets subject to significant non-repeat business.

Table 6 summarizes the evidence from those studies that have examined the effect of non-repeat business on the propensity to franchise. This table shows that the evidence on non-repeat is mixed. One explanation for this may be that franchisors find other ways to control franchisee free-riding, for example by using approved-supplier requirements or self-enforcing contracts. If so, the role of the franchisor in maintaining service quality and trademark

Table 6: The Effect of Non-Repeat Business on the Propensity to Contract Out

Author	Year	Data	Measure	% Contracted
Brickley & Dark	1987	1- Franchising Firms from All Sectors	Dummy Variable for Non Repeat Sectors	–*
		2- Outlets from 36 Franchising Firms in Various Sectors	Highway Dummy Variable	+*
Norton	1988	Restaurants and Motels by State & Sector	Tourism: Household Trips in the State	+* (in hotels)
Brickley, Dark and Weisbach	1991	1- State Level Sectoral Data	Non-Repeat Industry Dummy	+ (at means)
		2- Outlet Data from 36 Chains	Non-Repeat Industry Dummy	+ (at means)
Minkler	1994	Taco Bell Restaurants in Northern California and Western Nevada	Highway Dummy Variable	–
Lafontaine	1995	Fast Food in Pittsburgh and Detroit Metropolitan Areas	Highway Dummy Variable	+

Note: * indicates a result that is significant in the original study at the 0.05 level based on a two-tailed test.

reputation should be particularly important in sectors where most business is transient. This, in turn, brings us back to the issue of franchisor incentives in a double-sided moral-hazard model of franchise contracting. In fact, measures of the "value of the trade name" have been used in the literature to test both the notion that franchisors must be given more incentives to perform when the trade name is very valuable (see Table 5) and the notion that franchisee free-riding opportunities are greater under those circumstances. Furthermore, both sides of this coin lead to the same prediction — that chains will rely more on vertical integration when the trade name is very valuable — and are thus empirically indistinguishable. The results in Table 5 are consistent with this prediction, whereas the results in Table 6 overall do not support the non-repeat component of the free-riding model.

Product Substitutability

In some franchising industries, products are easily distinguishable from one another. For example, most customers have definite preferences between McDonald's hamburgers and KFC's chicken. There are, however, other industries in which the services that the agents provide are perhaps the only things that distinguish the output of one firm from that of another. Real-estate franchises, for example, fall in the latter group. Given that, across industries, there are varying degrees of differentiation among products that are provided within the industry, one can ask how these differences affect contract choice.

The situation just described is the converse of the spillover case. Specifically, one can rewrite the demand equation as

$$q = 1 - p + \delta \bar{p} + a + \varepsilon. \tag{18}$$

There are two differences between equations (16) and (18). First, $\bar{p}$ in (18) is the price charged by an outlet from a rival chain, whereas it was the price charged in another unit of the same chain in (16). Second, δ here represents the degree of product substitutability between the two chains. We assume that δ is positive, but less than 1 so that the own-price effect is greater than the cross-price effect. The principal now chooses price, p, and the share parameter, α, given rival choices, $\bar{p}$ and $\bar{\alpha}$.[34] With these modifications, the corresponding equation for the optimal contract is

$$\alpha^* = \frac{1}{(1+r\sigma^2)(2-\delta)-1} \qquad (19)$$

and $d\alpha^*/d\delta > 0$. In other words, as products become closer substitutes, the power of the agents' incentives should be increased. This is true because it becomes more important to induce the agent to promote the product so that sales will not be eroded by customers switching to rival brands. Indeed, one can interpret the substitution effect as yet another measure of the importance of the agent's effort. The higher the degree of substitutability, the harder is the agent's task of preventing the erosion of its sales. Therefore, the principal has an additional motive for emphasizing high-powered incentives relative to other objectives.

Note that in modeling competition, we have implicitly assumed that the random variables that are associated with own and rival demand are uncorrelated. If, however, these variables are correlated, and if the agent has private information about his own demand realization, the tendency towards separation is strengthened when competition increases.[35] Indeed, as shown in Gal-Or (1995), demand correlation is information that the principal can use to reduce the agent's informational rent and thus the need to integrate.

Given that most agency-theoretic models neglect the demand side of the market, it is not surprising that most empirical studies rely solely on attributes of the upstream firm and its outlets and ignore the firm's competitors. To our knowledge, Coughlan (1985) and Slade (1998b) are the only studies that have looked at contract choice as a function of the demand characteristics that agents face. Coughlan finds that firms are more likely to use a middleman (separation) to enter a foreign market if they sell highly substitutable products, and to sell directly (integration) if their product is more unique. Similarly, Slade relates outlet-level own and cross-price elasticities of demand to the contracts under which outlets operate. As the model predicts, she finds that higher cross-price elasticities are associated with higher-powered incentives for the agent.[36]

Strategic Delegation of the Pricing Decision

We have assumed thus far that, when prices are endogenous, the principal chooses the retail price herself. In reality, however, with franchising, whether traditional or business-format, the principal usually delegates the pricing decision to the agent.[37] We now examine the principal's incentive to delegate in a strategic setting.

When price is exogenous, it is possible to normalize and make no distinction between rewarding the agent on the basis of revenues or sales. With endogenous prices, in contrast, particularly when the agent chooses price, it is important to be more specific. We therefore adopt an alternative notation that conforms more closely with actual compensation schemes in franchise chains. We maintain the demand assumption of the previous subsection (i.e., $q = 1 - p + \delta \bar{p} + a + \varepsilon$), and assume that the business-format franchisee now pays the franchisor a royalty, ρ, per unit sold as well as a fixed franchise fee, F.[38] The retailer's surplus is then

$$(p-\rho)(1-p+\delta\bar{p}+a) - F - \frac{a^2}{2} - \left(\frac{r}{2}\right)(p-\rho)^2 \sigma^2. \qquad (20)$$

The agent now chooses effort, a, and price, p, to maximize this surplus, given rival choices, $\bar{p}$ and $\bar{a}$, where the rival is again a franchisee from another chain in the same industry.

The two first-order conditions for the maximization of (20) imply the retail reaction functions,

$$p = \frac{1+r\sigma^2\rho+\delta\bar{p}}{1+r\sigma^2}, \qquad (21)$$

which are clearly upward sloping. Furthermore, in a symmetric equilibrium, the retail price is

$$p_D^* = \frac{1+r\sigma^2\rho}{1+r\sigma^2-\delta}, \qquad (22)$$

where the subscript D stands for delegation.

Comparative statics results, with ρ exogenous to the retailer, yield $dp/dr < 0$, $dp/d\sigma^2 < 0$, $dp/d\delta > 0$, and $dp/d\rho > 0$. Finally, if the retailer is risk neutral or there is no risk, the equilibrium retail price is[39]

$$p_D^* = \frac{1}{1-\delta}. \tag{23}$$

We compare the delegated situation to the integrated, in which the retailer is a salaried employee, whose wage is F, and ρ is equal to 0. In this case, the manufacturer (who is, as always, assumed to be risk neutral) chooses the retail price p, given rival price $\bar{p}$, which is chosen by the rival manufacturer. In a symmetric equilibrium of the integrated game, the retail price is

$$p_I^* = \frac{1}{2-\delta}, \tag{24}$$

where the subscript I stands for integrated. Clearly, if the retailers are risk neutral, principals prefer the delegated situation. Indeed, since reaction functions slope up, when a principal increases the royalty rate to her franchisee, not only does her retailer raise price but also the rival retailer responds with a price increase. In equilibrium, prices and profits are higher as a consequence.[40]

Under agent risk neutrality then, delegation is a dominant strategy. However, as $r\sigma^2$ increases, the advantages of delegation fall. This occurs because the higher retail price is accompanied by an increase in the proportion of the franchisee's income that is variable, thereby increasing the risk that the retailer must bear, and the risk premium he therefore requires. At some level of risk and/or risk aversion, the retailer's need for compensation for bearing increased risk makes vertical separation unattractive, and the firm chooses to vertically integrate instead. On the other hand, the more substitutable the products of the competing chains (the higher is δ), the more firms benefit from delegation (franchising) and thus the more likely it will be chosen. Overall then, this model predicts that vertical separation will be preferred when products are highly substitutable and there is little risk or risk aversion.

One can test these hypotheses individually; in earlier subsections, we have discussed the relevant literature and main results. Alternatively, a joint test can be constructed from the observation that delegation is more apt to occur when reaction functions are steep, since the slope of the reaction functions is $\delta/(1+r\sigma^2)$. As with the product-substitutability model, however, these tests require information about each unit's competitors. Slade (1998b), who has such data, finds that delegation is more likely when rival reaction functions are steep, as predicted.[41]

It is interesting to note that once again we come face to face with the prediction that franchising should be discouraged by local-market risk. As we have already discussed, however, the data are inconsistent with this prediction.

Multiple Tasks

In many retailing situations the agent performs more than one task. For example, a service-station operator might repair cars as well as sell gasoline, a publican might offer food services as well as beer, and a real-estate agent might rent houses as well as sell properties. Generally, when this is the case, the optimal contract for one task depends on the characteristics of the others. See Holmstrom and Milgrom (1991 and 1994).

There are many possible variants of multi-task models. We develop a simple version that illustrates our point. Suppose that there are n tasks and that the agent exerts effort, a_i, on the ith task. Effort increases output according to the linear relationship

$$q = a + \varepsilon \quad \text{where} \quad \varepsilon \sim N(0, \Sigma), \qquad (25)$$

where q, a, and ε are vectors of outputs, efforts, and shocks, respectively, and Σ is the variance/covariance matrix of ε. The agent's cost of effort is given by $(a^T a)/2$, so the risk premium is $-(r/2)\alpha^T \Sigma \alpha$. First-order conditions for the maximization of the agent's certainty-equivalent income with respect to the vector of effort levels yield $a_i = \alpha_i$, $i = 1,...,n$.

The principal chooses the vector of commissions, α, to maximize the total surplus, which after substitution of the incentive constraint is

$$\alpha^T j - \frac{\alpha^T \alpha}{2} - \left(\frac{r}{2}\right)\alpha^T \Sigma \alpha, \qquad (26)$$

where j is a vector of ones. First-order conditions for this maximization can be manipulated to yield:

$$\alpha^* = (I + r\Sigma)^{-1} j. \tag{27}$$

In the special case where $n = 2$ and $\sigma_{11} = \sigma_{22} = \sigma^2$, equation (27) simplifies to

$$\alpha_i^* = \frac{1}{1 + r(\sigma^2 + \sigma_{12})}, \quad i = 1,2. \tag{28}$$

If one compares equations (6) and (28) it is clear that, when a second task is added, the power of the agent's incentives in the optimal contract falls (rises) if the associated risks are positively (negatively) correlated. This occurs for pure insurance reasons. In other words, positive correlation means higher risk, whereas negative correlation offers risk diversification for the agent.

In this simple model, tasks are linked only through covariation in uncertainty. There are, however, many other possible linkages. For example, the level of effort devoted to one task can affect the marginal cost of performing the other, and, when prices are endogenous, nonzero cross-price elasticities of demand for the outputs can link the returns to effort.

Slade (1996) develops a model that incorporates these three effects and shows that, if an agent has full residual-claimancy rights on outcomes for a second task, the power of incentives for a first task (here gasoline sales) should be lower when the tasks are more complementary. Her empirical application of the model to retail gasoline supports the model's prediction. Specifically, she finds that when the second activity is repairing cars, which is less complementary with selling gasoline than managing a convenience store, agent gasoline-sales incentives are higher powered.

Franchise Contract Terms

As noted in the introduction, much of the empirical literature on retail contracting has focused on the dichotomous choice between integration and separation rather than on the terms of the franchise contract. Some authors, notably Lafontaine (1992a and 1993), Sen (1993), Rao and Srinivasan (1995), Wimmer and Garen (1996), Gagné et al. (1997), and Lafontaine and Shaw (1999a), however, have examined factors that affect the share parameter, α, directly. Three principal conclusions arise from this set of studies. First, the effects of factors such as risk, the importance of the agent's or the principal's inputs, outlet size, and monitoring difficulty are consistent with those that we have discussed. In other words, factors that tend to increase the degree of separation also tend to increase the agent's share of residual claims. Second, these factors explain a much larger proportion of the variation in the extent of vertical integration than of the variation in share parameters.[42] Thus it appears that firms, in responding to risk, incentive, and monitoring-cost issues, adjust by changing how much they use franchising rather than by altering the terms of their franchise contracts. In that sense, the theoretical models seem to be missing some important aspects of the upstream/downstream relationship. Third, and finally, franchise fees are in general not negatively correlated with royalty rates, despite the fact that the standard principal-agent model suggests that they should be.[43] Instead, fixed fees tend to be set at levels that compensate the franchisor for expenses incurred in setting up a franchised unit.[44]

Lafontaine and Shaw (1999a), who have access to panel data on contract terms, show that these are not only the same for all franchisees that join a chain at a point in time, as established in the earlier literature, but that they are quite persistent over time as well. In fact, they show that firm fixed effects account for about 85% of the variation in royalty rates and franchise fees, and that a very small proportion of this firm-level heterogeneity is related to sectoral differences. They conclude that royalty rates are mainly determined by differences across firms, differences that likely arise from unobserved heterogeneity in production and monitoring technologies, as well as potential quality differences. In addition, the authors find that contract terms do not follow any systematic pattern up or down when they are adjusted, and that they do not vary in any obvious way as firms age or grow.

Finally, several studies examine the use of various franchise contract terms other than royalty rates and franchise fees.[45] For example, Dnes (1993) focuses on franchisor control of leases, and on non-compete covenants, tie-in clauses, and clauses governing the transfer of franchisee assets upon termination. He argues that these clauses act together to protect each party from the potentially opportunistic behavior of the other. Brickley (1997) finds evidence that franchisors impose restrictions on passive ownership, rely on area-development plans, and require mandatory advertising contributions more often when the potential for franchisee free riding is high. He also finds that these contract clauses are complementary. Finally, Mathewson and Winter (1994) show that certain contract clauses, especially exclusive territories and various forms of quantity forcing, occur together in franchise contracts.[46]

3. Further Comments

Having completed our survey of the factors that determine contract choice and contract terms in franchise markets, we are left with a number of loose ends. In this section, we address issues that we believe are important but that do not lend themselves to being integrated into the framework of section 2. In particular, we first discuss one of the most important consequences of franchising — its effect on the level of retail prices. We then consider the effect of franchising on firm performance, the reasons why franchisors choose to employ a standard set of contracts rather than fine tune each contract to the characteristics of the agent and the market, and the reasons why royalties are typically calculated as a percentage of sales rather than profits. Finally, we address the issue of asset specificity that we touched upon briefly in the introduction.

Prices at Delegated Outlets

In addition to considering when firms might want to use delegation or integration, empirical research on retail contracting has also been concerned with some consequences of this decision. One area that has received relatively more attention is the effect of contractual form on the final prices that consumers pay.

There are several reasons why prices might be higher at separated outlets. First, some transactions are more costly in a market than inside a firm. For example, contracts written with franchisees are often more complex and thus costlier to write and enforce than those written with employees. Second, because separation involves two firms rather than one, it can introduce an additional administrative layer. Third, when retailers have market power, double-marginalization (i.e., successive output restrictions) can arise. Fourth, the existence of spillovers such as those already described can lead franchisees to choose prices above those that maximize the chain's profits. Finally, as we showed above, in a strategic model of contracting, separation lowers retailers' perceived elasticities of demand and thus increases retail markups.[47]

Table 7 summarizes results from six studies that are relevant to this issue. Three deal with retail prices of gasoline in the U.S., another deals with prices charged by retailers of separated and integrated soft-drink bottlers, still another involves beer sold in public houses in the U.K., and the last two are concerned with fast-food franchising in certain U.S. submarkets.

Table 7: The Effect of Vertical Separation on Price

Author	Year	Data	Price Effect
Barron and Umbeck,	1984	Gasoline Service Stations in Maryland	+*
Muris, Scheffman and Spiller	1992	Prices of Retailers Served by Integrated or Separated Soft-Drink Bottlers	+*
Shepard	1993	Gasoline Service Stations in Mass.	+ (sign. for one product)
Slade	1998	Beer in the UK	+*
Lafontaine	1998	Fast-Food in Pittsburgh and Detroit Metropolitan Areas	+*
Graddy	1997	Selected Fast-Food Chains in New Jersey and Western Pennsylvania	+*

Note: * indicates a result that is significant in the original study at the 0.05 level, based on a two-tail test.

Barron and Umbeck (1984) and Slade (1998b) look at legally mandated changes in contractual arrangements (i.e., before and after studies). Muris, Scheffman, and Spiller (1992) also do a before-and-after study in that they focus on the temporal effect on retail prices of soft-drink manufacturers' decisions to buy back some of their bottlers. The other studies investigate the effect of contract type on prices in a cross section of contracts, though Lafontaine (1998) considers both longitudinal and cross-sectional patterns in her data. As predicted by theory, in all six studies, increases in the degree of vertical separation, whether voluntary or mandated, result in higher retail prices.

Franchising and Firm Performance

Another fruitful area of research is the effect of franchising, or of franchise-contract terms, on firm performance, where firm performance can refer to profitability, service quality, or survival. Shelton's (1967) analysis is a classic in this respect. He uses data on costs, revenues, and profits for outlets in a single chain to examine the effect of switching from franchising to company ownership and from company ownership back to franchising. He finds no tendency for revenues to differ according to regime. However, under company ownership, costs are higher, and thus profits are lower, than under franchising.

The main advantage of Shelton's study is that its within firm design holds most things constant as the mode of organization changes. Its main drawback, however, is that units in this chain were operated under company ownership only when there was no franchisee available or during a transition period. In other words, franchising was the preferred mode, and company ownership was a transitory phase. Consequently, company ownership was likely to be inefficiently implemented.

Still, Shelton's findings suggest that franchising was indeed more efficient for all units of the firm that he studied. Thus one might expect company-owned units to under perform in other settings. In a context where firms prefer to own and operate some of their units, Krueger (1991) finds that company employees are paid slightly more and face somewhat steeper earnings profiles than employees in franchised units. He argues that the lower powered incentives given to managers of company restaurants make it necessary to offer greater incentives to employees, in the form of efficiency wages and steeper earnings profiles. Thus, consistent with Shelton (1964), Krueger (1991) finds that costs are higher in company units.

As for service quality, Beheler (1991) assesses the effect of company ownership on the health-inspection scores of a sample of 100 fast-food restaurants from 14 chains operating in the St.-Louis metropolitan area. He finds that these scores are poorer for company-owned units.[48]

Turning to the effect of franchise contract terms on performance, Agrawal and Lal (1995) assess how royalty rates affect the level of services provided by franchisees, where these are measured by hours of work per dollar of sales. They find that higher royalty rates lead to lower franchisee services. At the same time, and consistent with a double-sided moral hazard model of franchising, they find that higher royalties lead to greater brand-name investment by franchisors. They measure this investment as a combination (the sum in this case) of four standardized variables, namely advertising expenditures per dollar of sales, the number of franchises in the chain, the number of full-time franchisor staff per dollar of sales, and the number of ongoing services provided by the franchisor. Finally, Lafontaine and Shaw (1998) examine the effect of initial contract terms on franchisor survival five years later. They find a positive relationship with both royalty rates and franchise fees. Only the latter, however, is significant, suggesting a limited role for royalty rates in affecting future performance.[49]

To summarize, the limited evidence concerning the effect of franchising on performance suggests that lower-powered downstream incentives, in the form of company ownership or of higher royalty rates, tend to lower (raise) franchisee (franchisor) performance. However, much more work is needed in this area before one can draw more definitive conclusions.

Within-Firm Contract Uniformity

Though our basic model does not highlight this, most theoretical contracting models imply that the principal should tailor the terms of the contract to suit the characteristics of the agent, the outlet, and the market. In other words, equation (1) is the output/effort relationship for a particular franchisee and franchisor pair, and for a particular local market. It is clear then that the optimal share parameter, α^*, should differ by outlet within a chain as well as across chains. Contracts that are observed in practice, in contrast, are remarkably insensitive to variations in individual, outlet, and market conditions. Indeed, most firms use a standard business-format franchise contract — a single combination of royalty rate and franchise fee — for all franchised operations joining the chain at a point in time. The same lack of variation is observed in traditional franchising, where a manufacturer often charges the same wholesale price to all of her leased operations.[50] When this is true, the only choice that the principal makes in the end is whether to franchise or to self operate. In other words, when the characteristics of individual units differ, the upstream firm chooses to vertically integrate those units with characteristics that require less high-powered incentives, and to franchise those that require more, which explains the focus in empirical work on the choice between integration and separation rather than on the terms of the contract.

Models that emphasize incentive issues for both parties — double-sided moral-hazard models — provide one possible explanation for this lack of contract fine tuning. These models recognize that, with most franchising arrangements, not only does the agent have to provide effort, but also the principal must maintain the value of the trade name, business format, and company logo. With moral hazard on the part of both parties, even when both are risk neutral, an optimal contract involves revenue sharing.[51]

In such a double-sided moral-hazard context, Bhattacharyya and Lafontaine (1995) show that, under specific assumptions concerning functional forms, the benefits of customizing contracts can be quite limited, if not zero. This implies that the optimal contract is insensitive to many relationship-specific circumstances.[52] In addition, their model might at least partially explain the persistence of uniform contract terms over time found by Lafontaine and Shaw (1999a). Indeed, in the Bhattacharyya and Lafontaine model, the terms of the optimal contract remain unchanged as the franchise chain grows.

Other reasons that have been advanced in the literature to explain the lack of customization involve the high costs of customizing, either the direct cost of designing and administering many different contracts, as in Holmström and Milgrom (1987) and Lafontaine (1992b), or the high potential for franchisor opportunism that arises when contracts can vary, as in McAfee and Schwartz (1994).

Whatever the reason for the lack of customization in franchise contracting, it remains that most of the empirical research has focused either on the discrete choice to operate a unit as a franchise or not (when the data consist of individual contracts) or on the fraction of a franchisor's units that are franchised (when the data are at the upstream firm level). One might therefore ask if the same factors that lead to granting higher-powered incentives in the fine-tuning case also lead to a higher fraction of franchised outlets in the uniform-contract case. We now construct a formal model in which this is the case.

Suppose that each outlet or unit is associated with some characteristic x that affects its profitability, and let the expected profitability of that unit depend on the power of the agent's incentives as well as on this characteristic. One can express this relationship as $E\pi(\alpha,x)$. We assume that a) the expected profit function is concave, and b) $E\pi_{\alpha x} > 0$. In other words, as x increases, the marginal profitability of higher-powered incentives also increases.[53]

With the fine-tuning model in which contracts are outlet specific, the principal's problem is to choose α_i to maximize $E\pi(\alpha_i, x_i)$ for each unit i, subject to the agent's incentive constraint. The first-order condition for this maximization can be solved to yield the optimal contract, $\alpha^*(x)$. Moreover, assumption b) guarantees that $d\alpha^*/dx > 0$.

Now suppose that fine tuning is sufficiently expensive so that the principal offers only two contracts, a franchise contract with $\alpha > 0$ and a vertical integration contract with $\alpha = 0$. Moreover, the power of incentives (α) is the same for all franchisees. If the principal has n units, one can order those units such that $x_1 \geq x_2 \geq .. x_n$. Now the principal's problem is to

$$\max_{\alpha, i^*} \left[\sum_{i \geq i^*} E\pi(\alpha, x_i) + \sum_{i < i^*} E\pi(0, x_i) \right]. \tag{29}$$

Given i^*, the optimal contract $\alpha^*(i^*)$ can be obtained from the first-order condition, $\sum_{i \geq i^*} E\pi_\alpha = 0$, and given α, the optimal i^* satisfies (i) $E\pi(\alpha, x_{i^*}) - E\pi(0, x_{i^*}) \geq 0$, and (ii) $E\pi(\alpha, x_{i^*-1}) - E\pi(0, x_{i^*-1}) < 0$.[54] In this uniform-contract situation, an exogenous increase in x at some of a firm's units leads to both higher powered incentives (higher α^*) and to a larger fraction of outlets franchised (lower i^*).

Why Royalties on Sales

With most variants of the model of section 2, price is normalized to one and there are no input costs other than agent effort. As a result, there is no operational difference between royalties on sales, input markups, and royalties on profits. Indeed, most models of retail contracting make no distinction among these possibilities. In reality, however, business-format contracts usually involve royalties on sales.[55] The puzzling issue then is why business-format franchise contracts systematically emphasize "sales sharing" rather than profit sharing. For example, in Lafontaine's (1992b) survey, 123 of the 127 franchisors who responded to this question indicated that they charged some form of royalties. Of these, 112 asked for a percentage of sales or revenues. Only two franchisors requested a proportion of profits, while another four were paid a proportion of gross margins.[56]

The traditional explanation for the use of sales rather than profit-based royalties is that the latter are too difficult to measure. For example, franchisees can pad their costs by including personal cars and salaries for family members, and cost padding can be difficult to observe or to contract upon. However, this measurement argument does not explain why franchisors do not collect a proportion of gross margins more often.

Rubin (1978) proposes a more substantive explanation for sales sharing: he argues that franchisee effort controls costs as well as stimulates demand. Franchisor effort, in contrast, only affects demand. Consequently, franchisees should be given full residual claimancy on cost reductions, whereas franchisors should be paid some proportion of sales so that they have incentives to maintain the value of the trade name.

Maness (1996) formalizes this argument by assuming that costs are noncontractible, and as such must be borne by the owner of the outlet. Thus the decision to franchise (to have the franchisee own the outlet) or operate directly (to have the company own the unit) hinges on which party is better at controlling unit costs. Furthermore, the sharing rule must allow the owner of the unit to cover the costs of operation and thus satisfy his or her individual-rationality constraint. Therefore, in contrast to say sharecropping, where the 50/50 sharing rule for output often applies, royalty rates in franchise agreements are low, typically between 5 and 10%.[57]

Asset Specificity

Asset specificity is an important area of the theoretical literature that we have, up to now, had little to say about. We made this choice because we believe that it is far less important for retail contracting than for the purchase and sale of intermediate inputs. As a result, we don't think it sheds much new light on the empirical regularities highlighted herein. Nevertheless, as this issue regularly surfaces in the literature, we discuss how we arrived at this conclusion.

The positive effect of unit size on company ownership has been interpreted by some, e.g. Brickley and Dark (1987) and Scott (1995), as evidence that franchisors find it more costly to rely on franchising when franchisees are required to make large relationship-specific investments. We, however, find no evidence that total investment relates positively to asset specificity in retail contracting. For example, the largest gasoline stations are high-volume self-service stations that are the least specialized. The owner of such a station, if terminated by one refiner, could easily obtain a contract with another. The value of his assets should therefore not be significantly lower outside of the relationship. The same is true in business-format franchising. Within this group, the hotel industry requires the largest absolute level of investment. This investment, however, is again not specific; hotel banners are routinely changed with little effect on property values. Our point is that overall investment is not a good measure of asset specificity.[58]

Furthermore, Klein (1995) notes that, from an incentive perspective, what matters is not the level of specific investment by franchisees, as these are sunk and should not affect behavior, but rather the rents or quasi rents that the franchisee can expect to lose if he is terminated.[59] Moreover, Dnes (1993) finds that franchisees' specific investments are protected by the terms of franchise agreements. More specifically, he argues that the contractual clauses that govern the transfer of franchisee assets upon termination are set such that

> *"if the franchisor withdraws from a contract and offers to buy assets (even if this follows the franchisee offering assets for sale), then the prices are governed by something other than just the franchisor's wishes,"* (p. 390)

be it arbitration or some notion of fair-market value. Presumably, units of franchisees who are terminated for disciplinary reasons are viable, and franchisors will want to buy the assets or allow other franchisees to do so. Consequently, upon termination, the current franchisee does not forego the rents that are attached to specific assets, and in that sense, these rents cannot play a self-enforcement role.[60]

On the other hand, other rents are lost by franchisees upon termination. In particular, the non-compete clauses that are found in most franchise contracts can make it difficult for franchisees to put the human capital they have accumulated within the chain to good use upon termination. Similarly, given that franchisees are often allowed to expand their business by owning additional outlets in a chain, whatever rents are associated with the right to purchase these extra units are foregone upon termination from a franchised system.[61] However, the value of such rents is not well captured by a measure of specific investments.

4. Final Remarks

Our survey of retail contracting under exclusive marks has highlighted the existence of many stylized facts and the robust nature of the evidence. Indeed, in almost every case where a factor is statistically significant, its effect on the power of agent incentives in real-world contracts is the same across studies. In other words, in spite of the fact that researchers assess different industries over different time periods using a number of proxies for a given factor, their empirical findings are usually consistent with one another.

The theories, on the other hand, are much more fragile. In fact, in order to obtain a tractable model, it is important to use simple specifications for agent utility and risk preference. The results of the model then can depend non-trivially on these assumptions. Furthermore, the way in which the unobservable risk factors interact with the tangible variables is also crucial, as we have demonstrated in our discussion of outlet size. Nevertheless, we hope that our attempt to organize the evidence in a unified framework will be helpful to theorists in that it gives them a set of stylized facts to explain. As for applied researchers, we hope to have provided them with a framework and a sense of where more empirical work would be most beneficial.

One theoretical prediction, however, is not fragile; it surfaces over and over again. We refer to the effect of risk on agent incentives. Whether one considers the simplest incentive/insurance model, or imbeds this model in one with endogenous prices and strategic delegation or one with multiple tasks and linked efforts, the theory predicts that more risky units should tend to be operated by the parent company. The evidence, however, strongly rejects this predicted tendency. We have suggested one possible explanation for the discrepancy between theory and evidence — endogenous output variability in a situation where agents have private information about local-market conditions. However, as shown in Allen and Lueck's (1995) survey of the sharecropping literature, a similar empirical finding surfaces in the sharecropping context, an area where exogenous output fluctuations are apt to dominate endogenous fluctuations. Given the central role that agent risk plays in the incentive-contracting literature, and given the strength of the empirical evidence, we believe that this puzzle in particular deserves further attention.

References

Agrawal, D. and R. Lal (1995), "Contractual Arrangements in Franchising: An Empirical Investigation," *Journal of Marketing Research*, 32, 213-221.

Alchian, A.A. and H. Demsetz (1972), "Production, Information Costs, and Economic Organization," *American Economic Review*, 62, 777-795.

Allen, D.W. and D. Lueck (1992), "Contract Choice in Modern Agriculture: Cropshare versus Cash Rent," *Journal of Law and Economics*, 35, 397-426.

Allen, D.W. and D. Lueck (1995), "Risk Preferences and the Economics of Contracts," *American Economic Review*, 85, 447-451.

Anderson, E. (1985), "The Salesperson as Outside Agent or Employee: A Transaction Cost Analysis," *Marketing Science*, 4, 234-254.

Anderson, E. and R.L. Oliver (1987), "Perspectives on Behavior-Based Versus Outcome-Based Salesforce-Control Systems, *Journal of Marketing*, 51, 76-88.

Anderson, E. and D. Schmittlein (1984), "Integration of the Sales Force: An Empirical Examination," *Rand Journal of Economics, 15*, 385-395.

Athey, S. and S. Stern (1997), "An Empirical Framework for Testing Theories about Complementarities in Organizational Design," MIT, mimeo.

Bates, T., (1998), "Survival Patterns Among Newcomers to Franchising," *Journal of Business Venturing*, 13, 113-130.

Barron, J.M. and J.R. Umbeck (1984), "The Effects of Different Contractual Arrangements: The Case of Retail Gasoline," *Journal of Law and Economics*, 27, 313-328.

Beheler, R.L. (1991), "Control in Various Organizational Forms: An Empirical Study of Company-Owned and Franchisee-Owned Units' Health Inspections," in J.R. Nevin (Ed.), *Franchising: Embracing the Future*, Fifth Annual Proceedings of the Society of Franchising, University of St-Thomas.

Bercovitz, J.E.L. (1998a), "An Analysis of Contractual Provisions in Business-Format Franchise Agreements," mimeo, Fuqua School of Business, Duke University.

Bercovitz, J.E.L. (1998b), "Franchising vs. Company Ownership," mimeo, Fuqua School of Business, Duke University.

Bhattacharyya, S. and F. Lafontaine (1995), "Double-Sided Moral Hazard and the Nature of Share Contracts," *RAND Journal of Economics*, 26, 761-781.

Blair, R.D. and D. L. Kaserman, (1994), "A Note on Incentive Incompatibility under Franchising," *Review of Industrial Organization*, 9, 323-330.

Blair, R.D. and F. Lafontaine (1999), "Will *Khan* Foster or Hinder Franchising? An Economic Analysis of Maximum Resale Price Maintenance," *Journal of Public Policy in Marketing*, 18, 25-36.

Bonanno, G. and J. Vickers (1988), "Vertical Separation," *Journal of Industrial Economics*, 36, 257-266.

Bradach, J.L. (1997), "Using the Plural Form in the Management of Restaurant Chains," *Administrative Science Quarterly*, 42, 276-303.

Brickley, J.A., (1997), "Incentive Conflicts and Contracting: Evidence from Franchising," mimeo, Simon Graduate School of Business, University of Rochester.

Brickley, J. and F. Dark (1987), "The Choice of Organizational Form: The Case of Franchising," *Journal of Financial Economics*, 18, 401-420.

Brickley, J.A., F.H. Dark, and M.S. Weisbach (1991), "An Agency Perspective on Franchising," *Financial Management*, 20, 27-35.

Carmichael, H.L. (1983), "The Agent-Agents Problem: Payment by Relative Output," *Journal of Labor Economics*, 1, 50-65.

Carney, M. and E. Gedajlovic (1991), "Vertical Integration in Franchise Systems: Agency Theory and Resource Explanations," *Strategic Management Journal*, 12, 607-629.

Caves, R.E. and W.F. Murphy (1976), "Franchising: Firms, Markets, and Intangible Assets," *Southern Economic Journal*, 42, 572-586.

Coase, R. (1937), "The Nature of the Firm," *Economica*, 4, 386-405.

Coughlan, A.T. (1985), "Competition and Cooperation in Marketing Channel Choice: Theory and Application," *Marketing Science*, 4, 110-129.

Crocker, K. J. and S. E. Masten (1996), "Regulation and Administered Contracts Revisited: Lessons from Transaction-Cost Economics for Public Utility Regulation," *Journal of Regulatory Economics*, 9, 5-39.

Crocker, K.J. and K.J. Reynolds (1992), "The Efficiency of Incomplete Contracts: An Empirical Analysis of Air Force Engine Procurement," *RAND Journal of Economics*, 24, 126-146.

Dant, R.P., P.J. Kaufmann and A.K. Paswan (1992), "Ownership Redirection in Franchised Channels," *Journal of Public Policy and Marketing*, 11, 33-44.

Dnes, A. W. (1992), *Franchising: A Case-study Approach*, Ashgate Publishing Ltd., Aldershot, England.

Dnes, A.W. (1993), "A Case Study Analysis of Franchise Contracts", *Journal of Legal Studies*, 22, 367-393.

Dnes, A. W. (1996), "The Economic Analysis of Franchise Contracts," *Journal of Institutional and Theoretical Economics*, 152, 297-324.

Elango, B. and V. H. Fried (1997), "Franchising Research: A Literature Review and Synthesis," *Journal of Small Business Management*, 35, 68-82.

Fudenberg, D. and J. Tirole (1984), "The Fat-Cat Effect, The Puppy-Dog Ploy, and the Lean-and-Hungry Look," *American Economic Review*, 74, 361-366.

Gal-Or, E. (1995), "Maintaining Quality Standards in Franchise Chains," *Management Science*, 41, 1774-1792.

Gagné, R., S.P. Sigué, and G. Zaccour (1998), "Droit d'entrée et taux de redevance dans les franchises d'exploitation au Québec," *L'Actualité économique*, 74, 651-668.

Garen, J.E. (1994), "Executive Compensation and Principal-Agent Theory," *Journal of Political Economy*, 102, 1175-1199.

Graddy, K. (1997), "Do Fast-Food Chains Price Discriminate on the Race and Income Characteristics of an Area," *Journal of Business and Economic Statistics*, 15, 391-401.

Grossman, S.J. and O.D. Hart (1986), "The Costs and Benefits of Ownership: A Theory of Vertical and Lateral Integration," *Journal of Political Economy*, 94, 691-719.

Hart, O. and J. Moore (1988), "Incomplete Contracts and Renegotiation," *Econometrica*, 56, 755-786.

Holmstrom, B. (1979), "Moral Hazard and Observability," *Bell Journal of Economics*, 10, 74-91.

Holmstrom, B. (1982), "Moral Hazard in Teams," *Bell Journal of Economics*, 13, 324-340.

Holmstrom, B. and P. Milgrom (1987), "Aggregation and Linearity in the Provision of Intertemporal Incentives," *Econometrica*, 55, 303-328.

Holmstrom, B. and P. Milgrom (1991), "Multitask Principal-Agent Analyses: Incentive Contracts, Asset Ownership, and Job Design," *Journal of Law, Economics, and Organization*, 7, 24-51.

Holmstrom, B. and P. Milgrom (1994), "The Firm as an Incentive System," *American Economic Review*, 84, 972-991.

Ichniowski, C., K.L Shaw and G. Prennushi (1997), "The Effects of Human Resource Management Practices on Productivity: Evidence from the Steel Industry," *American Economic Review*, 87, 291-313.

Jensen, M. and K. Murphy (1990), "Performance Pay and Top-Management Incentives," *Journal of Political Economy*, 98, 225-264.

John, G. and B.A. Weitz (1988), "Forward Integration into Distribution: An Empirical Test of Transaction Cost Analysis," *Journal of Law, Economics, and Organization*, 4, 337-355.

Joskow, P. (1988), "Asset Specificity and the Structure of Vertical Relationships: Empirical Evidence," *Journal of Law, Economics, and Organization*, 4, 95-117.

Kalnins, A. and F. Lafontaine (1999), "Incentive and Strategic Motives for Vertical Separation: Evidence from Ownership Patterns in the Texan Fast-Food Industry," mimeo, University of Michigan Business School.

Kaplan, S.N. (1994), "Top Executive Rewards and Firm Performance: A Comparison of Japan and the U.S.," *Journal of Political Economy*, 102, 510-546.

Kaufmann, P. J. and R. P. Dant (1996), "Multi-Unit Franchising: Growth and Management Issues," *Journal of Business Venturing*, 11, 343-358.

Kaufmann, P.J. and Lafontaine, F. (1994), "Costs of Control: The Source of Economic Rents for McDonald's Franchisees," *Journal of Law and Economics*, 37, 417-543.

Kehoe, M. R. (1996), "Franchising, agency problems, and the cost of capital," *Applied Economics*, 28, 1485-1493.

Klein, B. (1980), "Transaction Cost Determinants of 'Unfair' Contractual Arrangements," *American Economic Review*, 70, 356-362.

Klein, B. (1988), "Vertical Integration as Organizational Ownership: The Fisher Body - General Motors Relationship Revisited," *Journal of Law, Economics, and Organization*, 4, 199-213.

Klein, B. (1995), "The Economics of Franchise Contracts," *Journal of Corporate Finance*, 2, 9-37.

Klein, B., R. Crawford, and A. Alchian (1978), "Vertical Integration, Appropriable Rents, and the Competitive Contracting Process," *Journal of Law and Economics*, 21, 297-326.

Knight, F. (1921), *Risk, Uncertainty, and Profit*, Houghton Mifflin Pub., Chicago, USA.

Krueger, A. B., (1991), "Ownership, Agency and Wages: An Examination of the Fast Food Industry," *Quarterly Journal of Economics*, 106, 75-101.

Lafontaine, F. (1992a), "Agency Theory and Franchising: Some Empirical Results," *RAND Journal of Economics*, 23, 263-283.

Lafontaine, F. (1992b), "How and Why do Franchisors do What They do: A Survey Report", in P.J. Kaufmann (Ed.), *Franchising: Passport for Growth & World of Opportunity*, Sixth Annual Proceedings of the Society of Franchising, University of St-Thomas.

Lafontaine, F. (1993), "Contractual Arrangements as Signaling Devices: Evidence from Franchising," *Journal of Law, Economics, and Organization*, 9, 256-289.

Lafontaine, F. (1995), "Pricing Decisions in Franchised Chains: A Look at the Fast-Food Industry," NBER Working paper #5247.

Lafontaine, F. (1998), "Retail Pricing, Organizational Form, and the New Rule of Reason Approach to Maximum Resale Prices," mimeo, University of Michigan Business School.

Lafontaine, F. and Bhattacharyya, S., 1995, "The Role of Risk in Franchising", *Journal of Corporate Finance*, 2, 39-74.

Lafontaine, F. and P.J. Kaufmann (1994), "The Evolution of Ownership Patterns in Franchise Systems," *Journal of Retailing*, 70, 97-113.

Lafontaine, F. and Shaw, K.L., (1998), "Franchising Growth and Franchisor Entry and Exit in the U.S Market: Myth and Reality," *Journal of Business Venturing*, 13, 95-112.

Lafontaine, F. and Shaw, K.L., (1999a) "The Dynamics of Franchise Contracting: Evidence from Panel Data", *Journal of Political Economy*, forthcoming.

Lafontaine, F. and Shaw, K.L., (1999b), "Targeting Managerial Control: Evidence from Franchising," mimeo, University of Michigan Business School.

Lafontaine, F. and Slade, M.E. (1996), "Retail Contracting and Costly Monitoring: Theory and Evidence" *European Economic Review Papers and Proceedings*, 40, 923-932.

Lafontaine, F. and Slade, M.E. (1997), "Retail Contracting: Theory and Practice," *Journal of Industrial Economics*, 45, 1-25.

Lal, R. (1990), "Improving Channel Coordination through Franchising," *Marketing Science*, 9, 299-318.

Lazear, E. P. (1996), "Performance Pay and Productivity," NBER Working Paper #5672.

Leffler, K.B. and R.R. Rucker (1991), "Transaction Costs and the Efficient Organization of Production: A Study of Timber-Harvesting Contracts," *Journal of Political Economy*, 99, 1060-1087.

Lewin, S. (1998), "Autonomy, Contractibility, and the Franchise Relationship," mimeo, Iowa State University.

Lutz, N.A. (1995), "Ownership Rights and Incentives in Franchising," *Journal of Corporate Finance*, 2, 103-130.

Lyons, B.R. (1996), "Empirical Relevance of Efficient Contract Theory: Inter-Firm Contracts," *Oxford Review of Economic Policy*," 12, 27-52.

Maness, R. (1996), "Incomplete Contracts and the Choice Between Vertical Integration and Franchising," *Journal of Economic Behavior and Organization*, 31, 101-115.

Martin, R.E. (1988), "Franchising and Risk Management," *American Economic Review*, 78, 954-968.

Masten, S.E. (1984), "The Organization of Production: Evidence from the Aerospace Industry," *Journal of Law and Economics*, 27, 403-417.

Masten, S.E. (1985), "Efficient Adaptation in Long-Term Contracts: Take-or-Pay Provisions for Natural Gas," *American Economic Review*, 75, 1083-93.

Masten, S.E., (1998), "Contractual Choice," *Encyclopedia of Law & Economics*, B. Boukaert and G. De Geest (Eds.), Edward Elgar Publishing, forthcoming.

Mathewson, F. and R. Winter (1985), "The Economics of Franchise Contracts," *Journal of Law and Economics*, 28, 503–526.

McAfee, R.P. and M. Schwartz (1994), "Multilateral Vertical Contracting: Opportunism, Nondiscrimination, and Exclusivity," *American Economic Review*, 84, 210–230.

McGuire T.W. and R. Staelin, (1983), "An Industry Equilibrium Analysis of Downstream Vertical Integration," *Marketing Science*, 2, 161-192.

Michael, S.C. and H.J. Moore (1995), "Returns to Franchising, " *Journal of Corporate Finance*, 2, 133-156.

Milgrom, P. and J. Roberts (1992), *Economics, Organization and Management*, Englewoods Cliffs: Prentice-Hall Inc.

Minkler, A. (1990), "An Empirical Analysis of a Firm's Decision to Franchise," *Economics Letters*, 34, 77-82.

Minkler, A. and T. A. Park (1994), "Asset Specificity and Vertical Integration," *Review of Industrial Organization*, 9, 409-423.

Mirrlees, J.A. (1976), "The Optimal Structure of Incentives and Authority Within an Organization," *Bell Journal of Economics*, 7, 105-131.

Monteverde, K. and D. Teece (1982), "Supplier Switching Costs and Vertical Integration in the Automobile Industry," *Bell Journal of Economics*, 13, 206-213.

Muris, T.J., D.T. Scheffman, and P.T. Spiller (1992), "Strategy and Transaction Costs: The Organization of Distribution in the Carbonated Soft Drink Industry," *Journal of Economics & Management Strategy*, 1, 83-128.

Murphy, K. (1984), "Incentives, Learning, and Compensation: A Theoretical and Empirical Investigation of Managerial Labor Contracts," *RAND Journal of Economics*, 17, 59-76.

Norton, S.W. (1988), "An Empirical Look at Franchising as an Organizational Form," *Journal of Business*, 61: 197-217.

Ozanne, U.B. and S.D. Hunt, (1971), *The Economic Effect of Franchising*, U.S. Senate, Select Committee on Small Business, Washington, Gov. Printing Office.

Price, S. (1996), "Behind the veneer of success: Propensities for UK franchisor failure," *Small Business Research Trust Report*.

Rao, R.C. and S. Srinivasan (1995), "Why are Royalty Rates Higher in Service-Type Franchises," *Journal of Economics & Management Strategy*, 4, 7-31.

Rey, P. and J. Stiglitz (1995), "The Role of Exclusive Territories in Producers' Competition," *RAND Journal of Economics*, 26, 431-451.

Rindfleisch, A. and J.B. Heide (1997), "Transaction Cost Analysis: Past, Present and Future Applications," *Journal of Marketing*, 61, 30-54.

Romano, R. E. (1994), "Double Moral Hazard and Resale Price Maintenance," *RAND Journal of Economics*, 25, 455-466.

Rubin, P. (1978), "The Theory of the Firm and the Structure of the Franchise Contract," *Journal of Law and Economics*, 21, 223-233.

Scott, F.A. (1995), "Franchising vs. Company Ownership as a Decision Variable of the Firm," *Review of Industrial Organization*, 10, 69-81.

Sen, K. C. (1993), "The Use of Initial Fees and Royalties in Business Format Franchising," *Managerial and Decision Economics*, 14, 175–190.

Shane, S.A. (1996), "Hybrid Organizational Arrangements and their Implications for Firm Growth and Survival: A Study of New Franchisors," *Academy of Management Journal*, 39, 216-234.

Shane, S.A. (1997), "Organizational Incentives and Organizational Mortality," mimeo, Sloan School of Management, MIT.

Shane, S.A. and P. Azoulay (1998), "Contract Design and Firm Survival," mimeo, Sloan School of Management, MIT.

Shelanski, H.A. and Klein, P. G. (1995), "Empirical Research in Transaction Cost Economics: A Review and Assessment," *Journal of Law, Economics, and Organization*, 11, 335-361.

Shelton, J. (1967), "Allocative Efficiency vs. 'X-Efficiency': Comment," *American Economic Review*, 57, 1252-1258.

Shepard, A. (1993), "Contractual Form, Retail Price, and Asset Characteristics," *Rand Journal of Economics*, 24, 58-77.

Slade, M.E. (1996), "Multitask Agency and Contract Choice: An Empirical Assessment," *International Economic Review*, 37, 465-486.

Slade, M.E. (1998a), "Beer and the Tie: Did Divestiture of Brewer-Owned Public Houses Lead to Higher Beer Prices?", *Economic Journal*, 108, 1-38.

Slade, M.E. (1998b), "Strategic Motives for Vertical Separation: Evidence from Retail Gasoline," *Journal of Law, Economics, & Organization*, 14, 84-113.

Smith II, R. L. (1982), "Franchise Regulation: An Economic Analysis of State Restrictions on Automobile Distribution," *Journal of Law and Economics*, 25, 125-157.

Stanworth, J., 1996, "Dispelling the myths surrounding franchise failure rates - some recent evidence from Britain," *Franchising Research: An International Journal*, 1, 25-28.

State Oil Co. v. Khan, US Sup. CT, No. 96-871.

Stiglitz, J.E. (1974), "Incentives and Risk-Sharing in Sharecropping," *Review of Economic Studies*, 41, 219-255.

Thompson, R.S. (1994), "The Franchise Life Cycle and the Penrose Effect," *Journal of Economic Behavior and Organization*, 24, 207-218.

U.S. Department of Commerce (1998), *Franchising in the Economy, 1986-1988*, prepared by Andrew Kostecka.

Vickers, J. (1985), "Delegation and the Theory of the Firm," *Economic Journal*, 95, 138-147.

Williams, D.L. (1999), "Why do Entrepreneurs Become Franchisees? An Empirical Analysis of Organizational Choice," *Journal of Business Venturing*, 14, 103-124.

Williamson, O.E. (1971), "The Vertical Integration of Production: Market Failure Considerations," *American Economic Review*, 61, 112-123.

Williamson, O.E. (1979), "Transaction Cost Economics: The Governance of Contractual Relations," *Journal of Law and Economics*, 22, 233-262.

Williamson, O.E. (1983), "Credible Commitments: Using Hostages to Support Exchange," *American Economic Review*, 73, 519-540.

Williamson, O.E.(1985), *The Economic Institutions of Capitalism*, Free Press, New-York, NY.

Wimmer, B.S. and J.E. Garen (1997), "Moral Hazard, Asset Specificity, Implicit Bonding, and Compensation: The Case of Franchising," *Economic Inquiry*, 35, 544-554.

Appendix: Algebraic Derivations

In each case below, the agent (A) maximizes his certainty-equivalent income, $E(y) - (r/2)Var(y)$, whereas the principal (P) maximizes the expected total surplus — expected output minus the agent's cost of effort minus the agent's risk premium, $E(q) - a^2/2 - (r/2)Var(y)$. With one exception, noted below, the agent's compensation in each case is given by $S(q) = \alpha q + W$. The cases

Incentive Contracting and the Franchise Decision

differ according to the specification of the function that maps effort into output, $q = f(a, \varepsilon, \Theta)$.

Risk:

$$q = a + \varepsilon.$$

A: $$\max_{a} \left[\alpha a + W - \frac{a^2}{2} - \frac{r}{2}\alpha^2\sigma^2 \right]$$

The resulting first-order condition (foc) is: $a = \alpha$. Substituting the agent's effort choice into the principal's problem yields:

P: $$\max_{\alpha} \left[\alpha - \frac{\alpha^2}{2} - \frac{r}{2}\alpha^2\sigma^2 \right], \quad foc: \ 1 - \alpha - r\alpha\sigma^2 = 0,$$

$$\alpha^* = \frac{1}{1 + r\sigma^2},$$

$$\frac{d\alpha^*}{dr} = -\frac{\sigma^2}{\Gamma^2} < 0, \quad \frac{d\alpha^*}{d\sigma^2} = -\frac{r}{\Gamma^2} < 0, \quad \text{where} \ \ \Gamma = 1 + r\sigma^2.$$

Agent Effort:

$$q = \eta a + \varepsilon.$$

A: $$\max_{a} \left[\alpha \eta a + W - \frac{a^2}{2} - \frac{r}{2}\alpha^2\sigma^2 \right], \quad foc: \ a = \alpha\eta.$$

P: $$\max_{\alpha} \left[\eta^2\alpha - \frac{\alpha^2\eta^2}{2} - \frac{r}{2}\alpha^2\sigma^2 \right], \quad foc: \ \eta^2 - \alpha\eta^2 - r\alpha\sigma^2 = 0,$$

$$\alpha^* = \frac{\eta^2}{\eta^2 + r\sigma^2},$$

$$\frac{d\alpha^*}{d\eta} = \frac{2\eta r\sigma^2}{\Gamma^2} > 0, \quad \text{where} \ \ \Gamma = \left(\eta^2 + r\sigma^2\right)$$

Outlet Size:

$$q = \eta a + (\gamma + \varepsilon)k.$$

A: $\quad \max_{a} \left[\alpha(\eta a + \gamma k) + W - \dfrac{a^2}{2} - \dfrac{r}{2}\alpha^2 \sigma^2 k^2 \right], \quad$ *foc:* $a = \alpha\eta.$

P: $\quad \max_{\alpha} \left[\eta^2 \alpha - \dfrac{\alpha^2 \eta^2}{2} - \dfrac{r}{2}\alpha^2 \sigma^2 k^2 \right], \quad$ *foc:* $\eta^2 - \alpha\eta^2 - r\alpha\sigma^2 k^2 = 0,$

$$\alpha^* = \dfrac{\eta^2}{\eta^2 + r\sigma^2 k^2},$$

$$\dfrac{d\alpha^*}{dk} = -\dfrac{2\eta^2 r\sigma^2 k}{\Gamma^2} < 0, \quad \text{where } \Gamma = \eta^2 + r\sigma^2 k^2.$$

Costly Monitoring:

This result is derived in Lafontaine and Slade (1996).

Franchisor Effort:

$$q = \eta a + \theta b + \varepsilon.$$

This problem has two incentive constraints:

A: $\quad \max_{a} \left[\alpha(\eta a + \theta b) + W - \dfrac{a^2}{2} - \dfrac{r}{2}\alpha^2 \sigma^2 \right], \quad$ *foc:* $a = \alpha\eta.$

P: $\quad \max_{b} \left[(1-\alpha)(\eta a + \theta b) - \dfrac{b^2}{2} \right], \quad$ *foc:* $b = (1-\alpha)\theta.$

The franchisor determines α to maximize total surplus, given by $\eta a + \theta b - \dfrac{a^2}{2} - \dfrac{b^2}{2} - \dfrac{r}{2}\alpha^2\sigma^2$, subject to the two incentive constraints. After substituting, we have:

$$\max_{\alpha}\left[\alpha\eta^2+(1-\alpha)\theta^2-\frac{\alpha^2\eta^2}{2}-\frac{(1-\alpha)^2\theta^2}{2}-\frac{r}{2}\alpha^2\sigma^2\right],$$

foc: $\eta^2-\theta^2-\alpha\eta^2+(1-\alpha)\theta^2-r\alpha\sigma^2=0,$

$$\alpha^*=\frac{\eta^2}{\eta^2+\theta^2+r\sigma^2},$$

$$\frac{d\alpha^*}{d\theta}=-\frac{2\eta^2\theta}{\Gamma^2}<0,\quad \frac{d\alpha^*}{d\eta}=\frac{2\eta(\theta^2+r\sigma^2)}{\Gamma^2}>0,\quad \text{where } \Gamma=\left(\eta^2+\theta^2+r\sigma^2\right)$$

Spillovers Within the Chain:

$$q=1-p-\mu\bar{p}+a+\varepsilon.$$

where $\bar{p}$ is the price at another outlet in the same chain.

A: $\quad \max_{a}\left[\alpha(1-p-\mu\bar{p}+a)+W-\frac{a^2}{2}-\frac{r}{2}\alpha^2\sigma^2\right],\quad$ foc: $a=\alpha.$

The principal chooses $p=\bar{p}$ and α to

P: $\quad \max_{p,\alpha}\left[(1-(1+\mu)p+\alpha)p-\frac{\alpha^2}{2}-\frac{r}{2}\alpha^2\sigma^2\right],$

foc: $1-2(1+\mu)p+\alpha=0,\quad p^*=\frac{1+\alpha}{2(1+\mu)},$
$\quad p$

foc: $p-\alpha-r\alpha\sigma^2=0.$
$\quad \alpha$

Substituting for p yields:

$$\alpha^*=\frac{1}{2(1+r\sigma^2)(1+\mu)-1},$$

$$\frac{d\alpha^*}{d\mu} = -\frac{2(1+r\sigma^2)}{\Gamma^2} < 0 \quad \text{where } \Gamma = 2(1+r\sigma^2)(1+\mu) - 1.$$

Product Substitutability:

$$q = 1 - p + \delta\bar{p} + a + \varepsilon$$

where $\bar{p}$ is now the price at a rival chain.

A: $\quad \max_{a} \left[\alpha(1-p+\delta\bar{p}+a) + W - \frac{a^2}{2} - \frac{r}{2}\alpha^2\sigma^2 \right], \quad \text{foc: } a = \alpha.$

P: $\quad \max_{p,\alpha} \left[(1-p+\delta\bar{p}+\alpha)p - \frac{\alpha^2}{2} - \frac{r}{2}\alpha^2\sigma^2 \right],$

$$\text{foc: } 1 - 2p + \delta\bar{p} + \alpha = 0.$$

Using symmetry to set $p = \bar{p}$ yields:

$$p^* = \frac{1+\alpha}{2-\delta}.$$

$$\text{foc: } p - \alpha - r\alpha\sigma^2 = 0.$$

Substituting for p yields:

$$\alpha^* = \frac{1}{(1+r\sigma^2)(2-\delta) - 1},$$

$$\frac{d\alpha^*}{d\delta} = -\frac{(1+r\sigma^2)}{\Gamma^2} > 0, \quad \text{where } \Gamma = (1+r\sigma^2)(2-\delta) - 1.$$

Strategic Delegation of the Pricing Decision:

$$q = 1 - p + \delta\bar{p} + a + \varepsilon,$$

where $\bar{p}$ is again the price at a rival chain. In this case, the agent is compensated by residual claims after he pays a royalty ρ per unit to the franchisor, as well as a franchise fee F. Thus we have:

A: $$\max_{a,p}\left[(p-\rho)(1-p+\delta\bar{p}+a)-F-\frac{a^2}{2}-\frac{r}{2}(p-\rho)^2\sigma^2\right]$$

$$\text{foc: } a = p - \rho,$$

$$\text{foc: } 1 - 2p + \delta\bar{p} + a + \rho - r(p-\rho)\sigma^2 = 0.$$

Substituting for a yields:

$$p = \frac{1 + \delta\bar{p} + r\rho\sigma^2}{1 + r\sigma^2},$$

$$\frac{dp}{d\bar{p}} = \frac{\delta}{\Gamma} > 0, \quad \frac{dp}{d\delta} = \frac{\bar{p}}{\Gamma} > 0, \quad \frac{dp}{dr} = \frac{\sigma^2(\rho - 1 - \delta\bar{p})}{\Gamma^2} < 0,$$

$$\frac{dp}{d\sigma^2} = \frac{r(\rho - 1 - \delta\bar{p})}{\Gamma^2} < 0, \quad \frac{dp}{d\rho} = \frac{r\sigma^2}{\Gamma} > 0,$$

where $\Gamma = 1 + r\sigma^2$. Using symmetry to set $p = \bar{p}$ yields:

$$p_D^* = \frac{1 + r\rho\sigma^2}{1 - \delta + r\sigma^2}$$

$$= \frac{1}{1-\delta} \quad \text{when } r\sigma^2 = 0.$$

By contrast, under vertical integration, assuming that $a = 0$ and $\rho = 0$, we have

P: $\quad\max_{p}\left[(1-p+\bar{\wp})p-F\right] \quad$ foc: $1-2p+\delta\bar{p}=0$.

Setting $p=\bar{p}$ yields $p_I^* = \dfrac{1}{2-\delta} < 1 < \dfrac{1}{1-\delta} = p_D^*$. Thus $p_I^* < p_D^*$ when $r\sigma^2$ is small.

Multiple tasks:

$$q = a+\varepsilon, \quad \varepsilon \sim N(0,\Sigma), \quad C(a) = \dfrac{a^T a}{2},$$

where q, a, and ε are vectors, as is α. However, W remains a scalar.

A: $\quad \max_{a}\left[\alpha^T a + W - \dfrac{a^T a}{2} - \dfrac{r\alpha^T\Sigma\alpha}{2}\right], \quad$ foc: $a=\alpha$.

After substituting, we have:

P: $\quad \max_{\alpha}\left[\alpha^T j - \dfrac{\alpha^T\alpha}{2} - \dfrac{r\alpha^T\Sigma\alpha}{2}\right], \quad$ foc: $j-\alpha-r\Sigma\alpha = 0$.

Hence, $\alpha^* = (I+r\Sigma)^{-1}j$, where j is a vector of ones. When $n = 2$, this becomes

$$\alpha_i^* = \dfrac{1+r\sigma_{jj}-r\sigma_{12}}{(1+r\sigma_{11})(1+r\sigma_{22})-r^2\sigma_{12}^2}.$$

Setting $\sigma_{11} = \sigma_{22} = \sigma^2$ yields:

$$\alpha_i^* = \dfrac{1+r\sigma^2 - r\sigma_{12}}{(1+r\sigma^2)^2 - r^2\sigma_{12}^2} = \dfrac{1}{1+r(\sigma^2+\sigma_{12})}$$

so that

$$\dfrac{d\alpha_i^*}{d\sigma_{12}} = -\dfrac{r}{(1+r(\sigma^2+\sigma_{12}))^2} < 0.$$

Notes

[1] See, for example, Murphy (1984), Jensen and Murphy (1990), Kaplan (1994), and Garen (1996).

[2] For other areas, see e.g. Lazear (1996) on the effect of piece rates on production-worker productivity. For a broader discussion of the effect of human-resource management practices on production-worker productivity, see e.g. Ichniowski, Shaw and Prennushi, (1997).

[3] For example, see Monteverde and Teece (1982), Anderson and Schmittlein (1984), Masten (1984), Anderson (1985), Masten and Crocker (1985), Joskow (1988), Klein (1988), and Crocker and Reynolds (1992). For surveys of this empirical literature, see Shelanski and Klein (1995), Crocker and Masten (1996), and Rindfleisch and Heide (1997).

[4] For surveys of the franchising literature with a different emphasis, see Dnes (1996) and Elango and Fried (1997). For surveys with a broader contracting focus, see Lyons (1996) and Masten (1998).

[5] The distinction between these two types of franchising can be blurred sometimes because business-format franchisors can sell inputs to franchisees (e.g. Baskin-Robbins), and traditional franchisors offer training and ongoing business support to their dealers as well. See Dnes (1992, 1993) for more on this.

[6] See for example Rubin (1978), Mathewson and Winter (1985), Lal (1990), and Bhattacharyya and Lafontaine (1995). Also see Stiglitz (1974) for the earliest application of agency theory to explain the use and properties of another type of share contract, namely sharecropping.

[7] In business-format franchising, different franchisors choose different contract terms — different royalty rates and franchise fees — but a given franchisor offers the same terms to all potential franchisees at a given point in time. This makes the franchise versus company-operation dichotomy a meaningful one; if contracts were allowed to vary for each franchisee, then, assuming for simplicity that the company manager is paid a fixed salary, company ownership would be a limit case where the royalty rate is zero and the franchise fee negative. Of course, such a limit case would hardly ever be observed. In reality, the dichotomy involves more than just differences in the compensation scheme of the unit manager; it also involves differences in asset ownership and in the distribution of responsibilities between upstream and downstream parties. Similarly, in traditional franchising, while commission rates and fees can vary across a firm's agents, the distinction between integration and separation is well defined. This distinction again involves differences in the distribution of power between manufacturer and retailer. See, for example, Smith II (1982) and Slade (1998a).

[8] For example, Brickley and Dark (1987), John and Weitz (1988), Martin (1988), Norton (1988), Lafontaine (1992a), and Scott (1995).

[9] For example, Anderson and Schmittlein (1984), Barron and Umbeck (1984), Anderson (1985), Brickley and Dark (1987), Minkler (1990), Muris, Scheffman and Spiller (1992), Shepard (1993), Graddy (1995), Lafontaine (1995), and Slade (1996 and 1998b).

[10] Note that, as we assume below that the error term enters all of our functional forms in some additive way, our assumption that $\varepsilon \sim N(0,\sigma^2)$ also implies that q is normally distributed.

[11] We use the word linear here as has traditionally been done in the share-contract literature. The contracts, however, typically include a fixed component and are thus affine.

[12] See e.g. Lutz (1995) for a discussion of this issue in the context of franchising.

[13] For a possible explanation, see Holmstrom and Milgrom (1994).

[14] For possible explanations, see Holmstrom and Milgrom (1987), Romano (1994), and Bhattacharyya and Lafontaine (1995).

[15] In franchising applications, see Lal (1990) for an example of the first type of interpretation, and Bhattacharyya and Lafontaine (1995) for an example of the second.

[16] See Kaufmann and Lafontaine (1994) for evidence that there are rents left downstream at McDonald's. The authors argue that they serve an incentive role similar to that of efficiency wages. Michael and Moore (1995) find evidence that such rents are present in other franchised systems as well.

[17] The participation constraint is normally used to determine W, not α.

[18] On the relative merits of these two measures, see Lafontaine and Bhattacharyya (1995).

[19] See also Allen and Lueck (1992, 1995) and Leffler and Rucker (1991) for evidence that risk-sharing does not explain contract terms well in sharecropping and in timber harvesting respectively.

[20] See Lafontaine and Bhattacharyya (1995) for a formal model. Note that the positive relationship between incentives and output variability that they find depends on the form of the function that maps effort and the random variables into output.

[21] As shown below, if one assumes that k enters (9) only in an additive way, then changes in k have no effect on the optimal share parameter, α. If one assumes that k multiplies a, then its effect is the same as that of η in the previous subsection, and increases in k lead to higher values of α, the reverse of what we obtain with our formulation. With a combination of interactive terms with risk and franchisee effort, we would get two opposing effects, and the sign of the net effect would depend on the specific parameters of the problem.

[22] See Brickley and Dark (1987) for more on this argument, which they refer to as the "inefficient risk-bearing" argument against franchising.

[23] Consistent with the above evidence, on a sectoral basis, company units have higher sales (are larger) than franchised units (US Dept. of Commerce, 1988). Moreover, Muris, Scheffman and Spiller (1992) argue that the increase in the efficient size of bottling operations led soft-drink manufacturers to buy back several of their independent bottlers and enter into joint-venture agreements with many others.

[24] This result also depends on the assumption that information flows are superior within the firm.

[25] This subsection is based on Lafontaine and Slade (1996).

[26] The first-best effort level is defined as the level that the principal would choose if she were not constrained by incentive considerations in maximizing the total surplus.

[27] One alternative source of information that we do not consider arises when uncertainty is correlated across agents in a multi-agent setting. In that case, the optimal contract for agent i includes some measure of other agents' performance in addition to his own, as in Holmstrom (1982). Empirically, such relative-performance contracts are not used in franchising.

[28] The type of mechanism that we have in mind is sometimes called "behavior-based" compensation, as opposed to "outcome-based" compensation. See Anderson and Oliver (1987).

[29] The model is similar to Holmstrom and Milgrom (1991), who model multiple tasks and signals.

[30] In mapping our results from more or less sales-based compensation to more or less vertical integration, we are implicitly assuming that behavior monitoring takes place, and behavior-based compensation is used, inside the firm, but that sales commissions are not or are little used inside the firm. With complete separation, in contrast, the agent is the residual claimant, and there is no (or very little) behavior monitoring or behavior-based compensation. See Holmstrom and Milgrom (1991) for a discussion of these issues. See Bradach (1997) for descriptions of business practices in five franchised restaurant chains that suggest that these assumptions are realistic.

[31] See e.g. Rubin (1978), Mathewson and Winter (1985), Lal (1990) and Bhattacharyya and Lafontaine (1995) for more on this. Consistent with the argument that the franchisor must be given incentives in these cases, in the one case of a franchise agreement that does not involve any ongoing royalties or company ownership on the part of the franchisor, Dnes (1993) notes that "Franchisees (in this system) do complain of insufficient effort by the franchisor in supporting the development of their businesses." (p. 386; text in parentheses added).

[32] For a review of the empirical literature on the "ownership redirection hypothesis", according to which franchising is just a transitory phase for firms that face capital constraints, see Dant, Kaufmann and Paswan (1992). For more recent contributions, see also Lafontaine and Kaufmann (1994), Thompson (1994) and Scott (1995).

[33] For our current purposes, it is simpler to assume that the franchisor chooses price. There is some evidence that franchisors try to control franchisee prices, e.g. Ozanne and Hunt (1971), but rules against resale price maintenance have made this difficult up until recently, when the U.S. Supreme Court decided in *State* Oil v. *Khan* that maximum resale price maintenance would no longer be a *per se* violation of antitrust law. See Blair and Lafontaine (1999) for more on this decision and its likely impact on franchising, and Lafontaine (1998) for more on price controls in franchising. Note that the spillover problem is exacerbated when the franchisee chooses price. This situation can be modeled by changing the sign of δ in the demand equation in the next subsection.

[34] We continue to assume that the franchisor chooses price. In the next section, we relax this assumption.

[35] Here the increase in the cross-price elasticity is due to an increase in the number of competitors.

[36] When our evidence is from very few studies, we do not construct a table.

[37] US Antitrust laws prevent franchisors from enforcing specific prices in franchised units as these are

independent businesses under the law. Of course, this does not prevent franchisors from trying to affect franchisees' choice of prices indirectly, through advertising (Caves and Murphy, 1976) or other means. Moreover, as noted above, a recent Supreme Court decision *(State Oil v. Khan)* has transformed the *per se* status of maximum resale price maintenance to a rule of reason status, which opens the possibility that franchisors will control franchisee prices more in the U.S. in the future. See Blair and Lafontaine (1999) and Lafontaine (1998) for more on this.

38 With traditional franchising, ρ can be interpreted as the wholesale price that the retailer pays to the manufacturer for the product, and F as the fixed rent that he pays for the use of the retail outlet, which we assume is owned by the upstream firm. If there were no rent, or equivalent fixed payment, dealings between principal and agent would be arms length, and the principal would maximize the wholesale, not the total, surplus.

39 Most of the theoretical papers on this subject assume that there is no uncertainty and thus no moral hazard, e.g., McGuire and Staelin (1983), Vickers (1985), Bonanno and Vickers (1988), and Rey and Stiglitz (1995).

40 In the terminology of Fudenberg and Tirole (1984), this is a fat-cat game.

41 In her model, however, there is no risk and therefore no agency cost.

42 See Lafontaine (1992a) on this.

43 This prediction results from the fact that, in most theoretical models, the principal is assumed to extract all rent from the agent, an assumption that we have not exploited.

44 See Lafontaine (1992a), Dnes (1993) and Lafontaine and Shaw (1999a) on this issue.

45 In addition to those specifically mentioned, see Bercovitz (1998a).

46 See Athey and Stern (1997) for theoretical arguments as to why one might expect such complementarities.

47 See also Rey and Stiglitz, (1995).

48 Barron and Umbeck (1984) examine the effect of divorcement, or "forced franchising," of gasoline stations on hours of operation. They find that franchising leads to a reduction in hours, which corresponds to lower quality. This finding, however, as those related to pricing, mostly reflects the ease of setting and controlling hours of operation in company units. In other words, this result occurs because there is no agency problem with respect to hours of operation (or pricing) under vertical integration, but there is one under separation.

49 See also Shane (1997) on the effect of franchise contract terms on instantaneous survival, and Shane and Azoulay (1998) on the effect of exclusive territories on survival. For assessments of franchisor survival rates, see Price (1996), Shane (1996) and Stanworth (1996). For the effect of franchising on small-business survival, see Williams (1999), Bates (1998), and the references therein.

50 In the U.S., the Robinson-Patman Act requires wholesale-price uniformity, at least locally. This is not true, however, in Canada. Nevertheless, price uniformity across buyers is common there as well, e.g., in gasoline markets; see Slade (1996 and 1998a) on this. Also, the Robinson-Patman act does not explain contract uniformity in business-format franchising, as the Act applies to the sale of commodities, which do not include franchising rights. See McAfee and Schwartz (1994) as well as Bhattacharyya and Lafontaine (1995) for further arguments against legal constraints as the main source of contract uniformity in business-format franchising.

51 See e.g. Rubin (1978), Lal (1990), and Bhattacharyya and Lafontaine (1995). Carmichael (1983) has shown that with two agents or more, and moral hazard on the principal's side as well as the agents', the first best can be achieved with a contract based on relative outputs. However, we do not observe this type of contract in franchising. Why this is the case is beyond the scope of the present paper.

52 More specifically, Bhattacharyya and Lafontaine (1995) show that, when the production function is Cobb-Douglas and the cost-of-effort function is exponential, the optimal share parameter is independent of the scale of operation, and, as a result, of the level of demand and the degree of competition in the market. The share parameter is also independent of both parties' cost-of-effort parameters.

53 For example, x might be the importance of the agent's effort or the negative of the agent's degree of risk aversion.

54 We are assuming an interior solution witth $1 < i^* < n$. Assumption b) guarantees that the left-hand side of (i) is greater than the left-hand side of (ii) for any i.

55 In traditional franchise agreements, the franchisor sells a manufactured product to the franchisee who then

resells it. Assuming that the franchisee has little leeway on prices, input markups are equivalent to royalties on sales. See Dnes (1993) and Lafontaine (1993) and the references therein for more on this and on tying in business-format franchising.

[56] Of the remaining 5 firms, 4 charged a fixed amount per time period, and one did not answer this part of the question.

[57] For another argument on this issue, see also Lewin (1998).

[58] See Dnes (1993) and Wimmer and Garen (1996) for attempts to capture the part of total investment that is specific.

[59] While specific assets tend to generate rents, they are not necessary; downstream rents can also arise because franchisors choose to leave them with franchisees. See Kaufmann and Lafontaine (1994) and Michael and Moore (1995) for evidence that some franchisors choose to do this.

[60] Dnes argues that franchisees sustain a loss if they fail and their franchisor decides not to buy back their unit because it is not viable, and that this loss is larger the more specific assets are involved. He argues that such potential losses give franchisees incentives to get involved only if they truly are able to perform as they say they are. Thus he concludes that specific investments can serve a franchisee-screening function.

[61] See for example Kaufmann and Dant (1996) and Kalnins and Lafontaine (1999) on multi-unit ownership in franchised chains. Also see Bradach (1997) on the importance of additional units for franchisee growth and statements that refusing to grant extra units to franchisees serves a disciplinary role.

6 COOPERATIVE GAMES AND BUSINESS STRATEGY

H.W. Stuart, Jr.

The main purpose of this chapter is to demonstrate how cooperative game theory can be applied to business strategy. Although the academic literature on cooperative game theory is extensive, very little has been written from a strategy perspective. The situation with textbooks is similar. With few exceptions (e.g., Oster, 1994), strategy textbooks generally do not mention cooperative game theory. In fact, the most visible applications of cooperative game theory to strategy have been in the popular press, MacDonald's *The Game of Business*[1] (1975) and Brandenburger and Nalebuff's *Co-opetition* (1996), for example. The material in this chapter is taken from academic papers and teaching notes that use cooperative game theory for analyzing business strategy. Applying cooperative game theory to strategy often results in suggestive interpretations; this chapter will emphasize these interpretations since they are, arguably, a significant benefit of using cooperative game theory.

A cooperative game consists of a player set and a function specifying how much value any subset of players can create. This sparse formalism can be used to model situations in which there are no restrictions on the interactions between players. Players are free to pursue any favorable deals possible, and, in particular, no player is assumed to have price-setting power. For this reason, cooperative game theory can be considered a structural, rather than procedural, theory. It specifies the structure of the game: who the players are and what value they might appropriate. But it does not specify the procedures for creating and dividing value.

The structural approach of cooperative game theory has certain advantages for the study of business strategy. Since many business interactions are free-form in nature, this might not seem so surprising. But there are other benefits that this chapter will demonstrate. The effects of competition can be clearly identified in the analysis of a cooperative game. Although price-setting power is not assumed in a cooperative game, it can emerge from the structure of the game.[2] And even the process of formulating the game yields insights, as it often requires that basic business questions be answered.

The non-procedural nature of cooperative game theory may also be viewed as one of its limitations. In situations in which the players' interactions must follow well-defined rules, the free-form nature of a cooperative game probably will not be desirable. The question of uncertainty poses another limitation. To date, uncertainty has not been integrated into the application of cooperative game theory to business strategy, and attempts to do so have been informal at best.

Section 1 of this chapter reviews the definition of a cooperative game. For business strategy, cooperative games will be analyzed by methods that model unrestricted bargaining. With unrestricted bargaining, players are assumed to be actively involved in the creation and division of value. Unrestricted bargaining can be modeled by either the core of the game or the added value principle. The core is a more traditional method of analysis, but in business strategy, the added value principle often suffices. This section provides both definitions.

To examine actual cooperative games for business strategy, Section 1 uses the Supplier-Firm-Buyer game. This game is used to demonstrate how competition, reflected in players' added values, can either partially or completely determine the division of value. Further, the discussion of added value in these games introduces some of the basic business questions.

Business strategy is as much about creating a favorable game for yourself as it is about doing well in some existing game. Section 2 introduces a game form, called a biform game, for analyzing situations in which players have the ability to affect what business game they play.[3] Two examples are provided. First, a familiar monopoly example demonstrates both the approach and the relationship between monopoly power and competition. The second example is a biform model of spatial competition, in which socially-efficient, spatial differentiation can be a stable outcome.

1. The Game of Business as a Cooperative Game

Cooperative Games

A transferable-utility (TU) cooperative game is composed of a finite set N and a mapping $v: 2^N \to \Re$. The set N denotes the set of players, and the mapping v, the characteristic function, specifies, for each subset of the players, the value created by the subset. Thus, for any $S \subseteq N$, $v(S)$ is the maximum economic value that the players in S can create among themselves.

An outcome of a cooperative game $(N; v)$ is described by a vector $x \in \Re^{|N|}$. This vector specifies both the total value created and how it is divided. The component x_i denotes the value captured by player i. And the total value created, namely $x(N)$, is then $\sum_{i \in N} x_i$.

Definition: For any $S \subseteq N$, let $x(S) = \sum_{i \in S} x_i$. The *core* of a TU cooperative game $(N; v)$ is the set of outcomes satisfying $x(N) = v(N)$ and $\forall S \subseteq N, x(S) \geq v(S)$.

Definition: The *added value* of a coalition $S \subset N$ is defined to be the coalition's marginal contribution: $v(N) - v(N \setminus S)$. (The term $v(N \setminus S)$ denotes the value created by the coalition consisting of all players *except* the players in S.) The added value of a given player, say i, is therefore $v(N) - v(N \setminus \{i\})$. An outcome satisfies the *added value principle* if no player captures more than his or her added value, that is, if:
$$\forall i \in N, x_i \leq v(N) - v(N \setminus \{i\}), \text{ and if } x(N) = v(N).$$

The added value principle generates a superset of the core; that is, core outcomes must satisfy the added value principle. To see this, consider a given player i. The two conditions $x(N) = v(N)$ and $x(N \setminus \{i\}) \geq v(N \setminus \{i\})$ imply the added value condition $x_i \leq v(N) - v(N \setminus \{i\})$.

Both the core and the added value principle can be interpreted as modeling unrestricted bargaining. With unrestricted bargaining, all players are actively involved in seeking out favorable transactions, and no player is assumed to have any price-setting power. Further, any player or group of players is free to pursue a more favorable deal. For both the added value principle and the core, the mathematical statements generating these interpretations can be identified. With the added value principle, it is implied by the equation $x_i \leq v(N) - v(N \setminus \{i\})$. If a player were to receive more than its added value, namely $x_i > v(N) - v(N \setminus \{i\})$, then it would have to be the case that $x(N \setminus \{i\}) < v(N \setminus \{i\})$. But if this were so, the group $N\setminus\{i\}$ would prefer to actively pursue a deal on its own, thus obtaining $v(N\setminus\{i\})$ of value. In other words, if a given player were to receive more than its added value, then the other players would do better by excluding that player from the game.

With the core, a similar interpretation holds. But, whereas the unrestricted-bargaining interpretation of the added value principle considers only coalitions of the form $N\setminus\{i\}$, the interpretation for the core considers any coalition. If any coalition, say S, anticipated capturing less value than it could create on its own, namely $x(S) < v(S)$, then it would create and divide $v(S)$ of value on its own. This interpretation is consistent with the core condition that $\forall\, S \subseteq N, x(S) \geq v(S)$.

With any group of players free to create value on its own, Aumann (1985, pg. 53) states that the core captures the notion of "unbridled competition." The idea that a core analysis can model competition dates back to Edgeworth (1881). As Shubik (1959) discovered, Edgeworth's reasoning about "contracting" and "re-contracting" is consistent with the core conditions. But this "competition" does not necessarily determine a unique outcome. For a given player, the core will usually specify a range of values rather than a single amount. The minimum of the range is interpreted as the amount of value guaranteed the player due to competition. The difference between the minimum and the maximum is then interpreted as a residual bargaining problem. With an added value analysis, the interpretation is the same. If the added value principle implies a minimum amount for a player, then this is the amount due to competition. If the added value principle allows the player to capture more than this minimum, then the difference is a residual bargaining problem.

Supplier-Firm-Buyer Games[4]

Shapley and Shubik (1972) use two-sided assignment games to gain insights into buyer-seller markets with small numbers of players. Since business strategy is often concerned with small buyer-seller interactions, this suggests that assignment games might provide a natural starting point for applying cooperative game theory to business strategy. But since strategy focuses on the firm, it is convenient to start with a three-sided assignment game. Three sides are useful since the firm is both a buyer and a seller – a buyer with respect to its suppliers and a seller with respect to its customers.

Definition: A *three-sided assignment game* consists of three disjoint sets, N_1, N_2, N_3, and a three-dimensional assignment matrix, A. The matrix has dimensions of $n_1 \times n_2 \times n_3$, where $n_1 = |N_1|$, etc. (Alternatively, the assignment matrix is a mapping $A: N_1 \times N_2 \times N_3 \to \Re$.) The disjoint sets are interpreted as sets of players. A *matching* is a 3-tuple ijk consisting of a player i from set N_1, a player j from set N_2, and a player k from set N_3. Element a_{ijk} of the matrix A is interpreted as the value that can be created by the matching ijk. Two matchings $i^a j^a k^a$ and $i^b j^b k^b$ are *distinct* if $i^a \neq i^b, j^a \neq j^b, k^a \neq k^b$. In turn, an *assignment* of size r is a set of r distinct matchings.

The construction of the TU cooperative game for an m-sided assignment game is based on two principles. First, value creation is determined solely by distinct matchings, and second, the value created is taken to be as large as feasibly possible. Equations (1) – (3) below incorporate these principles.

Define the player set, N, to be $N_1 \cup N_2 \cup N_3$. The characteristic function v is defined by $v(\emptyset) = 0$, and, for all $T \subseteq N$,

$$v(T) = 0 \quad \text{if } T \cap N_m = \emptyset, \; m \in \{1, 2, 3\}, \tag{1}$$

$$v(\{i, j, k\}) = a_{ijk} \quad \forall i \in N_1, j \in N_2, k \in N_3, \text{ and} \tag{2}$$

$$v(T) = \max_{r, AS_r} (a_{i^1 j^1 k^1} + \ldots + a_{i^r j^r k^r}), \tag{3}$$

where $r \leq \min\{|T \cap N_1|, |T \cap N_2|, |T \cap N_3|\}$ and AS_r is the set of assignments of size r constructed from set T.

Note that equation (1) implies that no value is created if a matching is not possible. Equation (2) defines the value created by any given matching, and equation (3) states that the value created by a given coalition is computed by arranging the players into a collection of distinct matchings that yields the greatest value.

Definition: The *supplier-firm-buyer* game is a three-sided assignment game with the following restriction:

$$a_{ijk} = w_{jk} - c_{ij} \ \forall \ i \in N_1, j \in N_2, k \in N_3. \tag{4}$$

The set N_1 is interpreted as a set of suppliers, the set N_2 as a set of firms, and the set N_3 as a set of buyers. The term w_{jk}, represents buyer k's willingness-to-pay for transacting with firm j. Similarly, the term c_{ij}, represents supplier i's opportunity cost for transacting with firm j.

A supplier-firm-buyer game may always be analyzed with the core, as the core is always non-empty in these games (Stuart 1997b). But these games can be more immediately analyzed with added value analysis, as the examples that follow will demonstrate. Before considering the specific examples, consider the case in which there is just one supplier, one firm, and one buyer. The value created will be just $w - c$, namely the buyer's willingness-to-pay minus the supplier's opportunity cost. How will this value be divided? The answer is that any division of this value is possible. The added value of each player is equal to the total value created, so that there are no limits (other than the total value) on any player's value capture.

This situation is interpreted as one in which competition plays no role in determining the division of value. Since each player's added value is equal to the total value created, the residual bargaining problem is the "whole pie," and any division of value is possible. Figure 1 depicts a possible division of value. Bargaining between the buyer and the firm will lead to a price p_1 for the firm's product. Bargaining between the firm and the supplier will lead to a price p_2 for the supplier's resource. From the firm's perspective, this second price is a cost.

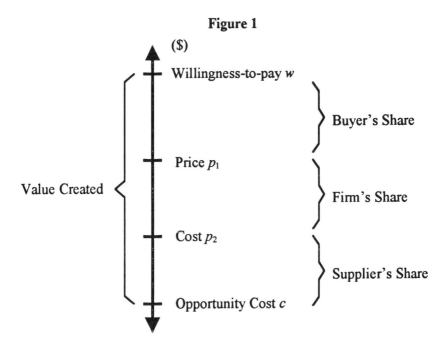

Figure 1

Examples

In the special case of one supplier, one firm, and one buyer, the division of value is completely indeterminate. In general, the division of value will not be completely indeterminate. The next three examples demonstrate the role of added values in determining the division of value.

Example 1. $N_1 = \{s_1, s_2\}, N_2 = \{f_1, f_2\}, N_3 = \{b_1\}$;
$w_{11} = 100, w_{21} = 150$;
$c_{11} = c_{12} = c_{21} = c_{12} = 10$.

Equations (1) through (4) specify how to construct the characteristic function for this example. In particular note that $v(N)$ equals 140. Since there is only one buyer, there can be only one matching. From equation (3), $v(N)$ is determined by the largest possible matching, namely the buyer with the second firm and either of the two suppliers.

Table 1 below provides the added value analysis for this example. For a given player, say player l, the guaranteed minimum is given by the quantity

$\max\{0, v(N) - \sum_{m \in N \setminus \{l\}} (v(N) - v(N \setminus \{m\}))\}$. This minimum derives from the fact that with the added value principle, no player can receive more than its added value. Thus, if every other player receives its added value, and if there is still some value remaining, then player l is guaranteed to receive this "leftover" value, namely $v(N) - \sum_{m \in N \setminus \{l\}} (v(N) - v(N \setminus \{m\}))$.

Table 1

Player l	$v(N)$	$v(M\{l\})$	Added Value $v(N) - v(M\{l\})$	Guaranteed Minimum
Buyer	140	0	140	90
Firm 1	140	140	0	0
Firm 2	140	90	50	0
Supplier i	140	140	0	0

In this example, there are two suppliers, two firms, and one buyer. Each supplier has an opportunity cost of $10 for providing resources to either firm. The buyer has a willingness-to-pay of $100 for the first firm's product, and a willingness-to-pay of $150 for the second firm's product. Each player's added value can be interpreted in terms of competition. Since only one supplier is required, and since the suppliers are identical, the added value of each supplier is $0. The second firm has added value of $50, but the first firm has no added value. With the second firm in the game, the first firm provides no additional benefit. The buyer has added value of $140. Without the buyer, no value is created, so the buyer could capture all the value. Could the buyer capture none of the value? The answer is no. Although the first firm has no added value, it does provide partial competition for the second firm. Consequently, the second firm can capture, at most, $50, thus guaranteeing that the buyer captures at least $90 of value. In summary, competition between the suppliers and partial competition between the firms guarantee $90 to the buyer. The remaining $50 is divided in a residual bargaining problem between the second firm and the buyer.

In the above analysis, added values significantly narrowed down the range of possible outcomes, but they still left residual value to be divided. In the next example, added values will completely determine the division of value. In Example 2, the sum of the added values equals the total value created. With the added value principle, this is a necessary and sufficient condition to

Cooperative Games and Business Strategy

uniquely determine the division of value. With the core, it is only a sufficient condition, provided the core is non-empty.[5]

Example 2. $N_1 = \{s_1, s_2, s_3, s_4\}$, $N_2 = \{f_1, f_2, f_3\}$, $N_3 = \{b_1, b_2\}$;
$w_{jk} = 100$, $j = 1,2,3, k = 1,2$;
$c_{ij} = 10$, $i = 1,2,3,4, j = 1,2,3$.

Table 2

Player l	$v(N)$	$v(N\{l\})$	Added Value $v(N)-v(N\{l\})$	Guaranteed Minimum
Buyer 1	180	90	90	90
Buyer 2	180	90	90	90
Firm j	180	180	0	0
Supplier i	180	180	0	0

In this game, the value created is $180 = 2 \times \$(100 - 10)$. The added value of each supplier and each firm is \$0. Each buyer has an added value of \$90, and since all the other players have zero added value, each buyer captures its added value. As before, there is a natural interpretation for this. There are an excess number of firms with respect to the buyers, and so each firm has zero added value. Also, note that although the firms are in a favorable position with respect to suppliers, this favorable position is not a sufficient condition for capturing value. In contrast, consider Example 3, in which the buyers have a higher willingness-to-pay for transacting with the third firm.

Example 3. $N_1 = \{s_1, s_2, s_3, s_4\}$, $N_2 = \{f_1, f_2, f_3\}$, $N_3 = \{b_1, b_2\}$;
$w_{jk} = 100$, $j = 1,2, k = 1,2$; $w_{3k} = 150$, $k = 1,2$;
$c_{ij} = 10$, $i = 1,2,3,4, j = 1,2,3$.

Table 3

Player l	$v(N)$	$v(N\{l\})$	Added Value $v(N)-v(N\{l\})$	Guaranteed Minimum
Buyer 1	230	140	90	90
Buyer 2	230	140	90	90
Firm 1,2	230	230	0	0
Firm 3	230	180	50	50
Supplier i	230	230	0	0

The game in Example 3 has much in common with the game in Example 2. The number of each type of player remains the same, and the sum of the added values equals the total value created: the value created is $230 = $(150 − 10) + $(100 − 10); the added values of the suppliers, the first firm, and the second firm are each $0; the third firm has an added value of $50, and each buyer has an added value of $90. As in Example 2, this is an example of perfect competition. But, unlike Example 2, one of the firms has positive added value, which it captures.[6] The source of this firm's added value is that it is "different" from its competitors. That is, it has a favorable asymmetry between itself and the other firms. In this example, the favorable asymmetry takes the form of buyers having a higher willingness to pay for its product. In moving from Example 2 to Example 3, the third firm established positive added value through a favorable willingness-to-pay asymmetry.

Alternatively, the third firm could have established positive added value through a favorable asymmetry on the supplier side. In such a case, the suppliers would have had a lower opportunity cost of providing resources to the third firm, as compared with providing them to the other two firms. In either case, a favorable asymmetry would have led to positive added value.

This focus on favorable asymmetries as a source of added value can prompt some of the basic questions which a (potentially) profitable business would want to answer. These questions arise by performing the following thought experiment.

Suppose, hypothetically, that a company were to close its business. If it has added value, it must be true that

(1) its buyers would then buy a product for which they had a lower willingness to pay (WTP) (or not buy at all), or
(2) its suppliers would incur a higher opportunity cost (OC) in supplying their resources to another business (or not supply at all), or
(3) both.

The logic behind this thought experiment is implied by this informal question. If a company disappeared from the market, and its buyers wouldn't care and its suppliers wouldn't care, why would it be making any money? In other words, without a favorable asymmetry, why would the company capture any value? Testing for such a favorable asymmetry is what prompts the basic

Cooperative Games and Business Strategy

business questions. For instance, the company should ask, if it were to disappear:

    Whom might its buyers buy from?
    Would its buyers have a lower WTP for this alternative?
    Why would its buyers have a lower WTP for this alternative?
    Whom might its suppliers sell to?
    Would its suppliers have a higher OC for this alternative?
    Why would its suppliers have a higher OC for this alternative?

These questions, though seemingly straightforward, actually require a good understanding of a business to answer properly. In particular, notice that answers to these questions require that the company understand:

    who its buyers are, and why they might prefer its products;
    who its suppliers are, and why they prefer doing business with it;
    whom else its buyers might want to buy from; and
    whom else its suppliers might want to do business with.

In short, the question of positive added value is the question of existence of a favorable asymmetry, which, in turn, is the question of whether a business is viable.

2. Choosing the Game

The previous section used the supplier-firm-buyer game to demonstrate how the structure of a business context could be modeled by a cooperative game. When using cooperative games, the term "structure" is well-defined: the players in the game and the value created by any group of these players. Therefore, whenever this structure is changed, the game also is changed. Thus, many a strategic decision is actually a decision about what game to be in. Examples include investing in a technology that reduces costs, finding ways to increase the willingness-to-pay for a product, changing production capacity, deciding to merge or integrate, and so on. In short, any decision that affects either the players in the game or the value created by any group of the players is a decision about choosing what business game to be in.

This section discusses biform games. A biform game is a hybrid game form designed to model situations in which players can choose what business game to play. Roughly speaking, a biform game is a non-cooperative game in which the consequences are cooperative games rather than specific payoffs.

Following a description of the formalism, two applications of a biform game will be presented: a monopoly model and a spatial competition model.

The Biform Formalism

The definition of a biform game starts with a strategic game form, that is:

(1) a finite set N of players, and
(2) for each player $i \in N$, a set A_i of strategies.

Let $A = \underset{i \in N}{\times} A_i$. Consider:

(3) a function $V: A \to \Re^{2^N}$ satisfying that for every a in A,
$V(a)(\varnothing) = 0$, and
(4) for each player i, a number α_i in $[0, 1]$.

A *biform game* is then a collection $\langle N; \{A_i\}_{i \in N}; V; \{\alpha_i\}_{i \in N} \rangle$.

The formalism of a biform game may be interpreted as follows. The players first make strategic choices from the strategy spaces A_i. Each resulting profile a in A of strategic choices induces a TU cooperative game $V(a) : 2^N \to \Re$, which is interpreted as a business game. For each player i, the number α_i is termed player i's *confidence index*. When a player's value-capture depends, partially or totally, upon a residual bargaining problem, the confidence index describes the extent to which player i anticipates that its appropriation of value will be in the upper, rather than lower, part of the residual problem.

Similar to a strategic-form, non-cooperative game, players simultaneously choose strategies in a biform game. But the consequence of a profile of strategies is a cooperative game, not a vector of payoffs. The analysis of a biform game therefore requires the specification of each player's preferences over different cooperative games. This specification is a three-step process, as described below.

For every profile a in A of strategic choices and resulting TU cooperative game $V(a)$,

(1) compute the core of $V(a)$, and, for each player $i \in N$,
(2) calculate the closed bounded interval of payoffs to player i delimited by the core,[7] and

Cooperative Games and Business Strategy

(3) evaluate the interval as an $\alpha_i : (1 - \alpha_i)$ weighted average of the upper and lower endpoints.

As in the discussion of the supplier-firm-buyer game, unrestricted bargaining is assumed and modeled by either the added value principle or the core. Step 1 uses the core, and in Step 2, the residual bargaining problem is calculated. (If using the added value principle, the determination of each player's range of possible value-capture replaces Steps 1 and 2.) The remaining task is to establish preferences over the residual bargaining problems for each player. This is Step 3.[8] Notice that with Steps 1 through 3, the consequence of a strategy profile now reduces to a vector of payoffs. This allows a biform game to be analyzed as a strategic-form, non-cooperative game.

A Monopoly Game[9]

A monopolist's capacity decision is one of the simplest examples of a strategic decision affecting the structure of the game. The following example presents a basic monopoly situation, with one seller and a finite number of buyers. The player set N is $\{s, 1, 2, \ldots, b\}$, where player s is the seller, and players $1, 2, \ldots, b$ are the buyers. The seller is the only player with a strategic decision to make, namely how much capacity to install. Thus, the strategy set A is $\{0, 1, 2, \ldots\}$, with typical element a. The seller has a constant cost-per-unit for installing capacity, namely k, and, for simplicity, a zero cost-per-unit for producing its product. Each buyer has a willingness-to-pay for only one unit of product. For a given buyer, say j, its willingness-to-pay is denoted by w_j, with $w_1 \geq w_2 \geq \ldots \geq w_b > k > 0$. The ordering of the buyers in terms of descending willingness-to-pay is without loss of generality. The condition $w_b > k$ ensures that it is socially optimal to install capacity for all the buyers.

The characteristic function for this example is defined by

$$V(a)(S) = \begin{array}{ll} 0 & \text{if } s \notin S, \\ -ka & \text{if } S = \{s\}, \\ -ka + \sum_{1 \leq j \leq R} \mathbf{I}_s(j) w_j & \text{otherwise;} \end{array} \quad (5)$$

where

$$\mathbf{I}_S(j) = 1 \text{ if } j \in S, \quad (6)$$
$$0 \text{ otherwise;}$$

$$R = \max \{r : \sum_{1 \leq j \leq r} \mathbf{I}_s(j) \leq \min\{a, |S| - 1\}\}. \quad (7)$$

Equations (5) through (7) merely state that if the number of buyers in a coalition exceed the capacity choice of the seller, then the buyers with the higher willingnesses to pay are assumed to be the ones who transact with the seller.

Given this model, the question is: how much capacity will the seller choose to install? The answer to this question depends upon how much value the firm will capture for each possible capacity choice. Using the added value principle to model unrestricted bargaining, the following propositions determine the players' value capture. (Brandenburger and Stuart (1996a) obtain the same results using the core. Their proof is easily adapted to prove the propositions below.)

Proposition 1.1. Suppose $a = 0$. Then with the added value principle every player receives 0.

Proposition 1.2. Suppose $0 < a < b$. Then with the added value principle:

(i) player s receives between $a(w_{a+1} - k)$ and $\sum_{1 \leq j \leq a}(w_j - k)$;

(ii) player j (for $j = 1, \ldots, a$) receives between 0 and $(w_j - w_{a+1})$;

(iii) player j (for $j = a + 1, \ldots, b$) receives 0.

Proposition 1.3. Suppose $a \geq b$. Then with the added value principle:

(i) player s receives between $-ka$ and $-ka + \sum_{1 \leq j \leq b} w_j$;

(ii) player j (for $j = 1, \ldots, b$) receives between 0 and w_j.

To interpret these results, first suppose that the seller chooses capacity sufficient to serve every buyer. What value might the seller capture? One answer would be that it depends on the price the seller sets. But this answer assumes that the seller has price-setting power. What is the basis for this assumption? Or as Kreps (1990, pp. 314-315) asks:

> "[H]ow do we determine who, in this sort of situation, does have the bargaining power? Why did we assume implicitly that the monopoly had all this power (which we most certainly did when

we said that consumers were price takers)? Standard stories, if given at all, get very fuzzy at this point. Hands start to wave ..."

Furthermore, why is it not true that the seller is involved with a collection of bilateral bargaining problems? If this is so, then the monopolist should not have any more power than the buyers. Proposition 1.3 implies just such a conclusion. If the monopolist has the capacity to serve the whole market, then it does not have any inherent "monopoly power." The cost of capacity, namely $-ka$, is a sunk cost incurred by the seller. The remaining value, namely $\sum_{1 \leq j \leq b} w_j$, is the sum of residual bargaining problems in which each buyer can capture between zero and its added value.

With capacity to supply the whole market, a monopolist does not have any monopoly power. But with under-supply, Proposition 1.2 suggests a different story. For concreteness, consider the example depicted in Figure 2.

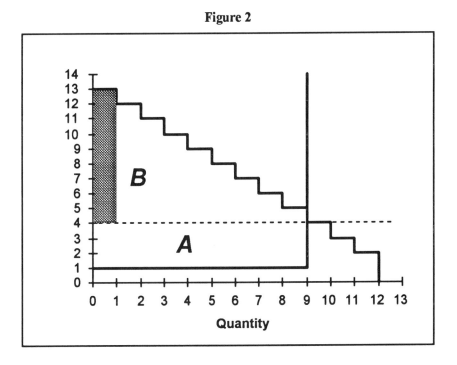

Figure 2

In this example, there are 12 buyers with willingness-to-pays ranging from $13 down to $2. The per-unit cost of capacity for the seller is $1, and the seller has chosen a capacity of 9 units. The added value of the first buyer, the buyer with a willingness-to-pay of $13, is $9. (In Figure 2, the first buyer's added value equals the shaded area.) Without this buyer in the game, the seller would instead transact with the just-excluded buyer, the buyer with a willingness-to-pay of $4. This would yield a loss in the value created of $9. Thus, the added value of the first buyer is $9, namely the value $w_j - w_{a+1}$ from part (ii) of Proposition 1.2. Notice what this implies for the seller. When the first buyer transacts with the seller, $13 of value is created. Since the buyer cannot receive more than its added value, the firm is guaranteed to capture at least $4 of value, namely w_{a+1}.

This reasoning can be repeated for the second through ninth buyers. From each buyer, the seller is guaranteed to receive at least $4 for a total of aw_{a+1}. Subtracting out the cost of capacity, namely ak, yields the term $a(w_{a+1} - k)$ in part (i) of Proposition 1.2. Region A depicts this value. Region B depicts the residual value. Similar to the case of full market supply, the seller still faces a collection of bilateral bargaining problems. But with under-supply, the size of these bargaining problems has been reduced, and a minimum price has *emerged*. With no *ex ante* assumptions about price-setting power, the monopolist now has the power to receive a price of at least $4.

With this model, the source of a monopolist's bargaining power can be interpreted as competition provided by just-excluded buyers. By limiting capacity, the monopolist creates excluded buyers. The just-excluded buyer provides competition among the buyers, reducing the added values of the included buyers. The reduction in buyers' added values guarantees value-capture to the firm, and a minimum price emerges.

Proposition 1 characterizes the consequences of the seller's capacity decision, but it does not identify the optimal choice of capacity. Due to the residual bargaining problems, the optimal choice will depend upon the seller's confidence index. Proposition 6.4 of Brandenburger and Stuart (1996a) shows that if $\alpha = 0$, the seller will choose a capacity equal to the quantity sold in a standard price-setting model. If $\alpha = 1$, the seller will choose capacity to serve every buyer, namely the quantity in the classic case of perfect price discrimination. For values between these two extremes, the optimal capacity choice is monotonically increasing in α.

A Spatial Competition Model [10]

The monopoly model provided an example of a biform game in which only one player had a strategic choice. In the spatial competition model that follows, there can be two or more firms, each having to make a strategic decision of where to locate. As in the monopoly example, buyers will be interested in obtaining only one unit of product.

In this model, the player set is the union of two disjoint sets, a set F of firms and a set T of buyers. For non-triviality, there are at least two firms and two buyers. Only the firms have strategic choices. Each player $i \in F$ has a strategy set A_i and a confidence index α_i. The sets A_i are compact, identical, and equal to a set $R \subset \mathfrak{R}^2$. Let $A = \underset{i \in F}{\times} A_i$, with typical element $a \in A$. The function V is defined by:

$$V(a)(N) = \sum_{j \in T} w - \min_{i \in F} c(a_i, b_j), \qquad (8)$$

and for $S \subset N$,

$$V(a)(S) = \sum_{j \in S \cap T} w - \min_{i \in S \cap F} c(a_i, b_j) \qquad (9)$$

if $S \cap F \neq \emptyset$ and $S \cap T \neq \emptyset$, and

$$V(a)(S) = 0. \qquad (10)$$

if $S \cap F = \emptyset$ or $S \cap T = \emptyset$.

The function c is a differentiable function from $\mathfrak{R}^2 \times \mathfrak{R}^2$ to $\mathfrak{R}^+ \cup \{0\}$.

An element $a \in A$ represents a choice of location for each firm. A given buyer, say j, is located at position $b_j \in \mathfrak{R}^2$. Buyer j is willing to pay $w - c(a_i, b_j)$ for the product from firm i. The function c may be interpreted as the buyer's transportation cost of transacting with the relevant firm. Thus, equations (8) and (9) state that the value created will be based upon each buyer purchasing from its "closest" firm, where the metric for "closeness" is transportation cost. Alternatively, each buyer may be interpreted as having a willingness-to-pay of w, with firm i incurring a cost of $c(a_i, b_j)$ to provide buyer j with one unit of product. (The choice of interpretation will not affect the analysis.) For simplicity, the firms have no cost of production.

Furthermore, it is assumed that any firm could feasibly supply the whole market: $w > c(a_i, b_j)$ for all $i \in F, j \in T, a_i \in A_i$. There are no assumptions about the distribution of the buyers, but since every firm is assumed to be able to supply all the buyers, the interpretation of the model is more reasonable if the distribution of the buyers is not too dispersed.

With this model, the central question is: where should each firm choose to locate? Answering this question requires virtually the same approach as in the monopoly example. The consequences of different choices of location must first be characterized, with unrestricted bargaining in the resultant cooperative games modeled by the added value principle. Then, given these consequences, optimal choices of location can be identified.

To analyze this biform game, it is convenient to define, given a profile of location choices, a buyer's closest firm and a firm's set of "local" buyers. For a given firm $i \in F$, let $T_i = \{j \in T : c(a_i, b_j) < c(a_k, b_j) \; \forall \, k \in F \setminus \{i\}\}$ denote its set of local buyers. (These are the buyers for whom firm i is strictly closer.) Note that a set T_i may be empty. For a given buyer $j \in T$, let $F_j = \{i \in F : j \in T_i\}$ denote the set containing the buyer's closest firm. Note that a set F_j is either a singleton set or the empty set.

Given an $a \in A$, the following proposition characterizes the added value principle for the resultant cooperative game $(V(a); N)$. Stuart (1998) proves a similar proposition using the core.

<u>Proposition 2</u>. In the game $(V(a); N)$, with the added value principle

(i) a player $k \in F$ receives between 0 and

$$\sum_{j \in T_k} \min_{i \in F \setminus \{k\}} c(a_i, b_j) - \min_{i \in F} c(a_i, b_j),$$

(ii) a player $j \in T$ receives between

$$w - \min_{i \in F \setminus F_j} c(a_i, b_j) \text{ and } w - \min_{i \in F} c(a_i, b_j).$$

Part (i) of this proposition states that the firm will capture an amount of value anywhere between zero and its added value. This added value is just its relative cost advantage with respect to its local buyers. (If a firm has no local buyers, its marginal contribution equals zero, and so it receives nothing.) Part (ii) states that each buyer may also capture value up to its added value. But, unlike the firms, a buyer is guaranteed a minimum amount of value. This value is equal to its added value minus the incremental cost of transacting with its second-closest firm. If a buyer does not have a unique, closest firm, i.e. $F_j = \emptyset$, then the buyer is guaranteed its added value.

As an example, Figure 3 depicts a two-firm case in which the buyers are uniformly distributed along a line. The horizontal axis represents location; the curve with the left-hand peak represents each buyer's willingness-to-pay for firm one's product; and the curve with the right-hand peak represents each buyer's willingness-to-pay for firm two's product. Region *R1* depicts the added value of firm one.[11] Without firm one, all the buyers would have to purchase from firm two, and the value created would correspond to regions *R2* and *R3*. Region *R1* must, therefore, be firm one's added value. By symmetric reasoning, region *R2* represents the added value of firm two.

Figure 3

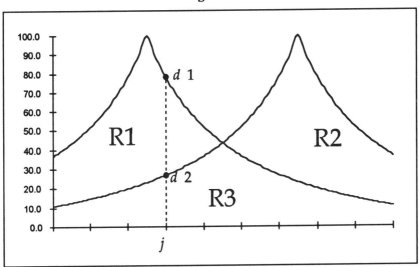

To relate Figure 3 to Proposition 2, consider the buyer labeled *j*. From part (ii) of the proposition, the value guaranteed to the buyer is $w - c(a_2, b_j)$, denoted by *d2* in the figure. The existence of guaranteed value-capture suggests the presence of competition, and this is indeed the case. Since firm two has capacity to supply all the buyers, it will surely have an excess unit to sell to buyer j. Although buyer j will prefer to transact with firm one, firm two's excess unit provides competition to firm one in its bargaining with buyer j. Further, since buyer j views firm two's product as inferior (since it is farther away), it only guarantees that buyer j captures some of its added value. At the other extreme, buyer j could capture up to its added value, $w - c(a_1, b_j)$, denoted by $d1$ in the figure. Part (i) of the proposition is almost immediate from part (ii). If each buyer captures its added value, the firms capture no value. If each buyer captures its minimum, then the firms capture their respective added values. If the quantity $d1 - d2$ is interpreted as a location advantage, then part (i) states that a firm may capture an amount ranging from zero to its relative location advantage.

Proposition 2 states that a firm will receive between zero and its added value. Since a firm evaluates this interval with an $\alpha : 1 - \alpha$ weighting, a given firm k will choose strategy a_k equal to

$$\arg\max_{a_k \in A_k} \left(\alpha_k \sum_{j \in T_k} \min_{i \in F \setminus \{k\}} c(a_i, b_j) - \min_{i \in F} c(a_i, b_j) \right),$$

given a choice $a_{-k} \in A_{-k}$ by all firms $i \in F \setminus \{k\}$.

With this best response function, a solution for this biform location model can be characterized. Let

$$f(a_k; a_{-k}) = \alpha_k \sum_{j \in T_k} \min_{i \in F \setminus \{k\}} c(a_i, b_j) - \min_{i \in F} c(a_i, b_j),$$

where a_{-k} is taken to be fixed. Then the following proposition provides a solution.

<u>Proposition 3</u>. (Stuart 1998): Suppose $\alpha_k > 0$ for all $k \in F$. There exists $a^* \in A$ such that

(i)   for all $k \in F$, $f(a_k^*; a_{-k}^*) \geq f(a_k; a_{-k}^*) \; \forall \; a_k \in A_k$, and

(ii) $V(a^*)(N) \geq V(a)(N) \; \forall \; a \in A$.

This last proposition states that the biform location model has a solution (part (i)) and that there exists a solution which is socially optimal (part (ii)). Thus, with unrestricted bargaining, socially-efficient spatial differentiation can be an optimal strategy for the firms.

A partial intuition for this result can be gained from two observations. First, each firm wants to maximize its marginal contribution, namely its added value (Proposition 2). In many contexts, this condition is sufficient for a stable, socially-efficient outcome. Although this sufficiency does not hold in general, (see, for example, Makowski and Ostroy (1995)), it does hold in the biform location model. The reason is due to the second observation: there is a kind of independence in this model. Specifically, given a firm k, the value of the game without that firm, namely $V(a)(N \setminus \{k\})$, does not depend upon firm k's choice of position. In other words, given that all the other firms choose positions a_{-k}, then $V(a_k, a_{-k})(N \setminus \{k\})$ has the same value for all a_k. With this sort of independence, individual maximization of added values leads to social efficiency.

3. Conclusion

Cooperative game theory is a structural, rather than procedural theory. It does not specify what actions the players can take, much less what they might do. At first glance, this might seem disappointing for business strategy, since business strategy is often concerned with what a firm does. Instead, the structural approach can be used to answer a broader question: is the firm (or will it be) in a favorable competitive environment? Section 1 of this chapter demonstrates how cooperative game theory can be used to answer this broader question. It uses Supplier-Firm-Buyer games to model business games and gain insights into the nature of competition.

Section 2 of this chapter addresses the "what might the firm do" question. Many business contexts are so complex that they resist specification of the players' actions. But the choice of what business to be is easier to make. Section 2 presents two examples of such choices: the capacity decision of a monopolist and the product positioning decisions of firms. Modeling these decisions does not, however, require that the structural approach of cooperative game theory be abandoned. Instead, the consequences of these decisions are complex business situations, which, with the biform game formalism, can be modeled as cooperative games. The answer to the "what might the firm do" question depends on an assessment of how favorable a competitive environment the firm finds itself in.

References

Aumann, R. (1985), "What is Game Theory Trying to Accomplish?" in Arrow, K. J. and S. Honkapohja (eds.), *Frontiers of Economics*, Oxford: Basil Blackwell, 28-76.

Brandenburger, A., and H. W. Stuart, Jr. (1996), "Biform Games," unpublished manuscript.

Brandenburger, A., and H. W. Stuart, Jr. (1996), "Value-based Business Strategy," *Journal of Economics & Management Strategy*, 5, 5-24.

Edgeworth, F. (1881), *Mathematical Psychics*. London: Kegan Paul.

Hart, O. and J. Moore (1990), "Property Rights and the Nature of the Firm," *Journal of Political Economy*, 98, 1119-1158.

Kreps, D. (1990), *A Course in Microeconomic Theory*. Princeton: Princeton University Press.

McDonald, J. (1975), *The Game of Business*. New York: Doubleday.

Makowski, L. (1980), "A Characterization of Perfectly Competitive Economies with Production," *Journal of Economic Theory*, 22, 208-221.

Makowski, L. and J. Ostroy (1995), "Appropriation and Efficiency: A Revision of the First Theorem of Welfare Economics," *American Economic Review*, 85, 808-827.

Milnor, J. (1954), "Games Against Nature," in Thrall, R., C. Coombs, and R. Davis, (Eds.), *Decision Processes*. New York: Wiley, 49-59.

Osborne, and A. Rubinstein (1994), *A Course in Game Theory*. Cambridge: MIT Press.

Oster, S.M. (1994), *Modern Competitive Analysis*. New York: Oxford University Press.

Ostroy, J. (1980), "The No-Surplus Condition as a Characterization of Perfectly Competitive Equilibrium," *Journal of Economic Theory*, 22, 183-207.

Roth, A. and X. Xing (1994), "Jumping the Gun: Imperfections and Institutions Related to the Timing of Market Transactions," *American Economic Review*, 84, 992-1044.

Shapley, L. and M. Shubik (1972), "The Assignment Game, I: The Core," *International Journal of Game Theory*, 1, 111-130.

Shubik, M. (1959) "Edgeworth Market Games," in Tucker, A. W. and R. D. Luce, eds., *Contributions to the Theory of Games*, Vol. IV (Annals of Mathematics Studies, 40). Princeton: Princeton University Press, 267-278.

Stuart, H. W., Jr. (1997), "Does Your Business Have Added Value?" unpublished manuscript.

Stuart, H. W., Jr. (1998), "Spatial Competition with Unrestricted Bargaining," unpublished manuscript.

Stuart, H. W., Jr. (1997), "The Supplier-Firm-Buyer Game and its m-sided Generalization," *Mathematical Social Sciences*, 34, 21-27.

Notes

[1] John MacDonald, a contemporary of von Neumann, was arguably one of the first to appreciate the relevance of cooperative game theory to business.
[2] Treating price as a consequence of the economic structure dates back at least to Edgeworth (1881). For a modern treatment and interpretation, see Makowski and Ostroy (1995).
[3] Biform games are defined in Brandenburger and Stuart (1996a). Related approaches can be found in Hart and Moore (1990), Makowski and Ostroy (1995), and Roth and Xing (1994).
[4] The material in this section is taken from Brandenburger and Stuart (1996b), Stuart (1997a), and Stuart (1997b).
[5] This fact is a discrete version of results in Ostroy (1980).
[6] See Makowski (1980) for a discussion of firm profitability under perfect competition.
[7] Formally, the projection of the core onto the ith coordinate axis.
[8] For an axiomatic treatment of this weighted average, see Proposition 5.1 of Brandenburger and Stuart (1996a).
[9] This material is taken from Brandenburger and Stuart (1996a).
[10] This material is taken from Stuart (1998).
[11] The area of $R1$ only approximates firm one's added value. It is not exactly equal due to the discreteness of the buyers.

7 RENEGOTIATION IN THE REPEATED AMNESTY DILEMMA, WITH ECONOMIC APPLICATIONS

Joseph Farrell and Georg Weizsäcker

1. INTRODUCTION

In many economic problems, efficient outcomes require that one party trusts a second party, who has a short-term incentive to behave opportunistically if trusted, while the first party will not "trust" if she expects such opportunism.[1] This problem is sometimes known as the "Trust Game" or as the "Amnesty Dilemma."

For example, a monopolistic seller of a good whose quality is unobservable to a buyer at the time of purchase typically has a short-run incentive to set low quality. Anticipating this, the buyer may not be willing to purchase the good. Similarly, the owner of a firm may decide not to employ a manager who, by choosing a high effort level, can generate an efficient outcome but who also has a short-run incentive to shirk. A related inefficiency arises in "sovereign" lending problems in which repayment cannot be externally enforced. Lending may not take place due to the incentive of the borrower not to repay, even if large mutual gains from successful lending are possible.

The standard subgame-perfect equilibrium theory of repeated competitive play would suggest that repetition could solve these problems if the parties (players) are sufficiently patient. If any of the games roughly described above is played infinitely often, all strictly individually rational and feasible outcomes can be sustained in subgame-perfect equilibria in which each player threatens to punish the other player if he or she deviates.

But many of the punishments threatened in such subgame-perfect equilibria may also harm the punishing player, and thus the players might (try to) renegotiate to Pareto-superior continuation equilibria after a deviation has occurred. Anticipating this renegotiation would, of course, weaken or remove the threat that is supposed to enforce efficient cooperation along the equilibrium path.

This renegotiation problem is particularly severe in trust problems like those described above, because in each of these situations the first player can only threaten to punish her opponent by putting less trust in him in future periods, and such a punishment constrains her own future payoffs. Hence, the requirement to use only punishments that do not encourage the players to renegotiate rules out these subgame-perfect equilibria.

We study this renegotiation problem by considering a simple game which serves as an "abstract" representation of the described situations, the Amnesty Dilemma. In the Amnesty Dilemma one player decides whether or not to trust her opponent who in turn chooses whether or not to play honestly.[2] Figures 1 and 2 depict the game in normal form and extensive form respectively.

		Player 2	
		Honest	Dishonest
Player 1	Trust	a, b	c, d
	Distrust	0, 0	0, 0

Figure 1: The normal-form Amnesty Dilemma, with $a > 0 > c$, $d > b > 0$.

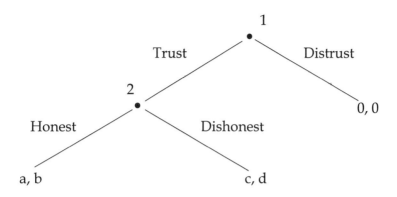

Figure 2: The extensive-form Amnesty Dilemma.

Since the second player has an incentive to cheat on the first player, the unique outcome of (either version of) the one-shot game that survives iterated elimination of weakly dominated strategies is for the first player not to trust and for the second player not to play honestly. The result is a strictly Pareto-inefficient outcome in which each player gets his or her minimax payoff, which we normalize to zero. In the normal-form version of the Amnesty Dilemma (Figure 1), this outcome is of course the unique Nash equilibrium outcome of the game. In the extensive-form version (Figure 2), it is the unique subgame-perfect equilibrium outcome.

By the Folk Theorem, if either version of the game is played infinitely often by sufficiently patient players, all strictly individually rational and feasible outcomes are sustainable in subgame-perfect equilibrium. In the next two sections we investigate to what extent the possibility of renegotiation constrains the set of sustainable outcomes. We then apply these abstract results to the economic problems described in the first paragraph.

Farrell and Maskin (1989) defined concepts of renegotiation-proof outcomes for subgame-perfect equilibria in repeated games (see equally Bernheim and Ray's (1989) "partial Pareto perfection"). For an equilibrium to be plausible when players can freely renegotiate after a defection it should, they argued, be *weakly renegotiation-proof* (WRP), meaning that none of its continuation equilibria strictly Pareto-dominate any others. Although this concept was originally developed for repeated normal-form games, it can equally be applied to repeated extensive-form games.

We study the WRP outcomes of both the repeated normal form and the repeated extensive form of the Amnesty Dilemma for the case of arbitrarily patient players.

In the normal-form game, strictly positive WRP payoffs always exist, but the extent of honesty is limited, i.e. the second player cheats at least with a certain positive probability. In particular, the pure action profile (Trust, Honest) cannot be sustained on the equilibrium path of any WRP equilibrium, however patient the players.

In the extensive-form game, as in the normal form, honesty is limited in any WRP equilibrium, as every WRP payoff vector in the extensive-form game is also a WRP payoff vector in the normal-form game. Nevertheless, trust occurs with a probability bounded away from zero in any WRP equilibrium with strictly positive payoffs, and the second player gets a large share of the total surplus.

In either form of the game, the maximum level of honesty that can be sustained in WRP equilibrium depends on the relative incentive of the second player to cheat in the stage game: The higher this short-term incentive is, the less cooperation can be achieved. This result is intuitive, but we remind the reader that it does not hold in the conventional subgame-perfect analysis.

Moreover, in the extensive-form game, if the short-term incentive to cheat is too large and the first player gets a sufficiently low utility if her opponent plays dishonestly, no WRP equilibrium with strictly positive payoffs exists, so no trust will occur if renegotiation is possible.

We then use these results to discuss some economic problems: the choice of product quality by a monopolist, shirking by a manager, and default on sovereign debt.[3] After showing how these problems may be illuminated using our analysis of the Amnesty Dilemma, we then show how particular features of those economic problems limit the applicability of the Amnesty Dilemma analysis. In particular, in the shirking manager case, the possibility of adjusting the wage depending on past performance changes the conclusions dramatically: A strongly perfect (and hence strongly renegotiation-proof) equilibrium exists for large enough discount factors, and indeed joint surplus can be maximized, which is not possible even in WRP equilibrium when the wage is fixed. In the product quality problem and in the problem of lending to a sovereign power, similar results hold if the price (the interest rate in the lending problem) can be made contingent on the outcomes in past periods.

In experimental economics, increasing attention has been paid to situations in which the agents can achieve Pareto-efficient outcomes only if subsequent players are trusted not to exploit their opponents. Berg,

Dickhaut, and McCabe (1995) conducted an experiment in which both players not only chose between two options (Trust v. Distrust, Honest v. Dishonest) but chose the precise level of trust and honesty by choosing the amount of money passed on to the other player. The authors observe that considerable trust occurs, but that on average an increase in the level of trust results in a slightly decreasing payoff for the first player: Slightly more than one half of the players who were trusted did not "reward" their opponents. Van Huyck, Battalio, and Walters (1995) confirmed this result in a similar experiment in which they also exogenously varied the total surplus (relative to the no-trust outcome) that could be achieved if the second player is trusted. They observe that the level of trust increases with the possible total surplus in the game. Güth, Ockenfels, and Wendel (1997) experimentally studied the Amnesty Dilemma (with two options for each player) in both the normal-form and the extensive-form representation. The results of both treatments show that, again, most subjects in the position of the first player trusted their opponents, but the trust was rarely rewarded.

At first glance, all of these observations are broadly consistent with our theoretical results; however, they can hardly be seen as strongly supporting the concept of WRP equilibria in the studied games, since in all three studies effects due to repeat play were at least partially ruled out by the experimental design.[4]

In the following two sections, we consider the "abstract" Amnesty Dilemma game and characterize the set of WRP payoffs for sufficiently patient players. The normal-form version of the game is analyzed in Section 2, the extensive-form version in Section 3. In Section 4, we analyze the provision of costly and unobservable product quality by a firm. Section 5 applies the previous results to the special case of a manager who is uniquely efficient in managing the principal's affairs, but is tempted to exploit his advantage by shirking. In Section 6, the "abstract" Amnesty Dilemma is applied to the problem of sovereign debt. We investigate the limits on the extent to which repeat lending can provide incentives to repay loans. Section 7 concludes.

2. RENEGOTIATION IN THE REPEATED NORMAL-FORM AMNESTY DILEMMA

The "abstract" Amnesty Dilemma, in its normal-form representation, is given by Figure 1. Player 1 must choose whether or not to trust player 2. Trust enables mutual gains (the outcome (a, b)) to be achieved, but it also tempts player 2 to be dishonest, leading to a payoff (c, d) that is best for him but worst for player 1, so the unique Nash equilibrium of the one-shot game is (Distrust, Dishonest) which gives each player zero. Our focus is on the extent to which better outcomes $v = (v_1, v_2)$ can be sustained by repeated play when players are patient.

We assume that the players play this stage game infinitely often and that they have a common discount factor $\delta < 1$. Also, we assume that the players can observe each other's private randomizations after each period, so each player can, in particular, react immediately to any deviation of his or her opponent from a specified mixed (not only pure) action.

We will use the following notation: In the stage game, for a given mixed action profile $a = (a_1, a_2) \in [0, 1]^2$, let $g_i(a)$ be the expected (per period) payoff for player i if a is played, $i \in \{1, 2\}$. We set (a_1, a_2) with $a_1 = a_2 = 1$ to be the pure action profile (Trust, Honest), so it follows e.g. that a_2 is the probability of player 2 choosing Honest and that $g_2(1, 1) = b$. Moreover, define $c_i(a)$ to be i's short-run best response payoff if j complies with playing a (i.e. $c_i(a) = \max_{a'_i} g_i(a'_i, a_j)$).

In the repeated game, a strategy for player i is a function that, for every date t and every history of the game up through date $t - 1$, defines a period-t action $a_i(t) \in [0, 1]$. For a strategy profile σ such that if both players adhere to it they will play a sequence of actions $\{a_1(t), a_2(t)\}_{t=1}^\infty$, define $g_i^*(\sigma)$ to be i's expected average payoff from playing σ, given by $g_i^*(\sigma) = (1 - \delta) \sum_{t=1}^\infty g_i(a_1(t), a_2(t)) \delta^{t-1}$.

Now consider WRP payoffs in the repeated game, i.e. payoffs that are sustainable in WRP equilibrium. As mentioned in the Introduction, a subgame-perfect Nash equilibrium σ is WRP if no pair of continuation equilibria of σ is strictly Pareto-ranked. (Hence, in a WRP equilibrium σ it is impossible that the players "agree on a different history", i.e. that they renegotiate at any time t to play a different continuation equilibrium of σ than the one prescribed by σ given the history up through $t - 1$.)

First notice that any feasible payoff vector $v = (v_1, v_2)$ with $v_1 = 0$ and $v_2 \geq 0$ is WRP for sufficiently patient players (in both the normal-form and the extensive-form game) since it can be sustained by an equilibrium prescribing to start with a mixed action pair a that gives the players per-period payoffs of (v_1, v_2) and to play (Distrust, Dishonest) forever after a player has deviated (yielding payoffs $(0,0)$). Since $v_1 = 0$, the only two continuation equilibria of this subgame-perfect equilibrium are not strictly Pareto-ranked, so it is WRP. However, such an equilibrium seems implausible for $v_2 > 0$ since player 1, by trusting her opponent, exposes herself to the possibility of being cheated without deriving positive utility from the equilibrium path, so one would expect her rather to choose Distrust from the beginning.

Also, any $v = (v_1, v_2)$ with $v_1 < 0$ cannot be WRP since it is not individually rational for player 1.

Lemma 1 *In both the repeated normal-form and the repeated extensive-form version of the Amnesty Dilemma, a feasible payoff vector $v = (v_1, v_2)$ is sustainable as a WRP payoff vector if $v_1 = 0$ for δ sufficiently large, and it is not sustainable in WRP equilibrium for any δ if $v_1 < 0$.*

By Lemma 1, we only have to consider strictly positive payoff vectors $v > 0$.[5] Theorem 1 in Farrell and Maskin (1989), applied to the repeated normal-form Amnesty Dilemma, implies that a payoff vector $v > 0$ is WRP for sufficiently large δ if and only if there exist action pairs $a^i = (a_1^i, a_2^i)$, for $i = 1, 2$, such that $c_i(a^i) < v_i$ and $g_j(a^i) \geq v_j$, where $j \neq i$. (The pair a^i can then be used to punish player i without giving j an incentive to renegotiate during i's punishment phase.)

The existence of an appropriate action pair a^1 to punish player 1 is immediate: For any $v > 0$ just set $a^1 = (1, 0)$, i.e. set a^1 to be the action profile (Trust, Dishonest). To determine whether there is a renegotiation-proof punishment for player 2, we simply apply the inequalities required by the theorem to the pair a^2, i.e. we want a^2 to satisfy

$$a_1^2 d < v_2$$

(since player 2 gets d if he is trusted, which occurs with probability a_1^2, and he cheats) and

$$a_1^2(a_2^2 a + (1 - a_2^2)c) \geq v_1$$

(where, analogously, the left-hand side is equal to $g_1(a^2)$). Such a pair a^2 exists if and only if there exists an action a_2^2 satisfying

$$v_1 < \frac{v_2}{d}(a_2^2 a + (1 - a_2^2)c),$$

which, since $a > c$, is true if and only if

$$\frac{v_1}{v_2} < \frac{a}{d}. \qquad (1)$$

Proposition 2 *For sufficiently patient players, a feasible payoff vector $v \geq 0$ can be sustained in a WRP equilibrium of the normal-form game if and only if (1) holds.*

Inequality (1) implies that any WRP equilibrium gives a relatively low utility to player 1, because she has less effective (that is, credible) means to punish her opponent if renegotiation is possible. This coincides with the intuition that player 2, who is in the more powerful position, should get a larger share of the total surplus. The set of WRP payoffs in the normal-form game is shown in Figure 3.

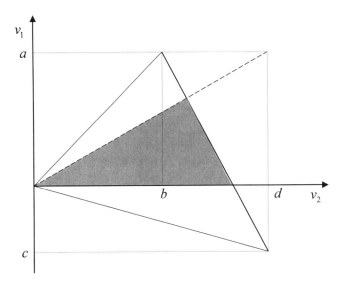

Figure 3: WRP payoffs in the normal-form game (shaded region).

3. RENEGOTIATION IN THE REPEATED EXTENSIVE-FORM AMNESTY DILEMMA

Now consider the extensive-form representation of the Amnesty Dilemma, which is depicted in Figure 2. In contrast to the normal form, here player 2 first observes his opponent's action, and chooses an action himself only if he is trusted in the current period. Equivalently, player 2's action is observable only if player 1 trusts him. Hence, player 2 has a stronger "bad" incentive in the extensive-form game: A deviation will be detected only in the case that it actually gives him a higher short-term payoff. (In the normal-form game, if player 1 plays mixed actions, player 2 is uncertain about whether or not cheating is "successful", whereas it is detected in any case.) As the analysis will show, it is therefore more difficult to effectively deter player 2 from cheating in the extensive-form Amnesty Dilemma, and the set of WRP outcomes is smaller than in the normal-form version of the game.

To simplify matters, we assume that the players have access to a publicly observable random device, so they can make the continuation of a punishment phase contingent on the realization of a random draw.[6] With this simplification we can restrict attention to a specific class of WRP equilibria if the players are sufficiently patient.[7]

Lemma 3 *For δ sufficiently close to 1, any payoff vector $v = (v_1, v_2)$ sustained by a WRP equilibrium can be sustained by a WRP equilibrium using a normal phase and two punishment phases as follows: (i) The players play fixed actions $(\tilde{a}_1(v), \tilde{a}_2(v))$ during the normal phase. (ii) For each player i, there is a fixed punishment action pair $a^i = (a_1^i, a_2^i)$ which is played during i's punishment phase. (iii) After each period of player i's punishment, play reverts to the normal phase with a fixed probability $p^i \in (0,1)$. (iv) Play starts in the normal phase. Whenever a player deviates, his or her punishment phase starts in the next period for sure. If both players deviate in the same period, player 2 is punished.*

Using Lemma 3, we know that for patient players any WRP payoff vector $v > 0$ can without loss of generality be sustained by an equilibrium characterized by inequalities (4) to (6) which are derived in the following:[8] In the normal phase, the action pair $(\tilde{a}_1(v) = \frac{v_1(d-b)+v_2(a-c)}{ad-bc}$, $\tilde{a}_2(v) = \frac{v_1 d - v_2 c}{v_1(d-b)+v_2(a-c)})$ is played, which is the unique action profile giving the players constant payoffs of (v_1, v_2). During player 2's punishment, a fixed action pair a^2 is played and play reverts to the normal phase with probability p^2, so player 2 has an average continuation payoff $\bar{g}_{2,2}^*$ satisfying

$$\bar{g}_{2,2}^* = (1-\delta)g_2(a^2) + \delta(p^2 v_2 + (1-p^2)\bar{g}_{2,2}^*),$$

which, using $g_2(a^2) = a_1^2(a_2^2 b + (1-a_2^2)d)$, can be written as

$$\bar{g}_{2,2}^* = \frac{(1-\delta)}{1-\delta+\delta p^2} a_1^2(a_2^2 b + (1-a_2^2)d) + \frac{\delta p^2}{1-\delta+\delta p^2} v_2. \quad (2)$$

To avoid renegotiation back to (v_1, v_2), the pair a^2 must also satisfy $g_1(a^2) \geq v_1$, or

$$a_1^2(a_2^2 a + (1-a_2^2)c) \geq v_1. \quad (3)$$

Combining (2) and (3) gives us the necessary condition

$$\bar{g}_{2,2}^* \geq \frac{(1-\delta)}{1-\delta+\delta p^2} \frac{a_2^2 b + (1-a_2^2)d}{a_2^2 a + (1-a_2^2)c} v_1 + \frac{\delta p^2}{1-\delta+\delta p^2} v_2. \quad (4)$$

If there is an action pair a^2 and a probability $p^2 \in (0,1)$ satisfying (4), then it is possible to find a strategy profile of the form described in Lemma 3 such that no renegotiation will occur after player 2 has deviated. But, of course, it is also necessary to deter player 2 from cheating. For the normal phase, it must be that[9]

$$(1-\delta)d + \delta \bar{g}^*_{2,2} \leq (1-\delta)(\tilde{a}_2(v)b + (1-\tilde{a}_2(v))d) + \delta v_2. \qquad (5)$$

To prevent cheating during player 2's punishment phase, we need that

$$(1-\delta)d + \delta \bar{g}^*_{2,2} \leq (1-\delta)(a_2^2 b + (1-a_2^2)d) + \delta(p^2 v_2 + (1-p^2)\bar{g}^*_{2,2}). \qquad (6)$$

Hence, (v_1, v_2) is a WRP payoff if and only if there exist $a^2 \in [0,1]^2$ and $p^2 \in (0,1)$ that satisfy conditions (4) to (6).

To find conditions under which these a^2 and p^2 exist, we first argue that a_2^2 necessarily exceeds player 2's normal phase action $\tilde{a}_2(v)$, i.e. player 2 has to play honestly during his punishment with a higher probability than during the normal phase. To see this, note that it must be that

$$a_1^2 d \leq v_2, \qquad (7)$$

because otherwise player 2 could profitably cheat forever. Combining this inequality with the no-renegotiation condition (3), we get

$$\frac{v_2}{d} \geq \frac{v_1}{a_2^2 a + (1-a_2^2)c},$$

or

$$a_2^2 \geq \underline{a}_2^2(v) \equiv \frac{v_1 d - v_2 c}{v_2(a-c)}.$$

Evidently, $\underline{a}_2^2(v)$ is strictly larger than $\tilde{a}_2(v)$ for any feasible v.[10] Figure 4 illustrates this observation in (v_1, v_2) space. For any given action pair $a = (a_1, a_2)$, the payoffs if player 2, instead of playing a_2, plays Dishonest for sure are $(a_1 c, a_1 d)$, lying on the straight line connecting the origin and (c, d). Therefore the probability a_1 is given by the distance from the origin to $(0, a_1 d)$ if d is normalized to 1. Similarly, if player 1 chooses Trust for sure, the resulting payoffs are $(a_2 a, a_2 b)$ and the probability a_2 is given by the distance between the origin and $(a_2 a, 0)$ if a is normalized to 1. In a given equilibrium sustaining v, the punishment action pair a^2 results in a per-period payoff vector $g(a^2)$ lying to the left of the dashed line, so $a_1^2 d \leq v_2$ and player 2 prefers playing his equilibrium strategy over cheating forever. The shaded region contains the payoffs $g(a^2)$ resulting from all possible action pairs a^2 that satisfy conditions (3) and (7). Since v results from playing the actions $(\tilde{a}_1(v), \tilde{a}_2(v))$ it must be that $a_2^2 > \tilde{a}_2(v)$.

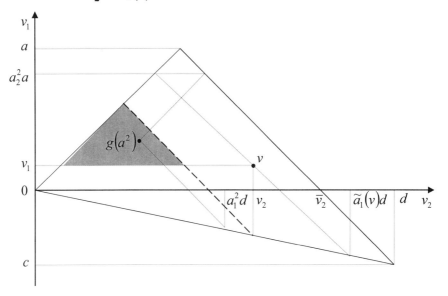

Figure 4: Possible range of g(a²) (shaded region).

Now rewrite (5) and (6) as

$$\bar{g}_{2,2}^{\hat{}} \leq v_2 - \frac{1-\delta}{\delta}\tilde{a}_2(v)(d-b) \tag{8}$$

and

$$\bar{g}_{2,2}^{*} \leq v_2 - \frac{1-\delta}{\delta}\frac{1}{p^2}a_2^2(d-b). \tag{9}$$

Using $p^2 < 1$ and the observation that $a_2^2 > \tilde{a}_2(v)$, it follows immediately that (8) holds whenever (9) holds.[11] So a feasible and strictly positive payoff vector $v = (v_1, v_2)$ is WRP for a sufficiently large δ if and only if there is a pair $a^2 \in [0,1]^2$ and a probability $p^2 \in (0,1)$ satisfying both (4) and (9). Combining these two inequalities and simple rearrangements show that this is possible if and only if there are an action $a_2^2 \in [\underline{a}_2^2(v), 1]$ and a probability p^2 that satisfy[12]

$$\frac{a_2^2 b + (1-a_2^2)d}{a_2^2 a + (1-a_2^2)c}v_1 + \frac{1-\delta+\delta p^2}{\delta p^2}a_2^2(d-b) \leq v_2. \tag{10}$$

Proposition 4 characterizes the set of WRP payoff vectors for sufficiently patient players in the repeated extensive-form Amnesty Dilemma.

Proposition 4 *For δ sufficiently close to 1, strictly positive WRP payoffs exist if and only if*

$$d(a-c) + 2c(d-b) > 0. \tag{11}$$

If this condition holds, any WRP equilibrium sustaining a payoff vector $v > 0$ must give player 2 an average payoff of at least $\underline{v}_2 = -c\frac{d-b}{a-c} > 0$. Subject to this, the set of strictly positive WRP payoffs is characterized as follows: (i) If $b < \frac{1}{2}d$, a feasible payoff vector $v = (v_1, v_2) > 0$ is sustainable as a WRP payoff if and only if $v_2 \geq \underline{v}_2$ and it satisfies

$$(4v_1 d - 2v_1 v_2 - 2v_2 c - v_2^2 \frac{a-c}{d-b})(a-c) - (v_1 - c)^2(d-b) < 0. \tag{12}$$

(ii) If $b \geq \frac{1}{2}d$, v is sustainable as a WRP payoff vector under the same conditions if $v_2 \leq \tilde{v}_2 = d - b^2\frac{a-c}{ad-bc}$; if $b \geq \frac{1}{2}d$ and $v_2 > \tilde{v}_2$, v is sustainable as a WRP payoff vector if and only if

$$av_2 - bv_1 - a(d-b) > 0. \tag{13}$$

In particular, the proposition shows that (10) can be satisfied only if inequality (12) holds and that in any WRP equilibrium with strictly positive payoffs player 2 gets an average payoff that is above some value $\underline{v}_2$. For some parameter values, $v_2 > \underline{v}_2$ may not be feasible, in which case strictly positive WRP payoffs fail to exist.

The set of WRP payoffs is depicted in Figures 5a and 5b (for parameters satisfying (11)). Figure 5a illustrates the case $b < \frac{1}{2}d$ (case (i) in Proposition 4), Figure 5b the case $b \geq \frac{1}{2}d$ (case (ii)).

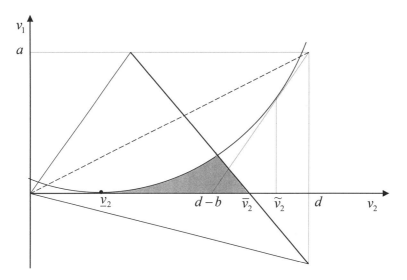

Figure 5a: b < ½ d.

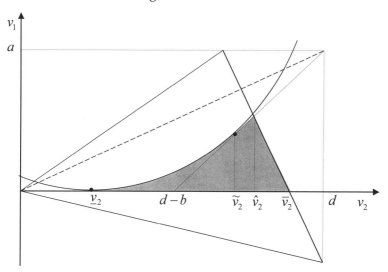

Figure 5b: b ≥ ½ d.

Figure 5: WRP payoffs in the extensive-form game (shaded regions). The dashed lines indicate the limits of WRP payoffs in the normal-form games.

The result that non-zero WRP payoff vectors exist only if condition (11) is satisfied says that if player 2's short-term incentive to choose Dishonest is is large (i.e. $(d-b)$ is large) and player 1 gets a low utility from the outcome (Trust, Dishonest) (i.e. $-c$ is large), then no trust will occur if renegotiation is possible.[13] This contrasts with the repeated normal-form Amnesty Dilemma (see Figure 2), where player 1 can always get positive utility in WRP equilibrium.

Proposition 4 also allows us to determine both the maximum extent to which honesty can occur and the maximum average utility player 1 can get in any WRP equilibrium with strictly positive payoffs in the repeated extensive-form version of the Amnesty Dilemma.[14] For simplicity, we only consider the more tractable case (ii).

It is straightforward that the set of Pareto-efficient outcomes of the stage game (the set of points on the line between (a,b) and (c,d)) is the set of payoff vectors (v_1, v_2) that satisfy $c \leq v_1 \leq a$ and

$$v_1 = a - \frac{a-c}{d-b}(v_2 - b). \tag{14}$$

For case (ii), we can therefore solve for the Pareto-efficient vector $\hat{v} = (\hat{v}_1, \hat{v}_2)$, where $\hat{v}_1$ denotes the lowest upper bound on player 1's utility in a WRP equilibrium (see Figure 5b), by substituting (14) into the following equation (from inequality (13)):[15]

$$av_2 - bv_1 = a(d-b).$$

The solution to these two equations is the vector $\hat{v}$ with

$$\hat{v}_1 = a\frac{b(a-c) + c(d-b)}{b(a-c) + a(d-b)}$$

and

$$\hat{v}_2 = \frac{b^2(a-c) + ad(d-b)}{b(a-c) + a(d-b)}.$$

The maximum level of honesty, i.e. the maximum relative frequency of player 2 choosing Honest, that is sustainable in case (ii) in a WRP equilibrium with strictly positive payoffs, denoted by $\bar{\lambda}$, can then be calculated as $\bar{\lambda} = \frac{d - \hat{v}_2}{d-b}$, or

$$\bar{\lambda} = \frac{b(a-c)}{b(a-c) + a(d-b)}. \tag{15}$$

Hence, the level of honesty that can be sustained in a WRP equilibrium is decreasing in the short-term incentive to cheat, $d - b$. This comparative-statics result also holds for the repeated normal-form Amnesty Dilemma (see Figure 2).

One might ask whether some WRP outcomes are more likely to be achieved than others. In particular, a WRP vector may be less plausible if it is strictly Pareto-dominated by another payoff that is also WRP (even though not another continuation payoff in the same equilibrium). Following this argument, Farrell and Maskin (1989) defined a WRP equilibrium to be *strongly renegotiation-proof* (SRP) if none of its continuation equilibria is strictly Pareto-dominated by any other WRP equilibrium. Unfortunately, our approach characterizing the WRP payoffs by the form given in Lemma 3 does not enable us to describe the set of SRP outcomes because we would have to consider the set of WRP equilibria (as opposed to the set of WRP payoff vectors).[16] Hence, Proposition 4 can be viewed as only containing negative results restricting the payoff vectors that can plausibly be achieved if renegotiation is possible.

4. PRODUCT QUALITY CHOICE

Consider the problem of a seller producing a good whose quality, $q \geq 0$, is unobservable to a buyer at the time of purchase. The seller's costs are $c(q)$ with $c(\cdot)$ strictly increasing. The buyer derives benefit $u(q)$ from quality q. The socially efficient quality is $q^* > 0$, which is assumed to generate strictly positive gains from trade: That is, q^* maximizes $u(q) - c(q)$ and $u(q^*) - c(q^*) > 0$. However, the seller has a short-run incentive to cut quality to zero (to save on costs), and at zero quality there are no gains from trade: $u(0) < c(0)$. In particular, we assume (to keep things tractable) that $u(0) = 0$. We suppose that, although quality is observable to the buyer *ex post*, it is never observable to third parties, and therefore no enforceable contract can make anything depend on realized quality.

Also, in this section we assume that the price p at which the transaction takes place is fixed for all periods, so it cannot vary with history. The only way for the buyer to deter the seller from cheating by setting low quality is then to buy with a lower probability following any cheating.

We restrict attention to prices $p \in (c(q^*), u(q^*))$, so this game (taking just two qualities, q^* and 0) is the extensive-form Amnesty Dilemma, with $a = u(q^*) - p, b = p - c(q^*), c = -p$, and $d = p - c(0)$.

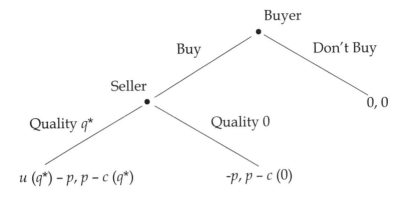

Figure 6: The product quality game.

By the Folk Theorem, the surplus-maximizing outcome (Quality q^*, Buy) can be enforced in subgame-perfect equilibrium for large enough δ. To do this, the buyer can play the strategy "buy as long as the seller has never produced quality less than q^*; thereafter, never buy" and the seller plays the strategy "produce q^* whenever the buyer buys as long as no player deviates; thereafter, produce $q = 0$". But this subgame-perfect equilibrium is not WRP: like the trigger-strategy equilibrium in the Prisoner's Dilemma, the punishment continuation equilibrium is strictly Pareto-dominated by the "normal" equilibrium. Moreover, as we saw above, this equilibrium path is not WRP, even with different punishments; this is quite different from the Prisoner's Dilemma.[17]

Instead, it turns out that, even for arbitrarily patient players, cooperation can be expected only if the gains from trade are sufficiently large and the price is set above some lower limit: Substituting for a, b, c, d in condition (11), we see that strictly positive WRP payoffs exist if and only if

$$u(q^*)(p - c(0)) - 2p(c(q^*) - c(0)) > 0, \qquad (16)$$

or, equivalently,

$$p(u(q^*) - 2(c(q^*) - c(0))) > u(q^*)c(0).$$

In particular, inequality (16) implies that $u(q^*) - 2(c(q^*) - c(0)) > 0$, so for any positive WRP payoffs to exist it must be that:

$$p > \underline{p} \equiv \frac{u(q^*)c(0)}{u(q^*) - 2(c(q^*) - c(0))}. \qquad (17)$$

Also, since $u(q^*) - 2(c(q^*) - c(0))$ is exactly the derivative of the left-hand side of (16) with respect to p, we know that non-zero WRP payoffs are more likely to exist if p increases. Substituting the upper bound $p = u(q^*)$ into (16) then yields the result that strictly positive WRP payoffs can exist only if

$$u(q^*) - c(q^*) > c(q^*) - c(0). \tag{18}$$

That is, purchases which result in any positive level of utility for the buyer can take place only if the gains from a successful trade with quality q^* are larger than the cost the seller can avoid by setting zero quality instead of quality q^*. If (18) fails, there can be no WRP trade, whatever the price.[18]

If condition (18) is satisfied, however, the surplus-maximizing outcome can (almost) be achieved in WRP equilibrium by setting p arbitrarily close to $u(q^*)$. To see this, first notice that if p is large then the set of WRP outcomes is given by clause (ii) of Proposition 4 because if we substitute for b and d the condition $b \geq \frac{1}{2}d$ becomes

$$p - c(q^*) \geq c(q^*) - c(0),$$

which always holds for p close to $u(q^*)$ if (18) holds. So for large p we can use the expression (15) to find the maximum sustainable level of honesty, i.e. the maximum frequency with which a high quality is chosen:

$$\bar{\lambda}(p) = \frac{u(q^*)(p - c(q^*))}{u(q^*)(p - c(q^*)) + (u(q^*) - p)(c(q^*) - c(0))}.$$

Since $\bar{\lambda}(p)$ approaches 1 as p goes to $u(q^*)$, the surplus-maximizing outcome (Quality q^*, Buy) can occur arbitrarily often on the equilibrium path of a WRP equilibrium.

The results are summarized in the following Proposition.

Proposition 5 *If the price paid when trade takes place is independent of history, no purchases will be made in WRP equilibrium if $u(q^*) - c(q^*) \leq c(q^*) - c(0)$. If this inequality does not hold and if the players are sufficiently patient, the WRP equilibrium that provides the most joint surplus (and will therefore be reached when ex ante side payments are possible) involves a price only slightly below $u(q^*)$ and provision of quality q^* almost all the time.*

Intuitively, setting a high price helps in two ways. First, it gives the seller a strong incentive to maintain the relationship, so that any threat of the buyer (even temporarily) not buying is at its most powerful. Second, it gives the buyer rather little at stake in maintaining the relationship, and thus makes her as willing as possible to impose a (temporary) end to the relationship.

If joint surplus is to be maximized, the buyer's share must be paid *ex ante*, in a lump sum at the beginning of the relationship (or otherwise independent of history). If, however, side payments are impossible, the buyer gets a relatively small share of the gains from trade, and the outcome and the price at which trade takes place will depend on the bargaining power of the agents. Assuming that one particular WRP payoff vector is chosen for a given p (e.g. $\gamma(\hat{v}_1, \hat{v}_2) + (1-\gamma)(0, \bar{v}_2)$ for some $\gamma \in (0, 1]$, see Figure 5b), it is possible to use the results of Proposition 4 to calculate the prices that are optimal for the buyer and the seller in this case. Notice that the seller always prefers a high price which is close to the surplus-maximizing $p = u(q^*)$, whereas the buyer does not generally prefer a price that is as low as possible.[19] The intuition for this is, as above, that the buyer's optimal choice of p is also motivated by the desire to have a more effective threat against the seller.

5. MANAGERIAL SLACK

Suppose that an owner of a firm (or other principal) identifies a uniquely qualified manager (or agent), who can create a value B for the owner at a personal "effort" cost $E < B$. (All these payoffs are relative to a normalization of zero for each player's best outside opportunities.)

For a variety of reasons, it may be impossible to make the manager's compensation depend on his current and even on his past performance; then the only sanction available to the owner is to fire the manager. We assume that it is at any point possible to re-hire him, so the owner can use limited punishments against the manager (this may make the analysis more applicable to repeated contracting than to employment). Suppose that the manager, when employed, is paid a wage $w \in (E, B)$ which is fixed in advance. For a given value of w, this game has the structure of the repeated extensive-form Amnesty Dilemma. In particular, this is a special case of the game analyzed in Section 4 if we substitute $p = w, u(q^*) = B, c(q^*) = E$, and $c(0) = 0$ (the product is re-interpreted as the manager's work and the quality as his working effort). Hence, we can immediately conclude from Proposition 5 that the manager can be employed in WRP equilibrium of the repeated game only if

$$B > 2E.$$

Also, if this condition holds, joint surplus can be maximized by choosing an efficiency wage w close to B.

Intuitively, if the short-run incentive to shirk is large, then no efficient trade can occur because the owner cannot credibly threaten to punish the manager hard enough to deter shirking.

However, the game between owner and manager (and likewise the game between buyer and seller in Section 4) only becomes the Amnesty Dilemma if the wage is independent of the game's history. If the wage can be changed after a defection, then the efficient outcome is SRP (and indeed *strongly perfect*):[20] A sufficient action pair a^2 to punish the manager is "owner pays $w^2 = 0$ and manager works".[21] Because this punishment is Pareto-efficient (indeed, continues to maximize joint surplus) there is no concern about renegotiation-proofness. And because the manager's cheating payoff is zero (his minimax payoff), the punishment is powerful enough to sustain cooperation for precisely the same values of the discount factor as will enable cooperation to be sustained in subgame-perfect equilibrium.

In fact, by allowing the wage to vary and assuming that the owner always employs the manager, we have changed the game. We now represent the stage game as given in Figure 7.

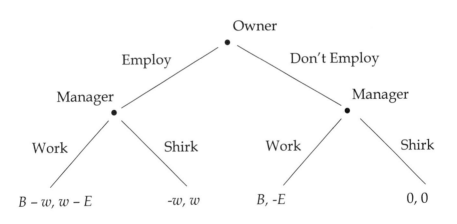

Figure 7: The owner-manager game with wage depending on history.

This game is an extensive-form version of the Prisoner's Dilemma. It is well established that in the conventional (normal-form) repeated Prisoner's Dilemma cooperation can be sustained in strongly perfect equilibrium, and this result immediately extends to the extensive-form version of the game: In such a cooperative equilibrium of the normal-form game, which prescribes sequences of Pareto-efficient outcomes after any history, there is no uncertainty about what the other player will play in the current period (in both punishment phases and in the normal phase), so no player can exploit the additional information he or she gets in the corresponding equilibrium of the extensive-form game, and the same punishment strategies can effectively be used in both representations of the game.

Proposition 6 *By making the manager's wage variable, the players can make the surplus-maximizing outcome into the equilibrium path of a strongly perfect (hence, strongly renegotiation-proof) equilibrium.*

The result also applies to the case of Section 4 if p is made to depend on history. In this context, we can give a simple economic interpretation of the punishment for the seller. We can regard him as continuing to do his equilibrium-path action (supplying q^* at a price p that lies strictly between $c(q^*)$ and $u(q^*)$) together with handing over money to consumers: Business as usual, but with a fine. Clearly, renegotiation is no problem when fines are available. In the owner-manager setup of this section, the same interpretation is straightforward: efficiency wages are not necessary if fines are possible.

The discussion in this section and Section 4 shows that it is important whether or not prices are fixed. With a fixed p (fixed wage w), non-zero WRP payoff vectors often fail to exist and efficiency is often unattainable, but if the price (wage) can vary with past quality (effort) choices then efficiency can easily be achieved. Therefore, one might suspect that choosing the price to depend on past quality is beneficial to both agents. However, there may be reasons not captured by our simple model to let prices be fixed over time: perhaps most obviously, the buyer's incentive to invoke punishment even if the seller has never in fact cheated.[22] Also, if *ex ante* side payments are impossible, our results suggest that whether or not the price of the good is fixed may depend on the bargaining power and on strategic actions of the agents: In an enriched model, the seller may want to commit to a fixed price in order to restrict the set of renegotiation-proof payoffs of the resulting game to payoffs that give him a large share of the total surplus.

6. LENDING TO A SOVEREIGN POWER

In this section, we analyze incentive problems associated with lending to a sovereign power. The problem is that there is no external enforcement of the contract and, therefore, repayment must be analyzed in game-theoretic terms. Intuitively, one might think that, at least for sufficiently high discount factors, repayment can be enforced by the threat of cutting off future loans. But, among other problems, this threat faces a credibility problem: It is not renegotiation-proof. In the following, we investigate renegotiation in the repeated lending game without external enforcement.

Here, player 1 is a lender, whose pure actions are Lend and Don't Lend; player 2 is a borrower, who observes whether or not a loan is made in the current period and chooses one of the two pure actions Repay and Default. For a fixed loan size L and interest rate r,[23] let $V(L)$ be the value to the borrower of being able to use the funds supplied in the loan L; $\bar{r}$ is the interest rate that the lender can obtain in an alternative use of the funds, e.g. by investing then in the capital market. This game is the extensive-form Amnesty Dilemma as depicted in Figure 3, with payoffs $a = (r - \bar{r})L, b = V(L) - (1+r)L, c = -(1+\bar{r})L$, and $d = V(L)$. Evidently, for an interesting problem, there must be values of L such that $V(L) > (1 + \bar{r})L$, and we can restrict attention to such L. Also, it is clear that the interest rate r has to lie above $\bar{r}$, so $a > 0$, if any lending is to take place.[24]

6.1 LENDING RELATIONSHIPS WITH TERMS INDEPENDENT OF HISTORY

As in the product-quality problem, we first study the case in which the terms-of-trade variables (r and L) are fixed by the players once and for all at the beginning of the repeated game. As we will see in the next paragraph, lending will occur in this case in WRP equilibrium only if the gains from trade are very large. In the next subsection we then investigate how the players may avoid this inefficiency by making the terms of trade depend on history.

If both r and L are fixed, we can use inequality (11) to derive that a WRP equilibrium with strictly positive payoffs (that is, in which lending occurs with positive probability) exists only if

$$V(L) > 2(1+\bar{r})L. \tag{19}$$

Notice that this condition does not contain the variable r, so the interest rate does not affect whether or not loans are made in WRP equilibrium.[25] Condition (19) can be rewritten as:

$$V(L) - L > \bar{r}L + (1+\bar{r})L.$$

Hence, lending occurs only if the return of the project for which L is used exceeds the return from investing L in the capital market by more than $(1+\bar{r})L$. Such a large (above-market) surplus seems unlikely in most contexts, especially for short-term loans (where returns are small compared to L).[26] Intuitively, it is extremely hard to sustain trust and honesty in the lending game because the short-term incentive of the borrower to default on the loan is very strong and cannot be weakened by changing the interest rate r: A defaulting borrower does not care about the interest rate.

However, as in the product-quality problem, the outcome (Lend, Repay) can occur almost all the time if (19) holds. For such a WRP equilibrium to exist, the interest rate r is set arbitrarily close to $\bar{r}$, so again player 1 must get approximately zero utility if full honesty (in terms of repayment) is to be achieved.

To obtain this result, first notice that by setting r arbitrarily close to $\bar{r}$, the game can be analyzed using the expressions for case (ii) of Proposition 4: The condition $b \geq \frac{1}{2}d$ becomes

$$V(L) \geq 2(1+r)L,$$

and this condition can be satisfied for small enough r if (19) holds. Hence, we can use condition (15) to see that the maximum sustainable level of honesty (repayment) is given by

$$\bar{\lambda}(r) = \frac{V(L) - (1+r)L}{V(L) - (1+\bar{r})L}.$$

Since $\bar{\lambda}(r)$ approaches 1 for r arbitrarily close to $\bar{r}$, the outcome (Lend, Repay) can occur almost always if (19) is satisfied. Notice that (Lend, Repay) is not the uniquely efficient or uniquely surplus-maximizing outcome of the one-shot game, because (in the one-shot game) repayment is simply a transfer.

Proposition 7 *In the lending game with fixed terms of trade, lending can occur in WRP equilibrium only if $V(L) > 2(1+\bar{r})L$. If this condition holds, the outcome (Lend, Repay) is sustainable almost all the time if r is chosen to be close to $\bar{r}$ and δ is arbitrarily close to 1.*

Although condition (19) seems very restrictive at first glance, it may still be satisfied in the relevant range of L if the size of the loan is chosen endogenously by the players. Consider the special case in which the borrower's benefit from the project financed by the loan is given by $V(L) = kL^\alpha$. The total surplus from lending is, in this example, equal to $kL^\alpha - (1+\bar{r})L$. Maximization of this expression over L yields the surplus-maximizing loan size

$$L^* = (\frac{1}{k\alpha}(1+\bar{r}))^{\frac{1}{\alpha-1}}.$$

This Pareto-optimal loan can, due to Proposition 7, be made in WRP equilibrium if L^* satisfies condition (19) (applied to the example), i.e. if

$$L^* < (\frac{2}{k}(1+\bar{r}))^{\frac{1}{\alpha-1}},$$

which is true if and only if $\alpha < \frac{1}{2}$.

6.2 TERMS OF TRADE THAT DEPEND ON HISTORY

Since no WRP equilibrium sustains positive amounts of lending for wide parameter ranges if r and L are independent of history, the players have a joint incentive to agree on a contract that prescribes varying terms of trade depending on past actions. Moreover, even if lending can occur with fixed r and L, we saw that the interest rate had to be arbitrarily close to $\bar{r}$ in order to achieve the outcome (Lend, Repay), so the lender gets almost none of the surplus of the game if side payments are not available. But the lending institutions (perhaps with justification in terms of longer-run considerations of efficiency) may want to implement (Lend, Repay) without making r so close to $\bar{r}$, so it is natural to ask whether this can be done by varying r and L following a default, in analogy to the method used in Section 5.

A change in the "price" (interest rate r) alone does not increase the set of sustainable loans L, because, as argued above, a defaulting borrower does not care about the interest rate. In the product-quality case, the price could be lowered to $c(0)$ as to make the seller's cheating payoff $c_2(a^2)$ small; here, mere changes in the interest rate do not reduce the borrower's cheating payoff.

In order to make $c_2(a^2)$ small without cutting off the relationship, it is necessary to reduce the loan size L following a default, say to L', where L' is small enough that defaulting on the small loan is worth less to the borrower than is getting and repaying the usual loan, i.e.,

$$V(L') < V(L) - (1+r)L.$$

This must be combined with sufficiently high post-default interest rates so that the lender's profit is not driven below her equilibrium value, i.e.,

$$(r' - \bar{r})L' \geq (r - \bar{r})L.$$

The resulting game is depicted in Figure 8.

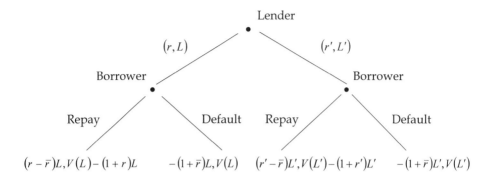

Figure 8: The lending game with terms of trade depending on history.

This game is again an extensive-form Prisoner's Dilemma, and by the argument given in Section 5, lending can occur in a strongly perfect equilibrium. Notice, in particular, that by choosing appropriate values of r' and L' we can sustain (Lend, Repay) for any values of r and L that make borrowing and repaying worthwhile for the borrower, i.e. for any (r, L) satisfying $V(L) > (1+r)L$ and $r > \bar{r}$.

Our interpretation of this result depends on the details of the time structure. If there is real meaning to the periods, then the need to reduce the loan size following defection means that, in contrast to the case of product quality where a fine-like punishment was available, punishment implies social inefficiency if the original loan size L is not larger than the surplus-maximizing loan. If, however, an additional artificial "punishment period" can be introduced, it can be used for a somewhat artificial play of the lending game as follows: The lender lends a very small amount L', and then immediately receives in return a large amount $(1+r')L'$, where, although L' is small, r' is so huge that $r'L'$ is large. A moment's thought will convince the reader that this curious arrangement is indistinguishable from a pure fine of $r'L'$, which is, as always, a renegotiation-proof punishment.

Proposition 8 *By making both r and L depend on history, lending and repaying the loan can occur in strongly perfect equilibrium if the players are sufficiently patient.*

7. CONCLUSIONS

For a very simple example of incentives for trust and betrayal, the Amnesty Dilemma, the paper investigates how difficult it can be to sustain honesty only to repeat play effects when renegotiation is possible. The reason for the largely negative results is that credible punishments may not be available because a punishment typically harms both players - in contrast to the Prisoner's Dilemma, in which Pareto-efficient punishments exist.

In the repeated extensive-form Amnesty Dilemma, if the short-term incentives to cheat are large and cheating hurts the trusting first party a lot, then no trust at all is sustainable in a renegotiation-proof equilibrium. In this sense the renegotiation-proof analysis may be more intuitive than is the subgame-perfect Folk-Theorem analysis, which claims that trust is always possible if discount factors are high enough.

From a theoretical point of view it is interesting that whether a game is played in its extensive-form or its normal-form representation can substantially change the set of WRP payoff vectors. In the extensive form of the Amnesty Dilemma, the second player always knows whether he is trusted and his action cannot be observed if no trust occurs. Hence, deterring him from dishonesty is harder than it is in the normal-form version of the game. However, this observation relies on the particular sequence of moves in the game: If, instead, the second player would first have to publicly commit to a level of honesty, cooperation could be sustained more easily. We also note that we made strong assumptions about observability of randomizations.

In the applications we examined, cooperation is possible if the players can let the variables of the stage game vary with past outcomes. Put differently, by making the terms of trade depend on history the players may be able to avoid playing the repeated extensive-form Amnesty Dilemma, and thereby achieve a cooperative outcome.

8. APPENDIX

We first state an additional lemma and then proceed to prove Lemma 3 and Proposition 4.

Lemma 9 *Let $(0, \bar{v}_2)$ be the individually rational payoff vector that is best for player 2. For δ sufficiently close to 1, if $v = (v_1, v_2) > 0$ is a WRP payoff vector, then the vector $v' = (v'_1, v'_2) = \alpha(v_1, v_2) + (1 - \alpha)(0, \bar{v}_2)$ is a WRP payoff vector for any $\alpha \in (0, 1]$.*

Proof:
First, notice that there is exactly one action pair $(\tilde{a}_1, \tilde{a}_2)$ such that $g_1(\tilde{a}_1, \tilde{a}_2) = v_1$ and $g_2(\tilde{a}_1, \tilde{a}_2) = v_2$ (see e.g. Figure 4). Likewise, there is exactly one pair $\tilde{a}' = (\tilde{a}'_1, \tilde{a}'_2)$ such that $g_1(\tilde{a}') = v'_1$ and $g_2(\tilde{a}') = v'_2$. Note also that $\tilde{a}'_1 \geq \tilde{a}_1$ and $\tilde{a}'_2 \leq \tilde{a}_2$ hold necessarily.

Define $\bar{a}$ to be the action pair yielding the payoffs $(0, \bar{v}_2)$.

Let σ be a WRP equilibrium sustaining v. To construct a WRP equilibrium σ' sustaining v', modify σ in the following way: For any action pair $a = (a_1, a_2)$ played at some date and history according to σ, multiply a_1 by $\frac{\tilde{a}'_1}{\tilde{a}_1}$ and a_2 by $\frac{\tilde{a}'_2}{\tilde{a}_2}$, and change the players' strategies to take these changed actions into account (e.g. if all actions $a_1 \neq \tilde{a}_1$ trigger a punishment by player 2 in σ, let all actions $a'_1 \neq \tilde{a}'_1$ trigger the corresponding punishment in σ'). Also, let σ' prescribe that if player 1 ever deviates then the pair $\bar{a}$ is played forever; if player 2 deviates after any prior deviation of player 1, then play reverts to the beginning of σ', so the continuation payoffs are (v'_1, v'_2).

For sufficiently large δ, player 1 will not deviate facing the punishment $\bar{a}$ and both players will not deviate after a prior deviation of her. So consider the decision problem of player 2 at any given date t and history up to $t-1$ (with no prior deviation of player 1), if σ is played, $a = (a_1, a_2)$ is the prescribed action pair at date t, player 2's average continuation payoff after he cheated in t is $\tilde{g}_2$, and his average continuation payoff after no deviation in t is $\hat{g}_2$. Suppose that player 2 is trusted in t (otherwise his decision problem is irrelevant). Since σ is a Nash equilibrium, it holds that

$$(1 - \delta)d + \delta \tilde{g}_2 \leq (1 - \delta)(a_2 b + (1 - a_2)d) + \delta \hat{g}_2. \quad (20)$$

This inequality implies that player 2 will not deviate in the proposed equilibrium σ' either: If we switch from σ to σ' all continuation equilibrium payoffs for player 2 are multiplied by some constant that is greater than 1, and we know that $\frac{\tilde{a}'_2}{\tilde{a}_2} a_2 \leq a_2$ (so $a_2 b + (1 - a_2)d$ increases) and $\tilde{g}_2 \leq \hat{g}_2$ (otherwise (20) were not true). Hence, an inequality corresponding to (20) holds for σ' and, by the one-stage-deviation principle, σ' is a subgame-perfect equilibrium.

Finally, if two continuation equilibria are Pareto-unranked, the two continuation equilibria resulting after any of the modifications made above are Pareto-unranked, too. This follows from the fact that the payoff vector $(0, \bar{v}_2)$ is Pareto-unranked with any other individually rational payoff vector. Hence, σ' is WRP if σ is.

Proof of Lemma 3:

In the proposed equilibrium, which we will construct below, let $\bar{g}^*_{i,j}$ be the average continuation payoff for i if play is in j's punishment phase (where both i and j could refer to the same player), so it holds that

$$\bar{g}^*_{i,j} = (1-\delta)g_i(a^j) + \delta(p^j v_i + (1-p^j)\bar{g}^*_{i,j}),$$

or $\bar{g}^*_{i,j} = \frac{(1-\delta)g_i(a^j)+\delta p^j v_i}{1-\delta+\delta p^j}$.

As the normal phase action pair set $(\tilde{a}_1(v) = \frac{v_1(d-b)+v_2(a-c)}{ad-bc}, \tilde{a}_2(v) = \frac{v_1 d - v_2 c}{v_1(d-b)+v_2(a-c)}$, which yields utilities (v_1, v_2). To punish player 1, set $a^1 = (1, 0)$ and define p^1 such that $\bar{g}^*_{1,1}$ is equal to zero. Then, player 1 will not cheat during the normal phase if δ is large. Since $(\bar{g}^*_{1,1}, \bar{g}^*_{2,1})$ is a weighted average of (v_1, v_2) and $(g_1(a^1), g_2(a^1)) = (c, d)$, player 2 is strictly better off during player 1's punishment then during the normal phase, so the two average continuation equilibrium payoff vectors $(\bar{g}^*_{1,1}, \bar{g}^*_{2,1})$ and (v_1, v_2) are Pareto-unranked. It remains to show that for any WRP equilibrium σ sustaining (v_1, v_2) we can construct a punishment of the proposed form for player 2 such that $\bar{g}^*_{2,2} < v_2$ and $\bar{g}^*_{1,2} \geq v_1$ and the resulting strategy profile is a subgame-perfect equilibrium.

Let σ^2 be the continuation equilibrium of σ that is worse for player 2, i.e. $g^*_2(\sigma^2) \leq g^*_2(\sigma^c)$ for any continuation equilibrium σ^c of σ. Such a worse continuation equilibrium σ^2 exists by Lemma 2 in Farrell and Maskin (1989) in the case of repeated normal-form games; their proof also applies to repeated extensive-form games. If there are several continuation equilibria that are worse for player 2, choose one that is best for player 1. Also, let $\hat{a}^2$ be the action pair in the first period of playing σ^2 and let $\hat{\sigma}^2$ be the continuation equilibrium after playing $\hat{a}^2$ in the first period of σ^2, so, for $i = 1, 2$,

$$g^*_i(\sigma^2) = (1-\delta)g_i(\hat{a}^2) + \delta g^*_i(\hat{\sigma}^2). \tag{21}$$

Choose $\hat{a}^2$ to be the action pair used to punish player 2 in the proposed equilibrium, i.e. set $a^2 = \hat{a}^2$.

Next, we will show that the following inequalities hold for $a^2 = \hat{a}^2$:

$$g_1(a^2) \geq v_1 \qquad (22)$$
$$g_2(a^2) < c_2(a^2) \leq g_2^*(\sigma^2) \leq v_2 \qquad (23)$$

Inequality (22) is shown in two steps. First, notice that $g_1(a^2) \geq g_1^*(\sigma^2)$ holds: Otherwise, equality (21) implies that $g_1^*(\hat{\sigma}^2) > g_1^*(\sigma^2)$. To avoid renegotiation from σ^2 to $\hat{\sigma}^2$, we then must have $g_2^*(\hat{\sigma}^2) \leq g_2^*(\sigma^2)$, which together with the previous inequality is a contradiction to the choice of σ^2. Second, $g_1^*(\sigma^2) \geq v_1$. Otherwise, we either have that $g_1^*(\sigma^2) < v_1$ and $g_2^*(\sigma^2) < v_2$ which would lead to renegotiation from σ^2 to σ, or we have $g_1^*(\sigma^2) < v_1$ and $g_2^*(\sigma^2) \geq v_2$ which violates the definition of σ^2. Hence, $g_1(a^2) \geq g_1^*(\sigma^2) \geq v_1$.

To see that $g_2(a^2) < c_2(a^2)$ note that in any WRP equilibrium sustaining $v > 0$, player 1 must trust her opponent with some probability in every period; otherwise the players could renegotiate and skip the period in which no trust occurs. Therefore, $g_2(a^2) < c_2(a^2)$ if $a_2^2 > 0$. But $a_2^2 = 0$ is impossible because we know $g_1(a^2) \geq v_1$.

Next, let $\tilde{g}_2$ denote player 2's average continuation equilibrium payoff after he has cheated in the first period of playing σ^2. Then, if $c_2(a^2) \leq g_2^*(\sigma^2)$ does not hold, the inequality $g_2^*(\sigma^2) \leq \tilde{g}_2$ implies that $(1 - \delta)c_2(a^2) + \delta \tilde{g}_2 > g_2^*(\sigma^2)$, and player 2 cheats in the original equilibrium. Hence, $c_2(a^2) \leq g_2^*(\sigma^2)$ is true.

Finally, suppose that $g_2^*(\sigma^2) \geq v_2$. First, consider a (v_1, v_2) that is close to $(0, \bar{v}_2)$ ($\bar{v}_2$ being the maximum individually rational per-period payoff player 2 can get) and rewrite v_2 as $v_2 = \bar{v}_2 - \varepsilon$. Since σ sustains (v_1, v_2), the equilibrium path of σ, which we denote by $\{a_1(t), a_2(t)\}_{t=1}^{\infty}$, must prescribe $a_1(t) \geq \tilde{a}_1(v)$ at least sometimes. (This can be seen in Figure 4: Since (v_1, v_2) is a weighted average of the equilibrium-path payoffs, at least some of these equilibrium-path payoffs have to lie on or above the straight dotted line through (v_1, v_2).) In the following, we will contradict this last statement. None of player 2's average continuation payoffs can be larger than $\bar{v}_2$ and (by $g_2^*(\sigma^2) \geq v_2$) any average continuation payoff for him cannot be worse than v_2, so we know from the fact that player 2 does not deviate in σ:

$$a_1(t)((1-\delta)d + \delta v_2) + (1 - a_1(t))\delta v_2 \leq \bar{v}_2$$

for all $t \in \{1, 2, ...\}$. Equivalently, using $v_2 = \bar{v}_2 \; \varepsilon$, it holds that

$$a_1(t)d \leq \bar{v}_2 + \frac{\varepsilon}{1-\delta}$$

for all $t \in \{1, 2, ...\}$ and some $\varepsilon > 0$. But, using the definition of $\tilde{a}_1(v)$, it is straightforward to see that $\tilde{a}_1(v)d$ is greater than and bounded away from $\bar{v}_2$ for small ε (see Figure 4), so for any given $\delta < 1$ and a sufficiently small ε it follows that $a_1(t) < \tilde{a}_1(v)$ for all t, which contradicts the statement above. Hence, $g_2^*(\sigma^2) \geq v_2$ cannot hold for (v_1, v_2) sufficiently close to $(0, \bar{v}_2)$. Now suppose $g_2^*(\sigma^2) \geq v_2$ is true for any WRP payoff vector $v > 0$. Then, by the construction of σ' in the proof of Lemma 9, it is also possible to sustain a vector $v' = (v_1', v_2') > 0$ arbitrarily close to $(0, \bar{v}_2)$ by a WRP equilibrium σ' which has the same property. This, as shown above, is a contradiction, so $g_2^*(\sigma^2) < v_2$ holds for any WRP vector $v > 0$ and the inequalities in (23) are true.

Now we construct player 2's punishment. Since (23) implies that $g_2(a^2) < g_2^*(\sigma^2) < v_2$, it follows from $\bar{g}_{2,2}^* = \frac{(1-\delta)g_2(a^2)+\delta p^2 v_2}{1-\delta+\delta p^2}$ that for a sufficiently large δ there is a (unique) $p^2 \in (0,1)$ which yields $\bar{g}_{2,2}^* = g_2^*(\sigma^2)$. Choose this p^2 as the probability to terminate player 2's punishment. Then, $\bar{g}_{2,2}^* < v_2$ and $\bar{g}_{1,2}^* \geq v_2$ hold. Also, player 2 does not deviate given the constructed strategies: In the normal phase, a deviation does not pay off for player 2 if

$$(1-\delta)d + \delta \bar{g}_{2,2}^* \leq (1-\delta)(\tilde{a}_2(v)b + (1-\tilde{a}_2(v))d) + \delta v_2,$$

which is true for large δ because $\bar{g}_{2,2}^* < v_2$ holds. During his punishment phase, player 2 will not cheat if

$$(1-\delta)d + \delta \bar{g}_{2,2}^* \leq (1-\delta)(a_2^2 b + (1-a_2^2)d) + \delta(p^2 v_2 + (1-p^2)\bar{g}_{2,2}^*),$$

or, equivalently,

$$(1-\delta)a_2^2(d-b) \leq \delta p^2(v_2 - \bar{g}_{2,2}^*),$$

which also holds for δ sufficiently close to 1. Moreover, no player deviates during his or her opponent's punishment, which completes the proof.

Proof of Proposition 4:
For δ arbitrarily close to 1, inequality (10) can be satisfied if and only if

$$\frac{a_2^2 b + (1-a_2^2)d}{a_2^2 a + (1-a_2^2)c} v_1 + a_2^2(d-b) - v_2 < 0, \qquad (24)$$

so we do not have to consider the probability p^2 for the case of sufficiently patient players. Let the left-hand side of (24) be denoted by $L(v, a_2^2)$.

For any $v > 0$, the action a_2^2 also has to satisfy the constraints $\underline{a}_2^2(v) \leq a_2^2 \leq 1$, so v is sustainable as a WRP vector if and only if it holds that $\min_{a_2^2 \in \{a_2: \underline{a}_2^2(v) \leq a_2 \leq 1\}} L(v, a_2^2) < 0$.

$L(\cdot)$ is convex in a_2^2, so the first-order condition of the unconstrained minimization problem yields the unconstrained solution

$$a_2^*(v) = \arg \min_{a_2} L(v, a_2) = \frac{1}{2}(\frac{v_2}{d-b} + \frac{v_1 - c}{a-c}).$$

Plugging $a_2^*(v)$ into (24) results, after straightforward algebra, in condition (12):

$$(4v_1 d - 2v_1 v_2 - 2v_2 c - v_2^2 \frac{a-c}{d-b})(a-c) - (v_1 - c)^2 (d-b) < 0$$

Define $L^*(v)$ to be the left-hand side of (12). The limit of the points satisfying condition (12), i.e. the set of points v with $L^*(v) = 0$, is given by the curved line in Figure 5. If we set $v_1 = 0$, the solution of the equation

$$L^*(0, \underline{v}_2) = 0$$

gives us the unique value $\underline{v}_2 = -c\frac{d-b}{a-c}$ at which the $L^*(v) = 0$ curve touches the horizontal axis in the figure.

Now consider the restrictions imposed on the set of WRP payoffs due to the constraints $\underline{a}_2^2(v) \leq a_2^2 \leq 1$. Using the expressions given above for $\underline{a}_2^2(v)$ and $a_2^*(v)$, it is straightforward to show that for a given v satisfying (12) the constraint $\underline{a}_2^2(v) \leq a_2^2$ binds (that is, $a_2^*(v) < \underline{a}_2^2(v)$) if and only if $v_2 < \underline{v}_2$. In particular, this constraint does not further restrict the set of WRP payoffs v with $v_2 \geq \underline{v}_2$. For the set of points v satisfying (12) and $v_2 < \underline{v}_2$ we can conclude immediately from Lemma 9 that v cannot be WRP: otherwise, all linear combinations of v and $(0, \bar{v}_2)$ are WRP, too, which is impossible due to condition (12) which implies that $(v_1, \underline{v}_2)$ is not a WRP payoff vector for any $v_1 \geq 0$ (see Figure 5).

The constraint $a_2^2 \leq 1$ binds for a feasible v satisfying (12) if and only if $a_2^*(v) > 1$. Since $L(\cdot)$ is convex in a_2^2, $L(v, a_2^2)$ is minimized over a_2^2 by setting $a_2^2 = 1$ in this case. Using $a_2^2 = 1$, condition (24) becomes

$$av_2 - bv_1 - a(d-b) > 0,$$

which is inequality (13). Let $M(v)$ denote the left-hand side of (13), so the limit of points satisfying (13) is the $M(v) = 0$ line, which is the straight line through the points $(0, d\ b)$ and (a, d) (see Figure 5). Any feasible non-zero payoff vector below the $M(v) = 0$ line can be sustained by a WRP equilibrium of the form in Lemma 3 with $a_2^2 = 1$.

To find conditions under which the constraint $a_2^2 \leq 1$ binds we now argue that the $L^*(v) = 0$ curve touches the $M(v) = 0$ line exactly once in a (not necessarily feasible) point $\tilde{v} = (\tilde{v}_1, \tilde{v}_2)$. First, the $L^*(v) = 0$ curve cannot cross the $M(v) = 0$ line due to the definition of $L^*(v)$ and $M(v)$: $L(v, a_2^2)$ at $a_2^2 = 1$ cannot be smaller than $L(v, a_2^*(v))$ for any v. Second, observe that if for a given point $\tilde{v}$ on the $M(v) = 0$ line it holds that $a_2^*(\tilde{v}) = 1$, then $\tilde{v}$ also satisfies $L^*(\tilde{v}) = 0$. Hence, the claim that the $L^*(v) = 0$ curve touches the $M(v) = 0$ line in $\tilde{v}$ is equivalent to requiring that $\tilde{v}$ solves the two equations $M(\tilde{v}) = 0$ and $a_2^*(\tilde{v}) = 1$. This system of equations has the unique solution $(\tilde{v}_1, \tilde{v}_2) = (a^2 \frac{d-b}{ad-bc}, d - b^2 \frac{a-c}{ad-bc})$. So indeed the $L^*(v) = 0$ curve touches the $M(v) = 0$ line exactly once in $\tilde{v} = (\tilde{v}_1, \tilde{v}_2)$.

The vector $\tilde{v}$ gives us the limit of payoff vectors v for which the constraint $a_2^2 \leq 1$ binds: Since $a_2^*(v)$ is strictly increasing in both v_1 and v_2 and it holds that $a_2^*(\tilde{v}) = 1$, it follows that for any $v > \tilde{v}$ we can sustain v as a WRP payoff vector if and only if (13) holds and v is feasible. Analogously, a given $v < \tilde{v}$ is sustainable in a WRP equilibrium if and only if both (12) and $v_2 \geq \underline{v}_2$ are satisfied and v is feasible. (Observe that the question whether $a_2^2 \leq 1$ binds is answered by now for all other v with $v_2 \geq \underline{v}_2$: If $v_1 > \tilde{v}_1$ and $v_2 \leq \tilde{v}_2$, or if $v_1 \geq \tilde{v}_1$ and $v_2 < \tilde{v}_2$, then both (12) and (13) hold; if $v_1 \leq \tilde{v}_1$ and $v_2 \geq \tilde{v}_2$, both conditions do not hold. See Figure 5.)

Finally, the question remains whether $a_2^2 \leq 1$ can bind for any *feasible* payoff vector v. By the observations in the previous paragraph, this is the case if and only if $\tilde{v}$ is feasible, which depends on the parameters of the game. Straightforward algebra characterizing the set of feasible outcomes (in particular, condition (14)) shows that $\tilde{v}$ is feasible if and only if $b \geq \frac{1}{2}d$.

Similarly, it follows that there are non-zero WRP payoff vectors if and only if $(0, \underline{v}_2)$ is feasible, which is equivalent to stating that condition (11) holds. Proposition 4 collects the results.

Notes

1. We thank the NSF and the Fulbright Commission, Bonn, for financial support.

2. This game is also known as the Trust Game.

3. For more examples and a discussion of economically relevant situations in which trust plays a central role see Williamson (1993) and the literature cited there.

4. We are not aware of experimental studies that systematically investigate the effects of repeating the Amnesty Dilemma or similar games.

5. The notation $v > 0$ is meant to be read as $v_i > 0$ for $i = 1, 2$. In constrast, $v \geq 0$ denotes $v_i \geq 0, i = 1, 2$, and $v_j > 0$ for some j. Observe that an action pair a satisfying $g_1(a) > 0$ and $g_2(a) = 0$ cannot exist, so $v_2 > 0$ whenever $v \geq 0$.

6. Using this assumption we do not have to consider restrictions on the set of WRP payoffs that are solely caused by the fact that optimal lengths of punishment phases can be non-integer.

7. The proofs of this section are relegated to the Appendix.

8. We only consider conditions that are due to player 2's incentives. Player 1 can easily be induced to comply by specifying a sufficiently strong punishment (see the proof of Lemma 3 in the Appendix). Hence, no extra conditions for player 1 are needed.

9. Conditions (5) and (6) are formulated for the case that player 2 is trusted in the current period (otherwise, his decision problem is irrelevant).

10. Notice that, since a_2^2 is a probability, the inequality $a_2^2 \geq \underline{a}_2^2(v)$ implies that for v to be WRP it must be that $\underline{a}_2^2(v) \leq 1$ or, equivalently, $\frac{v_1}{v_2} \leq \frac{a}{d}$, which is the non-strict version of condition (1). Hence, the limit of WRP payoffs in the extensive-form game cannot lie above the limit in the normal form. This is consistent with the intuition that the normal form makes it easier to discipline player 2 and, hence, easier to sustain cooperation.

11. This is intuitive because it simply says that player 2 willl not cheat during the normal phase if his punishment is severe enough to deter him from cheating during the punishment phase.

12. Necessity of $a_2^2 \geq \underline{a}_2^2(v)$ has been shown above. Conversely, if $a_2^2 \geq \underline{a}_2^2(v)$, then it also holds that $a_2^2 > \frac{v_1 - c}{a - c}$ (or, equivalently, $v_1 < a_2^2 a + (1 - a_2^2)c$) and the mixed action a_1^2 (which appears only in $\bar{g}_{2,2}^*$, see (2)) can always be chosen to be $a_1^2 = \frac{v_1}{a_2^2 a + (1-a_2^2)c}$. Using this a_1^2, (10) implies (4) and (9).

13. Notice that (11) always holds for $b \geq \frac{1}{2}d$, i.e. in case (ii) of Proposition 4.

14. From Figure 5 it is already obvious that, as in the normal-form game, player 2 must get to cheat a certain amount of the time.

15. This equation describes the limit of payoff vectors satisfying (13), which is in (v_1, v_2) space the straight line through $(0, d - b)$ and (a, d). See Figure 5b.

16. Note, however, that any feasible $v = (v_1, v_1)$ with $v_1 = 0$ (which is WRP for large δ by Lemma 1) is trivially SRP if no strictly positive WRP payoff vectors exist.

17. See Farrell and Maskin (1989) and van Damme (1989).

18. Strictly speaking, for this conclusion we need to assume that the buyer is not willing to agree to a WRP equilibrium of the kind described in Lemma 1, with $v_1 = 0$ and $v_2 > 0$ (which would only expose her to the risk of being cheated). In this and the following two sections we make this assumption.

19. For example, the constraint on p given in (17) may bind, i.e. $\underline{p} > c(q^*)$. Since for prices below $\underline{p}$ no trade can occur and no surplus is made, the buyer will then prefer some interior price.

20. In two-player games, a subgame-perfect equilibrium is strongly-perfect if all of its continuation equilibria are Pareto-efficient. See Rubinstein (1980).

21. The corresponding action pair for the product-quality problem is "buyer pays $c(0)$ and seller produces q^*". One might describe this as painfully rebuilding a reputation, although there is no private information in this game.

22. This is of course a generic concern in the theory of repeated games, but especially when (as is required for WRP equilibrium) the allegedly innocent party actually gains from claiming that the other has defected. One might want to limit this by ensuring that neither party can gain too much by claiming that the other has defected, but we do not explore this here.

23. The interest rate r that is charged on the loan should not be confused with the implicit utility-discounting interest rate that may be defined as $p \equiv (1/\delta) - 1$.

24. Kletzer and Wright (1998) study renegotiation-proof equilibria in a related model in which efficiency gains via lending arise due to the borrowing party's consumption smoothing motive (not, as in this paper, due to profitable projects that can only be undertaken if lending takes place). For a similar model and more bibliographical references concerning sovereign lending problems see also Kletzer (1994).

25. Certainly, however, the choice of r does influence the players' utilities if (19) is satisfied.

26. It is straightforward to show that $(1+\bar{r})L$ is exactly the minimum utility the borrower must get in any WRP equilibrium that sustain positive payoffs: Substituting the values for a, b, c, d into the expression for $\underline{v}_2$ (see Proposition 4) it follows immediately that $\underline{v}_2 = (1+\bar{r})L$.

References

Berg, J., J. Dickhaut and K. McCabe (1995); Trust, Reciprocity, and Social History; Games and Economic Behavior 10, 122-142.

Bernheim, B.D., and D. Ray (1989); Collective Dynamic Consistency in Repeated Games; Games and Economic Behavior 1, 295-326.

Farrell, J. and E. Maskin (1989); Renegotiation in Repeated Games; Games and Economic Behavior 1, 327-360.

Güth, W., P. Ockenfels and M. Wendel (1997); Cooperation Based on Trust: An Experimental Investigation: Journal of Economic Psychology 18, 15-43.

Kletzer, K. (1994); Sovereign Immunity and International Lending; in F. van der Ploeg (ed.), Handbook of International Macroeconomics; Oxford; Basil Blackwell.

Kletzer, K. and B. Wright (1998); Sovereign Debt as Intertemporal Barter; mimeo, University of California, Santa Cruz.

Rubinstein, A. (1980); Strong Perfect Equilibrium in Supergames; International Journal of Game Theory 9, 1-12.

van Damme, E. (1989); Renegotiation-Proof Equilibria in Repeated Prisoner's Dilemma; Journal of Economic Theory 47, 206-217.

van Huyck, J., R. Battalio and M. Walters (1995); Commitment versus Discretion in the Peasant-Dictator Game; Games and Economic Behavior 10, 143-170.

Williamson, O. (1993); Calculativeness, Trust, and Economic Organization; Journal of Law and Economics 36, 453-486.

8 REPUTATION AND SIGNALLING QUALITY THROUGH PRICE CHOICE

Taradas Bandyopadhyay,
Kalyan Chatterjee, and Navendu Vasavada

1. INTRODUCTION

There is a truism that people get what they pay for; in other words, the price is high because the quality of the item purchased is high. However, the truism need not always be true. Consumers who reason that high price automatically implies high quality are vulnerable to "cheating" by low-quality producers seeking to exploit this belief.

An example of the pitfalls of inferring quality from price, though in the opposite direction, appeared some years ago in the *Wall Street Journal* (1988). The newspaper mentioned Pathmark's premium all-purpose cleaner, whose chemical composition, according to the article, precisely duplicated that of the leading brand, Fantastik, but whose price was approximately half of the latter's. According to the report, the low price "discredited the intrinsic value of the product" in the eyes of the consumer. Pathmark chose to withdraw the product after its failure, rather than to raise the price, indicating that changing consumer perceptions about quality upward might be difficult if not impossible.

This chapter models price as a signal of quality, when quality is not discernable by inspection or by description, but is revealed (with noise) through use. In our model, information about quality is exchanged through a combination of the mechanisms in Kreps-Wilson (1982a) and Holmström (1983). There is both *strategic information transmission* through choice of price, and information obtained by experience of the good. Reputation plays a crucial role here, as in the literature beginning with Kreps and Wilson (1982a), but the reputation variable moves randomly, depending on the outcome of a consumer's experience, between pooling and partial separation.

Our model therefore adopts a somewhat different approach to the extensive literature that has addressed one or more of the issues we consider. Among these are the analyses of reputation and product quality by Allen (1984), Klein and Leffler (1981) and Rogerson (1983). Rogerson's paper, which is closest to ours in concept, analyzes the effect of word-of-mouth on the number of customers a firm can attract. In his model, there is sorting of firms by size and therefore quality. While such sorting takes place in our model as well, it is not the only means by which information is transmitted.

Milgrom and Roberts (1986) consider the other half of our model, namely the information revealed by price. Their model has several parameters, namely the cost of producing high quality items versus low quality items and the level of advertising. For some values of the costs, price is a sufficient signal of quality; other values (including the case where costs are equal for high and low quality goods) require a positive level of advertising in addition to price in order to achieve their objective. Our model deals with the case of equal costs, and we do not consider advertising at all. [1]

Other related work in the recent past is in Liebeskind and Rumelt (1989). These authors also consider a two-period model of an experience good. However, in their model, the seller chooses a price at the beginning of the game and is committed to it for two periods. Further, they do not consider the use of price as a signal, because they limit consideration to only those equilibria where price does not play a signalling role (that is, their equilibria are "robust" to interchanging the choices of price and quality). There also are no "types" in their model.

Wolinsky (1983) considers a model of price signalling where consumers have a positive probability of determining whether a firm is 'cheating' by asking for a high price for a low-quality good. He is able to show that price is then a perfect signal of quality. In our model, the probability of such detection is zero.

The main point of this chapter is, therefore, to illustrate the conditions under which price as an imperfect signal of the quality of an experience good can be sustained by reputation in the sense of Kreps-Wilson. In a single-period model, such an informative role for price is never possible in equilibrium. In the simplest finite-horizon model, with two periods, it is possible, but is one of several equilibria. We comment on the infinite-horizon case, but do not discuss it in detail, since in some sense the finite-horizon equilibria with signalling are harder to sustain than the infinite horizon ones.

In the rest of this section, we outline the remainder of the chapter in detail. Section 2 presents the basic model and calculates equilibria based on a restricted set of strategies. Section 3 removes these restrictions and discusses refinements to reduce the ensuing multiple equilibria. Without invoking a refinement, three types of equilibrium could exist, corresponding to a lemons market in which only low quality goods are produced, price being uninformative but experience leading to ex post sorting and the one we focus on where both price and experience are informative. Section 4 considers the case where buyers too have private information and Section 5 concludes.

2. THE BASIC MODEL AND A RESTRICTED ANALYSIS.

2.1 DESCRIPTION OF THE MODEL AND NOTATION

The basic model we use is as follows: There is a single, long-lived seller, facing one buyer in each of two periods. In each period, the seller has only one item to sell. If it is not sold in this period, the seller gets his per period reservation utility, here assumed to be zero.

The seller could be a high quality producer (type h) or a low quality one (type ℓ). A seller's type is private information, but the prior probability of its being h is commonly known to be π_1 at the beginning of period 1.

A type h seller has access to technologies H and L. Technology H produces stochastically better quality than technology L in the sense of first-order stochastic dominance.[2] Let realized quality in period t be $\tilde{x}_t$; with technology H, this quantity has a distribution $F_H(\cdot)$ and with technology L, $F_L(\cdot)$; the corresponding densities are $f_H(\cdot)$ and $f_L(\cdot)$, the expected values are $\overline{x}_H$ and $\overline{x}_L$ respectively. A type ℓ seller has access only to technology L at the beginning of the first period, but has an exogenously given probability $q > 0$ of learning the H technology by the beginning of the second period. Both types of seller have a production cost of 0 per item, irrespective of the quality.

In this and the next section, buyers are assumed to be identical with common reservation utility y. All agents are risk neutral, though this is not a crucial assumption. Future payoffs are discounted; the common discount factor is δ.

The realized qualities $\tilde{x}_t$ are conditionally independent and identically distributed across periods, given the technology that generates them.

At the beginning of the game, the seller, if he/she is type h makes a once-and-for-all choice of technology. Installing technology H entails an arbitrarily small but positive setup cost of $\varepsilon > 0$. (We will therefore consider the limiting equilibrium as $\varepsilon \to 0$; this is convenient to break ties and does not play a pivotal role in the model.) The type ℓ seller has no decision to make at this stage 0, but faces a similar choice at the beginning of period 2, if he learns the H technology in the first period.

After the choice of technology, when applicable, the seller makes a price offer[3] to the first buyer, who either accepts or rejects it. If the offer is accepted, the buyer experiences the realized quality of the good, x_1; his information is passed on to the second buyer by "word-of-mouth". If the offer is rejected, each player obtains his reservation utility. In the second period, there is an initial choice of technology, if the type ℓ seller has obtained the requisite know-how, followed by a seller offer and an acceptance or rejection by the (second) buyer. The game ends at the end of the second period; the second buyer experiences a realized quality of x_2.

In order to write down the payoffs to the players, we introduce some additional notation. We define

$$\pi'_t \equiv \text{the probability of } H \text{ technology prior to the price offer in period } t, \tag{7.1}$$

$$\pi''_t(p_t) \equiv \text{the probability of } H \text{ after the price offer in period } t, \tag{7.2}$$

$$\pi_{t+1}(x_t) \equiv \text{the probability of } H \text{ before any technology choice in } t+1. \tag{7.3}$$

Also let $\psi^i_{t+1}(\pi_{t+1}(x_t)) \equiv$ the equilibrium payoff to the seller using technology i in period t, in the subgame beginning in $t+1$.

The payoffs are given as follows; if the buyer in period t buys an item at price p_t, his expected payoff is

$$\pi''_t(p_t)\bar{x}_H + (1 - \pi''_t(p_t))\bar{x}_L - p_t, \tag{7.4}$$

and the seller's is

$$p_t + \delta E(\psi^i_{t+1}(\pi_{t+1}(\tilde{x}_t))), \tag{7.5}$$

where E stands for expectation.

In the next subsection, we consider a restricted version of the model in order to explore the main intuition behind our results. The main restriction is that the seller can choose one of two prices, a high price p_H and a low price p_L. This restriction is relaxed in Section 3.

2.2 EXOGENOUS PRICES

We now consider the model with two exogenously given prices, $p_H > p_L$, such that

$$\bar{x}_L - p_L = y, \tag{7.6}$$

and

$$\bar{x}_H - p_H > y. \tag{7.7}$$

(Note that y is the reservation utility of the buyer.)

We first consider the one-period game (or the basic model with $\delta = 0$).

Lemma 1 *Suppose that the type h seller chooses technology H. In any equilibrium of the resulting subgame, perfect separation of types through price offers is impossible.*

Proof. Suppose not, i.e. suppose that there exists an equilibrium with perfect separation of types. This can happen if:

1. Type h chooses p_H and type ℓ chooses p_L, or

2. Type h chooses p_L and type ℓ p_H.

In the first case, the buyer who receives an offer of p_H believes the good is of high quality and buys with probability one, from (7.7). Since $p_H > p_L$, type ℓ will then find it profitable to deviate, thus destroying the proposed equilibrium.

In the second case, type ℓ is clearly better off selling the good at p_L rather than not selling it at p_H and getting a zero payoff, and cannot therefore offer p_H in equilibrium. ∎

It is somewhat more interesting to verify that partially informative prices cannot be sustained in equilibrium either.

Lemma 2 *In any equilibrium of the single-period game, $\pi_1''(p) = \pi_1'$.*

Proof. For partial separation to occur, at least one type of seller must randomize. Suppose this is type ℓ. Thus, the type ℓ seller must be indifferent between asking for p_H and p_L. But since there is no difference in per period payoffs between type ℓ and type h, type h must be indifferent as well, at the stage where the price has to be chosen (after the technology has been chosen). However, the type h seller has to pay a positive ϵ to install type H technology. Therefore, this payoff is ϵ greater if he chooses type L rather than type H technology, given that he obtains no premium for type H technology. Therefore, $\pi_1' = 0$ and $\pi_1''(p_H) = 0$ as well, thus contradicting the premise. ∎

Note that, if ϵ were to be equal to zero, a potential signalling equilibrium could appear with type h choosing technology H and asking for p_H with probability one, while type ℓ randomized. This could occur even if both types were indifferent between p_H and p_L and is a consequence of allowing types to behave differently with identical payoffs.

We see also that the positive cost ϵ rules out all pooling equilibria (where the two types get identical payoffs) except the one where both types produce L goods and ask for p_L. This gives us Proposition 1.

Proposition 3 *In a single period game, the unique equilibrium of this model is for type h to choose to produce low quality (identical with type ℓ) and for all types of seller to charge p_L.*

The lemons problem is thus impossible to remove in a single period setup. We now turn to the two period setting and show that there could be signalling in the first period. Note that the one period game is not identical to the second period of the two period game, because the type h seller's decision to choose high or low quality is made once-and-for-all at the beginning of period one and is not repeated in period two. (In other words, the type h seller cannot switch from high to low quality in period 2, and will not choose to switch from low to high quality in any single period game.)

The lemons problem is attenuated in equilibrium in the two-period model in two ways. First, sorting may occur. That is, both types of seller pool, using the same pricing strategies, but the buyer's experience in the first period is more likely to be good with the high quality seller. Thus, the high quality technology will command a stochastically larger price in the second period and is therefore worth installing for the type h seller for small ϵ. Sorting is the phenomenon most frequently discussed in the previous literature on reputation and product quality. In addition, however, a second phenomenon, that of signalling through price, may occur. Signalling takes place through the mechanism of the high quality seller sacrificing expected profits in the first period (since a higher price decreases the probability of purchase) in order to increase the value of π_1'' in the "market". Given the value of π_1'', the expected incremental benefit, if a buyer purchases the item, is high for the high quality seller, because the buyer's experience is likely to be better with the high quality good than with the low quality good. The buyer's probability of purchase sustains this equilibrium by making type ℓ indifferent between demanding a high (pooling) and a low (revealing) price. Thus, the signalling is sustained in a complex way by the same features (monotonicity and first order stochastic dominance) that give rise to sorting.

We consider now the formal analysis of the two-period game beginning with the second period.

Before the price offers are made in the second period, the type ℓ seller obtains the type H technology with probability q and fails to obtain it with probability $1 - q$. The type ℓ seller then has the option to set up high quality technology at a cost of ϵ or continue using the low cost technology. As we saw in the previous analysis, the high quality and the low quality technologies must earn the same equilibrium payoff in the final period of the game. Therefore, we have

Lemma 4 *In the second period of the two period game, the type ℓ seller will produce a low quality good in equilibrium, whether or not he has obtained access to high quality technology at the beginning of the period.*

Remark 1 *If the game had continued beyond two periods, there could be an incentive for the type ℓ seller to use the high quality technology, as it became available. In this case, π_2' would be equal to $\pi_2 + (1 - \pi_2)q$ and the analysis could have proceeded in the sequel in essentially the same way as it now does.*

Given that the high type seller has chosen high quality technology and that the low type has low quality technology, we can characterize the second period pricing equilibria in the following proposition. We recall that π_2' is the probability that the seller is providing a high quality good. Later on we shall take into account explicitly the dependence of π_2' on realized first period quality x_1.

Proposition 5 *Let π_2' be the probability that the seller is providing high quality. Then the following equilibria can result in the second period of the two-period game:*

(i) The lemons equilibrium. Sellers of both types offer p_L and the buyer accepts. Any offer above p_L is deterred by an out-of-equilibrium conjecture by the buyer that the seller is of type ℓ.

(ii) A "trigger strategy" equilibrium. If $\pi'_2 \bar{x}_H + (1 - \pi'_2)\bar{x}_L - p_H \geq y$, both types of seller demand p_H. Otherwise, both demand p_L. The buyer accepts in either case. In this case, prices are completely uninformative to the buyer.

(iii) A trivial signalling equilibrium. Here $\pi''_2(p_L) = \pi^*$, where

$$\pi^* \bar{x}_H + (1 - \pi^*)\bar{x}_L - p_H = y, \tag{7.8}$$

provided $\pi'_2 < \pi^*$.

If $\pi'_2 > \pi^*$, p_H is asked for by both types of seller and price does not affect beliefs (also out of equilibrium).

If $\pi'_2 \leq \pi^*$, type h and type ℓ randomize so as to make $\pi''_2(p_H) = \pi^*$.

However, both type h and type ℓ are indifferent between p_H and p_L in this equilibrium.

Proof. It is easy to check that these are equilibria. To see that there are no others, we note that perfect separation is not an equilibrium, since a type ℓ seller can costlessly imitate a type h seller in a price offer. Pooling at p_H is optimal only if π'_2 is high enough as in (ii). Pooling at p_L is accounted for in (i) and (ii). Partial signalling is addressed in (iii). This exhausts the list of possibilities. ∎

The behaviour of players in stage one will depend upon the equilibrium to be followed in stage two. The multiplicity of equilibria in Proposition 2 thus poses a problem for determining what happens in the first period. We can see immediately, however, that the following must hold.

Lemma 6 *If the equilibrium in the second period is equilibrium (i) of Proposition 2, the unique first period response is for type h to choose technology L and for both types to demand p_L also in the first period.*

Proof. There is no second period difference in payoffs for the two types. The first period of the game then becomes similar to the one period game, which is solved in Proposition 1. ∎

Remark 2 *The lemons market is therefore an equilibrium outcome of the two period game as well. We shall show it is not the only one.*

First, we note that equilibria (ii) and (iii) of Proposition 2 give identical second period payoffs to the seller as illustrated in Figure 2.

That is, for $\pi'_2 > \pi^*$, the payoff is p_H and for $\pi'_2 \leq \pi^*$ it is p_L. The second period equilibrium expected payoff is therefore

$$E_h(\pi''_1(p)) = [p_H \cdot Prob.(\pi'_2(x_1) > \pi^* \mid H, \pi''_1(p))$$

$$+ p_L \cdot Prob.(\pi_2'(\tilde{x}_1) \leq \pi^* \mid H, \pi_1''(p))], \qquad (7.9)$$

for type h.

A similar expression can be derived for type ℓ. We denote ℓ's second period equilibrium expected payoff by $E_\ell(\pi_1''(p))$. We recall that the probability $\pi_1''(p)$ is the posterior probability of a high quality good after the price is observed by the period one buyer.

It is clear that, for the same $\pi_1''(p)$, E_h is strictly greater than E_ℓ. This is true, because

$$\pi_2'(x_1) = \frac{f_H(x_1) \cdot \pi_1''}{\pi_1'' f_H(x_1) + (1 + \pi_1'') f_L(x_L)}, \qquad (7.10)$$

is an increasing function of x_1 since the likelihood ratio f_L/f_H is decreasing. First-order stochastic dominance then is sufficient to assert that the probability that $\pi_2' > \pi^*$ is greater for type h than for type ℓ for the same value of $\pi_1''(p)$. We therefore have

Lemma 7 *For every value of π_1'', $E_h > E_\ell$, if the distributions $f_H(\cdot)$ and $f_L(\cdot)$ satisfy the monotone likelihood ratio property.*

We now turn to the analysis of the first period. The following proposition characterizes all the sequential equilibria of this game, given that second period equilibrium payoffs are given by (7.9).

Proposition 8 *If the second period equilibrium payoffs are given by (7.9), the equilibrium strategies in the first period of the two period game must be one of either (i), (ii), or (iii) below.*

(i) The ex-post sorting equilibrium. Seller type h chooses high quality technology. Sellers pool on offers, offering p_H if $\pi_1' \geq \pi^*$ and p_L if $\pi_1' < \pi^*$ where

$$\pi^* \bar{x}_H + (1 - \pi^*) \bar{x}_L - p_H = y. \qquad (7.11)$$

Buyers always accept equilibrium offers. If $\pi_1' > \pi^*$, an offer of p_L is also accepted. If $\pi_1' < \pi^*$, an offer of p_H is rejected. The buyer's conjectured probability remains equal to π_1' for all offers by the seller.

(ii) The signalling equilibrium. Seller type h chooses high quality technology and strictly prefers to offer p_H for any $\pi_1' \leq \pi^*$. Seller type ℓ randomizes between p_H and p_L if $\pi_1' \leq \pi^*$ so as to make $\pi_1''(p_H) = \pi^*$. The quantity $\pi_1''(p_L) = 0$ in this equilibrium. The buyer accepts p_L and accepts p_H in equilibrium with a probability α that makes the seller type ℓ indifferent between p_H and p_L.

(iii) The lemons equilibrium with only p_L being offered.

Remark 3 *We note that in equilibrium (i), the price offered in period one does not change the buyer's beliefs and is noninformative. However, the type h seller is willing to use type h technology, despite the positive setup cost, because he has a higher expected second period payoff than type ℓ. Thus, communicating higher quality is done entirely through realized quality in the first period and prices are uninformative about quality. Thus sorting takes place ex post, that is after first period performance.*

In equilibrium (ii), the type h player signals his type by asking for p_H, whatever the value of the prior probability π_1'. This, coupled with mixed imitation by the type ℓ player, induces an increase in probability to π^*, which has a beneficial effect on the second period expected payoff as well. Thus, both signalling and sorting take place. Equilibrium (iii) remains a possibility, sustained by particular buyer beliefs.

Proof. To check that (i) is an equilibrium, we note that given the buyer's beliefs and actions, a deviation of price upward by the seller would lead to a reduction in payoff by amount p_L. A reduction downward, when the equilibrium price is p_H, is clearly suboptimal since the buyer accepts equilibrium price offers, and since beliefs are unaffected. The buyer's action is clearly optimal given the seller's action.

The type h seller will choose type h technology because his payoff is $\delta \cdot (E_h - E_\ell)$ greater than the type ℓ seller's and this exceeds the incremental setup ϵ cost of using the high quality technology when ϵ is sufficiently small.

Equilibrium (ii) can be checked similarly. Note that α is chosen so as to make the type ℓ seller indifferent between asking for p_H and asking for p_L. Thus, when $\pi_1' \leq \pi^*$,

$$\alpha \left[p_H + \delta \{ p_H \ Prob._{\tilde{x}_1 | L}(\pi_2'(\tilde{x}_1) > \pi^*) + p_L \ Prob._{\tilde{x}_1 | L}(\pi_2'(\tilde{x}_1) \leq \pi^*) \} \right]$$

$$+ (1 - \alpha)\delta p_L = p_L + \delta p_L. \tag{7.12}$$

Note that the right hand side of equation (7.12) results from updating by the buyer based on the equilibrium being considered. Once p_L is observed, the probability of high quality drops to zero. A probability of zero cannot be changed by any observation $\tilde{x}$.[4] The second term on the left-hand side is the contribution to the expected payoff of a rejection of an offer p_H by the buyer. Since p_H has been offered and has been rejected, $\pi_2' = \pi^* = \pi_1''$. The second-period payoff to the seller with $\pi_2' = \pi^*$ is p_L.

From (7.12), we get

$$\alpha = \frac{p_L}{p_H + \delta(p_H - p_L)\text{Prob.}_{\tilde{x}_1|L}(\pi_2'(\tilde{x}_1) > \pi^*)}. \tag{7.13}$$

Since, by the monotonicity of the likelihood ratio, $\pi_2'(x)$ is monotonically increasing in x, we can apply first-order stochastic dominance to obtain

$$\text{Prob.}_{\tilde{x}_1|H}(\pi_2'(\tilde{x}_1) > \pi^*) > \text{Prob.}_{\tilde{x}_1|L}(\pi_2'(\tilde{x}_1) > \pi^*). \tag{7.14}$$

Applying this to the analogue of equation (7.12) for the high quality type,

$$\alpha \cdot \left[p_H + \delta(p_H - p_L)\text{Prob.}_{\tilde{x}_1|H}(\pi_2'(\tilde{x}_1) > \pi^*) \right] \tag{7.15}$$
$$> \alpha \left[p_H + \delta(p_H - p_L)\text{Prob.}_{\tilde{x}_1|L}(\pi_2'(x_1) > \pi^*) \right] = p_L.$$

This implies that the type h seller is strictly better off than the type ℓ seller and will therefore choose the type H technology despite the cost ϵ (provided ϵ is sufficiently small).

It can similarly be checked that the other players' strategies are best responses to one another. Pooling at p_L, the lemons equilibrium (iii), is sustainable by a buyer conjecture that any price demand above p_L implies a type ℓ seller with certainty.

It remains to be shown that this exhausts the possible equilibria. It is clear again that perfect separation with type h asking p_H and type ℓ asking p_L is not an equilibrium, because the buyer would then accept p_H always and this would lead to imitation by ℓ. For the same reason, randomization by type h and a pure strategy of playing p_L by type ℓ is also ruled out in equilibrium. Therefore, the proposition is a complete description of the equilibria. ∎

Despite the restrictive assumption of only two permitted prices, the equilibria of this section generalize to the more appealing models of Sections 3 and 4. The main result is that even in a model with the deck stacked against signalling, with no differentially costly opportunities or wasteful expenditure, a noisy signalling equilibrium appears. The intuition behind this result is simple. The experience of buyers is able to sustain signalling, even if the buyer is unable to detect ever whether he or she has bought a good of type H or type L; that is, even with the distributions of $\tilde{x}$ having identical supports for low and high type sellers. Further, this is essentially a multiperiod phenomenon, since in a single period only the lemons equilibrium survives.

The multiplicity of equilibria is not addressed in this section. In Sections 3 and 4 we shall discuss arguments for choosing an equilibrium. The important thing in this section is to note that a noisy signalling equilibrium is possible with minimal assumptions.

2.3 GENERALIZATIONS

We discuss the following possible extensions: (i) many buyers and sellers in each period; (ii) infinite horizon models; (iii) long-lived buyers; and (iv) buyers with private information about reservation prices.

(i) Many buyers and sellers: Instead of a single seller and a single buyer in each period, we could think of n sellers, each of whom is randomly matched with m buyers, where m is the number of items the seller has to sell. The buyers, if they decide to buy, all observe the same realization of $\tilde{x}_t$ in period t for a given seller, so that quality describes a batch rather than an individual item. Before committing to purchase, a buyer may decide to search at some positive cost. But if all sellers adopt the same equilibrium strategy, the result of Diamond (1971) holds and the buyer will never search, since this will lead to a net expected loss of the search cost. Thus, for example, a buyer who observes a seller offering p_H knows that other sellers will be doing the same, so that search is unprofitable. If the seller is an ℓ type and randomizes, with p_L being the outcome of the randomization, the buyer knows that even if she observes p_H elsewhere, her probability that the seller is H will increase just sufficiently to leave her indifferent. A positive search cost will tip the scales in favor of accepting the current offer. Thus, at least at first sight, extending the model in this way to many buyers and sellers does not yield any new results. It could, however, be that some buyers who are not matched with a seller in any period could begin a bidding war, thus lowering price. This possibility is not taken into account here.

(ii) <u>Infinite Horizon Models:</u> The use of an infinite horizon, rather than a two period model may complicate the exposition, but should not affect the main results. In equation (7.9), second period payoffs can be replaced by the expected future value for low and high types. With *stationary strategies* (depending on the probability that a type is h) and discounting, equilibrium in the infinite horizon model will be of the same form as the equilibria here.

(iii) <u>Long-lived buyers:</u> Suppose that both the buyer and the seller live for two periods so that a seller would be negotiating with the same buyer. Would this cause any difference in the results? Intuitively, one might think of value of information considerations leading the buyer to purchase even though the payoff from buying is below her reservation utility. Presumably the value of information should make up the difference. This might lead to an upward adjustment in price or a decrease in seller L's probability of quoting p_L. However, in this model, if the buyer is indifferent between buying and not buying in the second period, she is unable to make use of this information. Our conjecture is that there will be no difference between a sequence of buyers and a single long-lived buyer under the condition of a single buyer reservation price. The situation may change with buyer private information.

(iv) <u>Buyer private information:</u> This eliminates the need for buyer mixed strategies and is discussed in Section 4. There may be value of information considerations entering more explicitly with buyer private information.

3. ENDOGENOUS PRICES

In this section, we relax the assumption of fixed prices made in the last section.

Keeping the definition of p_L the same, that is, $p_L = \bar{x}_L - y$, we note that in a single period game, the lemons equilibrium, with a price p_L being charged by both types, will be the only equilibrium. In a two-period model, the lemons outcome will remain an equilibrium, sustained by the buyer's conjecture that any price above or below p_L signals type ℓ with probability one. There will also be equilibria of the type: both seller types demand p^* and the demand is accepted for any $p^* \leq \pi_1'\lambda + p_L$, where $\lambda = \bar{x}_H - \bar{x}_L$. If any price higher than p^* is asked for, the buyer conjectures the seller to be type ℓ with probability one and rejects the offer. Otherwise, the buyer's belief is unchanged. Any price below p^* is accepted. These can be considered to be in the same class as the lemons equilibria and will be ruled out by almost any refinement.

It is also clear that there is a continuum of two-price signalling equilibria of the kind described in the previous section, sustained by the buyer conjecture that any price offer other than these two equilibrium prices signals type ℓ with probability one. The problem seems to be, therefore, one of identifying a plausible set of signalling equilibria rather than of demonstrating that these are possible.

We therefore begin by considering the second period of a two-period game. Once again, second period payoffs must be identical for sellers with H and L technologies, therefore a type ℓ seller who obtains the high quality technology at the beginning of the second period will choose not to install it at incremental positive cost.

We now show that the lemons equilibrium will not survive in the second period of a two period game, unless $\pi'_2 = 0$, if out-of-equilibrium conjectures are restricted by using the Farrell-Grossman-Perry refinement. (See Farrell (1983) and Grossman and Perry (1986a).)

Suppose that $\pi'_2 > 0$ and that a price

$$p' = \pi'_2 \bar{x}_H + (1 - \pi'_2)\bar{x}_L - y$$

is offered.

The buyer may then conjecture that <u>both</u> types h and l have deviated and, based on the prior probabilities, decide to accept the offer. But acceptance would ensure that both types are strictly better off using p' than using $p_L = \bar{x}_L - y$. Thus, the conjecture is self-consistent and destroys the equilibrium.

Lemma 9 *The refined equilibrium second period payoff is equal to*

$$p' = \pi'_2(\bar{x}_H - \bar{x}_L) - y \tag{7.16}$$

$$= \pi'_2 \lambda p + p_L, \tag{7.17}$$

where we define $\lambda = (\bar{x}_H - \bar{x}_L)$ and p_L is defined as before.

Remark 4 *This equilibrium can be sustained by sellers of both types pooling on the price or by sellers of both types randomizing (when indifferent) with different probabilities.*

Proof. The proof is immediate, given the discussion. ■

We now turn our attention to the first period of the two-period game and assume that the high type has chosen high quality technology. Let the probability of high quality be π_1'.

Given the Farrell-Grossman-Perry refinement and Lemma 9, the lemons outcome is no longer an equilibrium (in the refined sense).

Among the pooling equilibria, this leaves the ex post sorting equilibrium. It is easy to check that the following is true:

Lemma 10 *The following is an equilibrium in the first period, given that the second period equilibrium payoffs are given by (7.16).*

- 1. The type h seller chooses high quality technology.
  2. Seller types pool on price with $p_1' = x_1'\gamma + p_L$ being the price demanded.
  3. The buyer accepts an offer at or below p_1' and rejects any offer above p_1'.

The argument we shall now make is as follows:
Suppose some $\hat{p} > p'$ is observed and define

$$\hat{\pi} = \frac{\hat{p} - p_L}{\lambda}. \tag{7.18}$$

If there exists a $\hat{\pi}$ and a possible mixed strategy best response $\alpha(\hat{p})$ of the buyer such that

(1) type h is strictly better off by choosing $\hat{p}$ rather than p', given the buyer's actions

(2) type ℓ is indifferent between choosing $\hat{p}$ and p',

(3) There exist strategies for types h and ℓ, such that, after Bayesian updating by the buyer,

$$\hat{\pi}_1''(\hat{p}) = \hat{\pi},$$

then the conjectures $\hat{\pi}_1''(\hat{p})$ are self-consistent and destroy the existing equilibrium with pooling price offer p'.

In order for such $\hat{\pi}, \hat{p}$ to exist, we have to impose an additional condition on $F_H(\cdot)$ and $F_L(\cdot)$. The condition is somewhat reminiscent of Spence's marginal cost condition for a signalling equilibrium as we shall see shortly. The value of $\hat{\pi}$ is also bounded by an expression that may be less than one, depending on the value of the parameters of the problem. We now turn to developing these conditions.

Consider type ℓ in the first period of the two period game. Given π'_1, the pooling equilibrium offer is

$$p' = \pi'_1 \lambda + p_L, \qquad (7.19)$$

and the expected payoff to L is

$$p' + \delta\gamma \left[\int_{-\infty}^{\infty} \pi'_2(\tilde{x}_1, \pi'_1) f_L(\tilde{x}_1) d\tilde{x}_1 \right] + \delta p_L. \qquad (7.20)$$

If an offer of $\hat{p}$ leads the buyer to change his or her beliefs to $\hat{\pi}$, then the expected payoff to type ℓ is

$$\alpha(\hat{p}) \left[\hat{p} + \delta\gamma \int_{-\infty}^{\infty} \pi'_2(\tilde{x}_1, \hat{\pi}) f_L(\tilde{x}_1) dx_1 + \delta p_L \right]$$

$$+ (1 - \alpha(\hat{p})) [\delta\gamma\hat{\pi} + \delta p_L]. \qquad (7.21)$$

The α that will make type ℓ indifferent can be found by equating expressions (7.20) and (7.21) and solving for α to obtain

$$\alpha = \frac{p' + \delta\gamma \left[\int \pi'_2(\tilde{x}_1, \pi'_1) f_L(\tilde{x}_1) d\tilde{x}_1 \right] - \delta\gamma\hat{\pi}}{\hat{p} + \delta\gamma \left[\int \pi'_2(\tilde{x}_1, \hat{\pi}) f_L(\tilde{x}) d\tilde{x}_1 \right] - \delta\gamma\hat{\pi}} \qquad (7.22)$$

It is clear that $\alpha \leq 1$ for $\hat{\pi} \geq \pi'_1$, with equality if $\hat{\pi} = \pi'_1$, since the expected second period probability π''_2 is increasing in π'_1. Of course, the value of α is the maximum of the expression (7.22) and 0. For type ℓ's indifference to hold, we must have $\hat{\pi}$ low enough so that the numerator of (7.22) is non-negative.

In order to economize on notation we rewrite equation (7.22) as

$$\alpha = \frac{e'_L}{\hat{e}_L}, \qquad (7.23)$$

where

$$e'_L = p' + \delta\gamma \left[\int \pi'_2(\tilde{x}_1, \pi'_1) f_L(\tilde{x}_1) d\tilde{x}_1 \right] - \delta\gamma\hat{\pi}, \qquad (7.24)$$

and

$$\hat{e}_L = \hat{p} + \delta\gamma \left[\int \pi'_2(\tilde{x}_1, \tilde{\pi}_1) f_L(\tilde{x}_1) d\tilde{x}_1 \right] - \delta\gamma\hat{\pi}. \qquad (7.25)$$

We define e'_H and $\hat{e}_H$ analogously, with $f_H(\cdot)$ replacing $f_L(\cdot)$ in equations (7.24) and (7.25).

Then we need

$$\alpha \hat{e}_H > e'_H, \qquad (7.26)$$

for the reference pooling equilibrium to be destroyed, or

$$\alpha = \frac{e'_L}{\hat{e}_L} > \frac{e'_H}{\hat{e}_H}, \qquad (7.27)$$

or

$$\frac{\hat{e}_H}{e'_H} > \frac{\hat{e}_L}{e'_L}. \qquad (7.28)$$

If $\hat{\pi} = \pi'_1$, then, of course (since $\hat{e}_H = e'_H$ and $e_L = e'_L$ by definition),

$$\frac{\hat{e}_H}{e'_H} = \frac{\hat{e}_L}{e'_L}.$$

As $\hat{\pi}$ increases above π'_1, $\hat{e}_H$ should increase faster than $\hat{e}_L$, for the desired result to hold.

We write this as condition (S); namely, the expected payoff to the type h seller (who has chosen technology H) increases with π at a faster rate than the corresponding payoff to the L type seller[5].

Note that this can be interpreted as asserting that the marginal benefit from signalling a certain level $\hat{\pi}$ is higher for the high quality type of seller. This condition does not, however, suffice to obtain a fully informative signalling equilibrium. Some probability of perfect verification of type seems to be necessary to obtain that in our framework. We therefore have

Proposition 11 *Given condition (S) and other earlier assumptions, there exists a $\hat{p} > p'$ and a self-consistent buyer conjecture $\hat{\pi}(p)$ such that the type h seller prefers to deviate from the pooling equilibrium offer p' and the type ℓ seller is indifferent. Therefore, the pooling equilibrium is not a perfect sequential equilibrium.*

Proof. See preceding discussion. ∎

It now remains to construct a signalling equilibrium based on the conjectures contained in equation (7.18). Such a conjecture will not be Bayesian consistent if the pooling price p' or any lower price is charged, since the high type will charge a higher price with probability one in equilibrium. Therefore, a price of p' or lower should signal a low type and this will lead to buyer rejection if the price is above p_L.

The signalling equilibrium, therefore, has two equilibrium price offers p_H and p_L, with p_L defined as before and p_H defined as

$$\overset{argmax}{p} \alpha(p)[p + \delta\gamma \int_{-\infty}^{\infty} \pi_2'(\tilde{x}_1, \pi_1''(p))f_H(\tilde{x}_1)dx_1 + \delta p_L]$$

$$+(1-\alpha(p))[\delta\gamma\pi_1''(p) + \delta p_L], \quad (7.29)$$

subject to

$$\alpha(p)[p + \delta\gamma \int_{-\infty}^{\infty} \pi_2'(\tilde{x}_1, \pi_1''(p))f_L(\tilde{x}_1) + \delta p_L]$$

$$+(1-\alpha(p))[\delta\gamma\pi_1''(p) + \delta p_L] = p_L + \delta p_L \quad (7.30)$$

and

$$\alpha(p) \geq 0. \quad (7.31)$$

The last condition gives

$$p_H \leq \frac{1+\delta}{\delta}p_L. \quad (7.32)$$

This means that the high price charged cannot be too high in equilibrium in comparison with the low one.

Let

$$\pi_1''(p_H) = \pi_1^*. \quad (7.33)$$

Proposition 12 *Suppose $\pi_1' \leq \pi_1^*$. Then there exists a signalling equilibrium described as follows:*

1. Type h chooses technology H and chooses price p_H in the first period. type ℓ randomizes between p_H and p_L so as to keep the buyer indifferent between buying and not buying. The price p_H is given by (7.29) to (7.32) and p_L has been defined earlier as equal to $\bar{x}_L - y$.

2. The buyer accepts p_L with probability one and all prices p above p_L with a probability $\alpha(p)$ with α chosen in $(0,1)$ to keep type ℓ indifferent between p and p_L.

3. Both types pool in the second period on $p_L = \pi_2' \gamma + p_L$ and this is accepted by the second period buyer.

The equilibrium in (1) to (3) is sustained by the following buyer conjecture for any p:

$$\pi_1''(p) = \frac{p - p_L}{\gamma}. \tag{7.34}$$

Under this conjecture, the buyer is indifferent between buying and not buying for any price p. The buyer then buys with probability α so as to keep player L indifferent between asking for p and p_L.

Proof. For any price p above p_L,

$$\pi_1''(p) = \frac{p - p_L}{\gamma}.$$

Thus, the buyer is indifferent between buying and not buying and randomizing, buying with probability $\alpha(p)$. This keeps seller type ℓ indifferent between p and p_L. First-order stochastic dominance implies that

$$\int \pi_2'(\tilde{x}_1, \pi_1''(p)) f_H(\tilde{x}_1) d\tilde{x}_1 > \int \pi_2'(\tilde{x}_1, \pi_1''(p)) f_L(\tilde{x}_1) d\tilde{x}_1.$$

This, in turn, implies that type h strictly prefers $p(> p_L)$ to p_L.

Type h then chooses his best price p_H (given by (7.29) to (7.32)). This ensures that (p_H, p_L) are equilibrium prices. The buyer's responses are also clearly optimal given his or her beliefs. ∎

The next question to ask is whether this refinement survives the Farrell-Grossman-Perry condition. We note first that there is another class of signalling equilibria in this model. In these equilibria, the prices charged are p_L and some price p_d less than p_L. The buyer always accepts both p_L and p_d. The conjecture $\pi''(p)$ <u>decreases</u> to zero as p_d <u>increases</u> to p_L and is chosen to make the type ℓ seller indifferent between asking for p_L and asking for p_d. Once again, type h then prefers p_d.

It is difficult to compare type h payoffs in these equilibria with the one in the previous proposition. The essential intuition is the same. Signalling involves sacrificing present payoffs for future payoffs; since type

h has higher future payoffs, it prefers doing this more than type ℓ. Sacrificing current payoffs could be through a reduction in the probability of purchase, as before, or in a reduction in price as in the (p_d, p_L) equilibria.

However casual empiricism suggests that high prices are usually regarded as signals of quality rather than low prices, so that the equilibria of Proposition 5 appear closer to real world behavior than the ones discussed in the preceding two paragraphs. We therefore impose condition (ND), namely,

$$\pi_1''(p) \text{ is nondecreasing in } p, \text{ for all } p. \tag{ND}$$

Lemma 13 *Under condition (ND) (and earlier assumptions), and if π_1' is sufficiently small relative to π_1^*, the equilibrium described in Proposition 5 cannot be destroyed by any self-consistent conjectures off the equilibrium path.*

Proof. Consider a deviation p^d. If $p^d < p_L$, $\pi_1''(p_d) = 0$, by condition (ND). Then it is clearly not a best response for either type h or type ℓ to deviate.

Suppose p^d is in (p_L, p_H). Then $\pi_1''(p^d)$ can satisfy one of the three conditions below. Either

$$\pi_1''(p^d) < \frac{p^d - p_L}{\gamma}, \tag{7.35}$$

or

$$\pi_1''(p^d) > \frac{p^d - p_L}{\gamma} \tag{7.36}$$

or

$$\pi_1''(p^d) = \frac{p^d - p_L}{\gamma}. \tag{7.37}$$

Equation (7.37) is the conjecture used in the equilibrium and we know that a deviation is not optimal. If (7.35) describes the buyer's belief, the buyer does not buy and both types of seller get zero first period payoffs. The seller's second period payoffs would clearly be better with $\pi''(p) = \frac{p - p_L}{\gamma}$ and even this is dominated by the equilibrium payoff from p_H for type h. Therefore type h does not deviate with this buyer belief.

Therefore, the only self-consistent conjecture satisfying (7.35) would be $\pi_1''(p^d) = 0$. This cannot destroy the equilibrium.

If the buyer's belief is given by (7.36), then the buyer will accept the price with probability one. Then the deviation p^d will be strictly preferred by type ℓ whose equilibrium payoff involves $p_L(< p^d)$ being accepted with probability one and with $\pi_1''(p^d) = 0$. It might or might not be preferred by H. Therefore,

$$\frac{p^d - p_L}{\gamma} < \pi_1''(p^d) \leq \pi_1' \leq \pi_1^*(p_H). \tag{7.38}$$

From (7.38), the second period payoffs from the deviation will clearly be less than the second period payoffs from the equilibrium (since second period payoffs are monotone increasing in π_1''). The first period payoffs might favor the deviation from equilibrium. However, we know that type h strictly prefers p_H (leading to $\pi_1'' = \pi_1^*$) to p_L (with $\pi_1'' = 0$) despite the fact that a buyer always accepts p_L. By continuity, if $\pi_1''(p^d)$ is sufficiently close to 0, the price p_H will be preferred. A sufficient condition for this, according to (7.38), is for π_1' to be small relative to π_1^*. If this happens, h will not deviate and $\pi''(p^d) = 0$ thus deterring ℓ from deviating as well.

The intuition behind this last result is plausible. When the prior is low, the high type most prefers to signal and the benefits from so doing ensure that the signalling equilibrium survives the refinement.

It should be pointed out that in an infinite horizon model with δ close to one, present payoffs will be small compared to the future and the robustness of the signalling equilibrium to FGP deviations can be more easily ensured. Thus, an infinite horizon model should support our conclusions more strongly.

We conclude from this section then that endogenizing the prices charged in equilibrium is possible and that the basic result of the model of the previous section survives in this more general context.

4. BUYERS WITH PRIVATE INFORMATION

Assume now that we still have one buyer per period, but that the buyer obtains private information about a reservation utility $\tilde{y}_t$ in period t, where the y_t are identically and independently drawn from a known distribution. To avoid triviality, suppose that the common knowledge probability distribution of $\tilde{y}_t$ is such that a buyer would be willing to buy a known high quality good at some price, no matter what his reservation utility.

We ask the question: which of the equilibria described in the previous section survive in this new context?

First, the signalling equilibrium survives with the following redefinitions. Let the pooling second period price be $\psi_2(\pi'_2)$. Given this, let the optimal first period price for a known type ℓ seller be denoted by p_L. For any price p above p_L, define $\pi''_1(p)$ as that posterior that leads to a buyer probability of purchase that makes type L indifferent between demanding p and demanding p_L. Note that the buyer probability of purchase is not obtained from a mixed strategy, since the buyer has private information too but from a natural calculation. If, for example, the distribution function of $\tilde{y}$ is $G(\cdot)$, then for any price p this probability is $G(\pi''_1(p)\bar{x}_H + (1 - \pi''_1(p))\bar{x}_L - p)$. To maintain the analogy let us call this $\alpha(p)$.

With π''_1 determined so as to keep type ℓ indifferent, the type h seller determines the optimal p as before. We call this p_H, in analogy with equations (7.29) to (7.32). Then we have

$$[p_H + \delta\gamma E_L \psi(\pi'_2(\tilde{x}_1 \mid \pi''_1)] G(\pi''_1(p_H)\bar{x}_H + (1 - \pi''_1(p_H))\bar{x}_L - p_H) + \delta\psi(\pi''_1)(1 - G)$$

$$= p_L \cdot G(\bar{x}_L - p_L)(1 + \delta). \qquad (7.39)$$

This is the analogue of the indifference equation that determined α in Sections 2 and 3.

The pooling equilibrium of the last section, which was sustained by probabilities π''_1 and π'_1, whatever the value of p, cannot survive under the same assumptions. The reason is that type h and type ℓ, with different expected future benefits, would come up with different optimum prices corresponding to a probability π'_1. (Note that this will not happen in the second period, since the payoffs to H and L technologies are identical in any one period game.)

Pooling could be sustained at the lemons equilibrium, where both types quote p_L and $\pi''_1 = 0$. This would be sustained by out-of-equilibriun beliefs that ascribed the same probability of a high type no matter what the price demanded.

Pooling at the high type's optimal first period price given π'_1 is also an equilibrium for high π'_1 if any price below this is considered to imply $\pi''_1 = 0$ and if the low type prefers the pooling price and its associated probability to revealing himself.

The equilibria in the model of this section are therefore similar to those in Sections 2 and 3 even though buyer mixed strategies are not used.

(Note, however, that signalling requires the low type seller to randomize on price; this is a consequence of the assumption of two types.)

An interesting extension to this model would be to have a long-lived buyer as well as a long-lived seller. The buyer's reservation utility $\tilde{y}_t$ could, as in this section, be randomly drawn every period. With a positive expected future value, the buyer would be more willing to purchase a good in the first period, since there would be an informational component as well as the utility obtained from the good itself. Our conjecture is that this might raise the equilibrium prices and increase the difference between p_H and p_L, but will not materially affect the equilibria possible in the model.

If the buyer's $\tilde{y}$ is drawn at the beginning of the first period and remains the same in the second period, the issue becomes more complicated because both the buyer and the seller will now be signalling to each other through their actions and sellers might want to experiment with price demands.

5. CONCLUSIONS

In this paper, we have sought to analyze the intuitive notion that prices often serve as a signal of product quality. We have shown that even in a model with no opportunities for single period signalling, an equilibrium exists in which price partially signals quality. We show that this possibility is sustained by the multiperiod structure of the model and that such signalling might occur in both finite and infinite horizon versions of the model. The price signalling equilibrium rests on conjectures that are intermediate between those that generate sorting purely based on observed quality and those that generate a lemons market.

We have also shown that these equilibria will have analogues in models where the buyer does not use mixed strategies. The assumption of two types of seller makes it necessary, however, for the low type seller to use mixed strategies in equilibrium. A continuum of types of seller model would have been significantly more complicated than the one in this paper, because of the heavy use made here of the monotone likelihood ratio property. We did not feel that the additional insights obtained would compensate for the increase in algebra.

We feel that, though there are other potential applications of this framework, the model is particularly appropriate for the use made of it here–namely the use of price as an imperfect signalling device without assuming that quality can be perfectly verified with some probability.

References

Akerlof, G. A. (1970): "The Market for Lemons," in the *Quarterly Journal of Economics*, Vol. 89, pp. 488-500.

Allen, Franklin (1984): "Reputation and Product Quality," *Bell Journal of Economics*, Vol. 15, pp. 311-327.

Banks, Jeffrey S. and Joel Sobel (1987): "Equilibrium Selection in Signalling Games," *Econometrica*.

Bhattacharya, S. (1979): "Information, Dividend Policy, and the 'Bird in Hand' Fallacy," *Bell Journal of Economics*, Vol. 10, pp. 259-270.

Bhattacharya, S. (1979): "Delegated Portfolio Management," *Journal of Economic Theory*, Vol. 36, pp 1-25.

Chatterjee, Kalyan and Navendu Vasavada (1984): "Signalling Through Price Choice in a Repeated Principal-Agent Problem with Asymmetric Information About Agency Quality," Mimeo.

Chatterjee, Kalyan and Larry Samuelson (1988, April): "Bargaining Under Two Sided Incomplete Information: The Unrestricted Offers Case," *Operations Research*, Vol. 36, pp. 605-618.

Cho, I.K. and David M. Kreps (1987): "Signalling Games and Stable Equilibria," *The Quarterly Journal of Economics*, Vol. 102, pp. 179-221.

Cothren, Richard and Mark A. Loewenstein (1989): "A Sequential Bargaining Fame with Asymmetric Information: The Market for Lemons Reconsidered," Mimeo, Department of Economics, VPI, Blacksburg, VA.

Diamond, Peter A. (1971): "A Model of Price Adjustment," *Journal of Economic Theory*, Vol. 3, pp. 156-168.

Farrell, Joseph (1983): "Communication in Games I: Mechanism Design without a Mediator," Mimeo, Department of Economics, MIT Cambridge, MA.

Ferguson, T.S. (1967): *Mathematical Statistics: A Decision Theoretic Approach,* New York: Academic Press.

Gibbons, R. (1985): "Optimal Incentive Schemes in the Presence of Career Concerns," Mimeo, Department of Economics, MIT, Cambridge, MA.

Grossman, Sanford and Motty Perry (1986a): "Perfect Sequential Equilibria," *Journal of Economic Theory,* Vol. 39, pp. 97-119.

Grossman, Sanford and Motty Perry (1986b):" Sequential Bargaining under Asymmetric Information," *Journal of Economic Theory,* Vol. 39, pp. 120-154.

Holmström, Bengt R. (1983): "Managerial Incentive Problems: A Dynamic Perspective," Mimeo, Yale University School of Organization and Management.

Klein, Benjamin and Keith B. Leffler (1981): "The Role of Market Forces in Assuring Contractual Performance," *Journal of Political Economy,* Vol. 89, pp. 615-637.

Kreps, D.M. and R.B. Wilson (1982a): "Reputation and Imperfect Information," *Journal of Economic Theory,* Vol. 27, pp. 253-279.

Kreps, D.M. and R.B. Wilson (1982b): "Sequential Equilibria," *Econometrica,* Vol. 50, pp. 863-894.

Liebeskind, Julia and Richard P. Rumelt, (1989), "Markets for Experience Goods with Performance Uncertainty," RAND Journal of Economics, Vol. 20, No. 4, Winter, pp. 601-621.

Milgrom, Paul and John Roberts, (1986): "Price and Advertising Signals of Product Quality", the *Journal of Political Economy,* Vol. 94, No. 4, pp. 796-821.

Roell, Ailsa (1987): "Signalling, Taxes and the Transaction Cost of Takeovers," Mimeo, London School of Economics.

Rogerson, William P. (1983): "Reputation and Product Quality," *Bell Journal of Economics,* Vol. 14, pp. 508-516.

Ross, S. A. (1977): "The Determination of Financial Structure: The Incentive Signalling Approach," *Bell Journal of Economics,* Vol. 8, pp. 23-40.

Spence, A. M. (1973): *Market Signalling,* Boston, MA: Harvard University Press.

Wolinsky, A. (1983): "Prices as Signals of Product Quality," *Review of Economic Studies,* pp. 647-658.

Wilson, R. (1985); "Reputation in Games and Markets," in A. Roth (Editor), *Game Theoretic Models of Bargaining.* " New York: Cambridge University Press.

Wall Street Journal, November 15, 1988

Notes

1. An earlier version of our paper discussed the informational value of plush offices for accounting firms, which might be considered a form of advertising, but the current version considers a stripped-down model that concentrates on the role of reputation rather than advertising. Of course, these models are all related, as in some sense all signalling models are variations of Spence (1973).

2. This is also implied by the assumption, made later, that the likelihood ratio $\frac{f_L(x)}{f_H(x)}$ is decreasing in x.

3. Note that the price cannot be contingent on the experienced quality in this model; this requires a mechanism to generate a commitment to such a contingent contract from the seller as well as the absence of moral hazard on the part of the buyer, who is the only one who actually experiences the good.

4. We assume this support restriction holds.

5. This seems, in general, to be an independent condition. Apart from the case where output has two values, we have not been able to derive it from the MLRP.

9 GAME THEORY AND THE PRACTICE OF BARGAINING

Kalyan Chatterjee

The interest of game theorists in bargaining and negotiation is of long standing. Among the early contributors to the study of bargaining were Howard Raiffa (1953), John Harsanyi (1956) and, of course, John Nash (1950, 1953). While Raiffa explicitly labeled his work "arbitration," thus emphasizing its normative aspects, Nash appeared to be seeking a solution that would describe actual outcomes of the negotiating process. His axiomatic framework can be interpreted as setting out the principles that describe a class of bargaining processes. A later authoritative work on the axiomatic literature by Roth (1979) adopts a similar interpretation.

Nash saw the need for explicit modeling of bargaining procedures or extensive forms. This approach to modeling laid out the actual sequence of possible decisions that negotiators would have to make during the course of the process, including responding to and making offers and, in some models, deciding when to make offers and to whom. In such a description the decision to cooperate or to sign a binding contract would appear as a choice for the players in some institutional setting. The axioms would then be justified if the equilibrium of some reasonable game corresponded with the outcome implied by the axioms – thus, the so-called Nash program of research.

The last twenty years have seen an enormous volume of work on game-theoretic models of bargaining. Most of this has sought to concentrate on extensive forms that represent important aspects of the real world processes of trading, rather than on axioms characterizing the outcome. The strategic use of proprietary information, the role of time preference, the influence of the competitive environment and of alternative trading opportunities have all been modeled and discussed. Bargaining research has also provided a basis for the interaction between theory and experiment in investigating fundamental behavioral questions of self-interest and fairness as driving forces in human action.[1]

Despite the excitement this work has generated, a growing number of individuals are critical of its achievements. They include pioneers such as Raiffa and influential theorists such as David Kreps. In *Game Theory and Economic Modeling* (1991), Kreps writes "bargaining is an extremely difficult topic because in many settings it runs right up against the things game theory is not good at." (p. 92). Binmore, Osborne and Rubinstein (1992) assert that "it ... seems premature to advocate any of the proposed resolutions of the problem of bargaining under incomplete information for general use in economic theory" (p. 210-211).

This discontent among theorists with the literature is paralleled by skepticism in other fields, often due to unfamiliarity with the highly technical aspects of the new work. At the same time, the importance of studying negotiation in the real world has become well recognized. Most major business schools and law schools now have courses on applied negotiations, and many use Raiffa's *The Art and Science of Negotiation*. Best-selling practical books on bargaining such as Herb Cohen's, *You Can Negotiate Anything*, form the bases for popular short executive courses.

The aim of this chapter is to explore what the recent advances in game theory have to say to the practitioner, especially in view of the doubts expressed by many theorists themselves. The chapter is organized as follows. Section 1 contains a discussion of the main themes of the recent research and the implications of these results for practice. Section 2 lays out some of the criticisms of this literature – criticisms made of game theory in general. Section 3 addresses these criticisms by examining how we should interpret game theory models, and section 4 concludes.

The contention in this chapter is that theory provides insights that are more detailed and nuanced than those of common sense, and it is this refinement of common sense that makes it valuable. One caveat: There is no attempt to survey the enormous literature in this area. The focus is on particular models

Game Theory and the Practice of Bargaining 275

and results. Admittedly, the presentation is condensed and, therefore, complete derivations are not included. References are provided for those readers who wish to verify these results for themselves. Here the results and the models are taken for granted and the emphasis is on what, if anything, they have to say for the practice of bargaining.

1. Models and Practice

One indication of the importance of understanding negotiation is the proliferation of negotiation courses, both in universities and in special executive programs. There are also several books on negotiation written by practitioners. One example is Herb Cohen's, *You Can Negotiate Anything*. (Cohen, a lawyer, is frequently called on as a consultant on negotiation by both the government and the private sector.) I shall select a few of the insights of this book and compare them with the results of the non-cooperative models of the last twenty years.

Cohen's work has prescriptive and descriptive aspects, and the descriptive aspects are used to generate prescriptions for action. The main feature of the descriptive analysis is the investigation of what constitutes power in negotiation. The more *powerful* one is, the closer the negotiated outcome is to one's most desired agreement. Cohen identifies the following factors as contributing to power:[2]
1. *Competition*. You are more powerful if you have fewer competitors and if your opponent has many competitors.
2. *Risk-taking behavior.*
3. *Legitimacy and recognized authority.*
4. *The power of commitment.*
5. *Expertise in the subject area of the negotiation.*
6. *Knowledge of the other party's needs.*

Cohen also discusses tactics, ways of making the most of the power one has. First, one should recognize that most things in the real world are negotiable. The tactics are designed to ensure successful conclusion of an agreement (for example by developing a negotiating style that engenders trust) and to capture more of the gains from trade in the event of an agreement. (This might mean committing to a first-and-final offer or inducing the other party to invest in the relationship so that he will be inclined to reach an agreement at any cost or using any number of other strategies detailed in the book.) Some of

Schelling's (1960) classic insights come through in this work, including the times when weakness is strength and the importance of making ultimatums credible.

Academic research deals with many of the same issues but in different ways. For example, the concept of "power" is not used as an explanatory variable, an input to the analysis. Rather, the elements of competition, risk preference, commitment and so on are modeled and the effect of each individually on the negotiated outcome is derived. A more desirable outcome for a party could be said to display greater "power" for that party, but this term is simply a convenient shorthand. There is no wrangling about definitions of the word. One can conceive of the power of different players as determined by their personal and environmental characteristics, i.e. the "parameters" of the model. The discussion of negotiation tactics, in a stylized form, then becomes the determination of the best way for an individual to play the bargaining game, given the parameters and the method of play of other active decision makers. The game-theoretic models assume that the methods of play are rational, and the use of Nash equilibrium is justified by positing mutual knowledge of rationality and common knowledge of beliefs by the players. As demonstrated by Aumann and Brandenburger (1995), these are sufficient conditions. There is also work on bounded rationality in the game-theoretic literature, but as yet it has had a limited impact on bargaining theory.

Let us now consider how game-theoretic models attempt to explain the causes of impasse or delay, and the factors that lead one bargainer to do better than another. The first factors we consider are the bargainers' utility functions, the degree of competition, incomplete information, and reputation. These could all be thought of as inherent bargaining characteristics. However, a bargainer is often able to affect values of these characteristics by his or her actions, and this leads us into questions of tactics (though not of questionable tactics) and of considerations beyond those necessary for a single game.

Impatience and Sequential Bargaining

The easiest way to highlight the effects of utility functions and of competition is to consider the simplest bargaining problem: two players can share a "pie" of unit size, provided they agree how to divide it. Later we shall consider the case where the two players bargain in the shadow of an external market (or other outside alternatives). A natural process to consider is the "alternating-offers" extensive form, where a player opens the bargaining by making an offer, which the other player can accept or reject. A rejection leads to a counter-offer by the recipient and so on. Such a process of haggling is studied by Rubinstein (1982) and Stahl (1972). Right away, we notice something that may not have been obvious without the formal statement of the procedure. Alternating offers imply a minimal degree of commitment to an offer; a proposer cannot rescind it unless it is rejected. A procedure that allows players to withdraw accepted offers would be expected to have different properties. Also, bargaining in good faith requires a recipient to come up with an offer once he or she has rejected an earlier offer. Again, such expectations of good faith are not common to all situations, and it is important to take this into account when formulating analyses and recommendations. Interestingly, research studies on deal making between computers as in Rosenschein and Zlotkin (1994) relies on specifying the procedure exactly before the computers begin to "bargain." Thus, the properties of different procedures are potentially of benefit to those constructing such computer programs.

To gauge the effects of the players' preferences on the allocation of the pie, we appeal to two versions of Rubinstein's (1982) alternating offers procedure. Suppose that the players prefer more of the pie to less and also discount future consumption at different rates. Then, Rubinstein shows that the advantage will lie with the more patient player. However this is in a specific offer/counter-offer extensive form. The patient player may not be able to capitalize on his or her power if, for instance, there is no commitment to an offer until the other party has an opportunity to respond.[3] The important aspect of this is not just recognizing what constitutes power but also recognizing the conditions that make it valuable. Similarly, a variant of the same Rubinstein model with an exogenous probability of termination allows us to make precise a popular intuition about risk aversion. An individual who is less risk averse than his bargaining partner does better in the bargaining.

The Rubinstein result is one of the most cited in the theory of bargaining. However, under different assumptions the finding of a unique solution (subgame perfect equilibrium) does not hold. For example, if offers are simultaneous rather than sequential, so that both sides make offers, it has been shown that every individually rational outcome and any length delay is an equilibrium in addition to the Rubinstein outcome. If offers were on a grid, that is if there were a smallest unit of the pie that could not be further subdivided and the discount factor were sufficiently high, the uniqueness of the Rubinstein solution would again collapse. These are not deficiencies of Rubinstein's model or approach. If players do not take into account the strategic effects of a smallest unit of pie in their analysis, then a canonical assumption of infinite divisibility is justified. The Rubinstein outcome then gets chosen, even though it is one of multiple equilibria in the "true" game.[4]

The Effect of Competition

Bargaining theory has much to say about the role of competition. The competitive landscape influences a bargainer's "Best Alternative to a Negotiated Agreement" – BATNA as Raiffa calls it or the status quo point in Nash's view. If one bargainer has many good alternatives to the current negotiation, his payoff if the current negotiation fails will not be terrible and this might increase his willingness to break off rather than to continue talking. Binmore, Rubinstein and Wolinsky (1986) have examined this intuition carefully in the context of the Rubinstein model, and Binmore, Shaked and Sutton (1988) have tested it experimentally. It is not automatic that a better BATNA will increase your payoff in the current negotiation. For example, if you are already getting 50 percent of the pie, it is not much use going to your counterpart and saying, "You'd better increase your offer, because I have received an attractive outside offer of 40%." On the other hand if the outside option is forced on you by exogenous breakdown, then this *will* matter. In a given situation, we need to ask which of two possibilities is true. Can you choose when to leave a negotiation? Or will you be forced to take your outside option if breakdown occurs exogenously? Without careful attention to bargaining theory, such a question would not arise. Indeed, the question is also of practical importance. In real-world negotiations, an individual can gain a bargaining advantage if he can commit to take an outside option in the event of an impasse.

The role of competition can be modeled in two other ways. One approach is to formulate a model with several buyers and sellers, where a player can choose to terminate a negotiation and go to someone else. Here again, the institutional detail is important and should determine whether a negotiator is able to use his competitive power. If there is a particularly productive player and the others can bid for her against each other, this player can usually do very well for herself (unless the other parties are able to collude). However, if the negotiation has to be bilateral and each player has to reject an offer before moving on to another partner, the powerful player may not do very well. See

Chatterjee, Dutta, Ray and Sengupta (1993) and Selten (1981) for alternative views on a particular game with one strong and two weak players, and Binmore (1985) and Chatterjee and Dutta (1994) for discussions of public offers and telephone bargaining and the different outcomes that may result.

It is also useful to consider another strategy a player might follow even if he or she has no current alternatives to the negotiation he is in. This is to go out and search for such alternatives in a bid to strengthen his bargaining position. Chikte and Deshmukh (1987) consider a case in which negotiations are temporarily suspended as players look for alternatives. If an alternative is found, a player has to decide whether or not to take it. If he rejects it, he can return to the original negotiation. Lee (1994) and Chatterjee and Lee (1998) model a somewhat different setup, where a player can return to the existing negotiation with the search result in hand and ask for a new offer. The offers resulting from the search are not modeled as coming from strategic players, but as draws from a probability distribution of offers. (While this is a shortcoming, it makes the models tractable.) These two papers show that enhancing a player's ability to search (i.e. lowering his search costs) will not necessarily lead to a better offer from his current partner. Beyond a certain point, a decrease in a buyer's search cost makes it unattractive for the seller to make an immediately acceptable offer; the seller would rather take his or her chances and let the buyer come back with an outside option that the seller can then match. However, if the buyer is unable to communicate his outside option credibly to the seller (for example she alone knows how much she liked the colleagues at that alternative job), matching is no longer possible and the seller protects herself by making lower demands on average, to the advantage of the *buyer*. These findings are far from obvious and not simply a matter of common sense (though after some reflection they are not at odds with one's intuition).

Private Information and Bargaining Inefficiency

Considerable research has been devoted to the strategic use of private information. Indeed, this literature looks at two issues raised earlier. First, one can use the notion of a bargainer holding better private information to represent superior expertise. Second, private information illustrates the tradeoffs between creating value and claiming a bigger share of the total value, to use the language of Lax and Sebenius (1986). Simple games of pie sharing are necessarily "win-win", by which I mean a mutually beneficial agreement is available and, this is commonly known to the players. (The alternative interpretation of "win-win," that exchanging smiles and good cheer is enough to overcome any conflict, is one I find to be far too optimistic.) But private information raises the possibility that there is no pie at all. There might be a surplus (or a zone of agreement), but there is also the chance that the negotiators would be better off walking away.

To consider the effect of private information, consider negotiations between a single buyer and single seller, each with privately known valuations for the good or service to be transacted. This negotiation is no longer a "pure bargaining problem." In addition to the aspect of how to divide the pie, players must try to determine whether there is a positive surplus. Myerson and Satterthwaite (1983) showed that this dual task leads inevitably to ex post inefficiency, that is players are not able to realize all the available surplus, all of the time. Chatterjee and Samuelson (1983) explored a simultaneous-offer bargaining game in which the tradeoff was displayed explicitly in equilibrium. In a setting of incomplete information, the buyer's optimal strategy is to submit an offer that is less than his value, while the seller makes a demand that is greater than her cost. Thus, the game may end with incompatible offers and a disagreement, even when a zone of agreement exists. In recent years, there has been a great deal of research on "double auctions" as market institutions.

Under private information, the inability of bargainers to attain all mutually beneficial agreements gives rise to phenomena such as "cheap talk." Farrell and Gibbons (1989) show that negotiators can gain (at least in particular cases) by expressing keenness about coming to an agreement. This flies in the face of much of the conventional bargaining wisdom. In the usual view, eagerness to complete a deal is tantamount to a unilateral concession and therefore frowned upon as a tactic. By contrast, bargaining theory holds that making a unilateral concession can be a sound tactic if it signals that mutually beneficial agreements are possible.

Bargainers' incentives to explore each other's preferences lead to the jockeying for position that Raiffa (1982) calls the negotiation "dance." A further strand of analysis has sought to model this negotiation dance, when there is private information on one or both sides. This work has usually been technically difficult. Moreover, it has been criticized as a series of thought experiments having more to do with the technical issues of refining equilibrium than with real-world bargaining. However, three main types of results have been obtained. Consider a seller who has incomplete information about the value of the item to the buyer. The seller makes an initial price offer that the buyer can accept or reject. If the offer is rejected, time passes before the seller makes a subsequent offer. If this offer is accepted, player profits are discounted due to the time delay. Fudenberg, Levine and Tirole (1985) and others have characterized the seller's equilibrium sequence of offers, showing that these prices decline over time as the game continues. The seller initially names high prices to take advantage of the chance that the buyer has a high value. An initial rejection signals a lower buyer value, implying a lower second-stage offer, and so on. However, if the seller can choose how fast to make offers, she would want to accelerate the process of offer and acceptance, and this acts against her ability to charge high prices. The persistence of delay and impasse must therefore be due to constraints on how fast the uninformed party can make concessions. The uninformed player uses a sequence of offers to find whether the informed player is of high value (and therefore suffers a high cost of delay). Delay could also be used by the informed player to signal toughness as in Cramton (1992) and Admati and Perry (1987).

A related but somewhat different approach to the negotiation dance is to view it as a concession game. Each party limits its price concessions (even risking disagreement) to convince the other of its "toughness," before one party is the first to give in, thereby revealing its "true" nature. Chatterjee and Samuelson (1987) examine such a model, with each bargainer being one of two "types," *hard* or *soft*, and playing one of two actions at each stage, tough (not concede) or weak (concede). Abreu and Gul (2000) generalize this approach to allow for a finite number of types, and prove uniqueness of a concession kind of equilibrium. Chatterjee and Samuelson show that the gains from trade go to the player whose initial "index of strength" is high.[5] This initial index depends on the player's patience (as in Rubinstein) but also on the starting probability that one side is a tougher bargainer than the other. The player who is more impatient or has a lower probability of toughness concedes first with certainty. Thus establishing a public reputation for toughness (it must bee public, otherwise it is of no use) can be beneficial in negotiations. (In some settings, however, such a reputation could drive away potential negotiating partners; this aspect has not been covered in the reputation models discussed, but is possible to incorporate in principle.)

Another stream of research in incomplete information has considered quality uncertainty. In many settings, the seller and buyer valuations have common elements. For instance in a corporate acquisition, a target firm possessing a superior technology (or a strong brand) will create additional value for the potential acquirer. Samuelson (1984), Evans (1989), and Vincent (1989) show that bargaining impasses frequently arise in such situations, even if the buyer obtains some synergistic value from the purchase (i.e. the buyer's value for the firm is always greater than the seller's). The impediment to trade stems from adverse selection or the "buyer's" curse – that is, a buyer's price offer is most likely to be accepted by low quality sellers. It turns out that the buyer's value must be substantially greater than the seller's if a mutually beneficial negotiation is to succeed.

What do we learn from all this? Qualitatively, the aggressive use of private information to seek a bigger share can lead to disagreement and inefficiency. Thus, there is no separation here between achieving the joint surplus and sharing it. In short, a negotiator must balance these twin elements. In turn,, a unilateral concession or an expression of keenness might help, by convincing the other party that there is a surplus to share, or might hurt by ceding too much to the other. Second, a "tough" bargaining stance (or the perception of being tough) can be a source of power and bargaining advantage (unless it deters bargaining altogether). Third, the quality of a bargainer's information is important. If your opponent knows everything you do and then some, your potential gains from the negotiation will be severely limited. While some of these game-theoretic results conform to conventional bargaining wisdom, others (the persistence of inefficiency, the benefits from "cheap talk" and the consequences of informational asymmetries) are not obvious. In general, thinking about the other side's preferences and information is crucial.

Incomplete Contracts

The framework of incomplete contracts and property rights developed in the last decade by Grossman, Hart and Moore has proved to be of surprising explanatory power in a whole variety of contexts. See Grossman and Hart (1986), Hart (1995), and Hart and Moore (1990). Incomplete contracts arise in situations where *ex ante* contracts must perforce leave actions and payoffs unspecified in some contingencies. (These provisions will be negotiated later if the contingencies actually occur.) The reasons for contract incompleteness might be because such contingencies are hard to describe in advance or because such contracts cannot be verified and enforced, or for other reasons. Relying on incomplete contracts leads to a "hold up" problem. A market participant who has to make an *ex ante* investment (for example, a seller who researches a buyer's special needs in order to provide a customized product) does not obtain the full marginal benefit of the investment and, therefore, underinvests relative to the optimum level.

Grossman, Hart and Moore show that the firm can come closer to the optimum investment level by the appropriate assignment of property rights to productive assets and residual control rights. The key to this argument is that ownership of an asset gives the holder the right to exclude the other party from its use, and thereby reduces the non-owner's bargaining power in the negotiation following the occurrence of an unspecified contingency. The idea of incomplete contracts is fundamental to the theory of the firm. Indeed, it is what distinguishes the firm from a set of bilateral or multilateral arms-length contracts. The result is crucially dependent on the bargaining power of participants in "thin" markets. A recent study by Chatterjee and Chiu (1999) considers a model in which these competitive negotiations determine the outside options and market power and goes on to derive results that are somewhat different from those predicted by Hart and Moore. Nonetheless, the framework developed by Hart and Moore has proved to have a surprisingly broad range of applications. Continuing the explicit analysis of firms' market settings and firms' outside options would increase the value of this contracting approach.

Much of the debate about the Grossman-Hart-Moore framework is now foundational in nature. What explains contract incompleteness? What can and cannot be done in dealing with "indescribable states of nature" or with contingencies that are observable but not contractible? Though these questions might be one step removed from the thrust of applied modeling, it is interesting that one approach to this issue, of de Meza and Lockwood (1998), also relies on models of bilateral search and bargaining in thin markets.

Multi-person Bargaining

In many settings, a bargainer faces a number of partners with whom he or she can negotiate, simultaneously or in sequence. One important example is the negotiations between the United Auto Workers union and the major automobile manufacturers in the United States – a case analyzed by Marshall and Merlo (1999) and Banerji (1999). The UAW faces a series of negotiations with opponents of unequal sizes and economic strengths and has the choice to negotiate simultaneously or sequentially (and, if sequentially, in what order). Banerji's work concentrates on two different models: one in which only the wage is negotiated and employment is chosen by the firms following the negotiations, and the other in which the union negotiates the total wage bill and level of employment with each firm. The first model is driven by the fact that the auto companies compete with each other, and that a high wage agreed on by one of them increases the profits for the others (other things equal). Therefore, later bargainers have more to give away, and this results in upward pressure on negotiated wages in the later bargains. The second model is similar to a strategic bargaining version of the Marshall-Merlo model, which uses Nash bargaining. In Marshall and Merlo's model of "pattern bargaining," the negotiated outcome in the first encounter affects the status quo point for future negotiations with other firms.

In a different context, Winter (1997) considers a pair of parties negotiating with respect to two issues, one of which is much more important than the other. He concludes that it is better for the parties concerned to tackle the more important issue first, so as to reduce the incentive for early posturing to get future advantage. In ongoing work, Chatterjee and Kim (1999) focus on a context in which a single buyer has to buy an object from each of two separate sellers. The buyer's value for one seller's object is significantly higher than its value for the other seller's object. The values are linearly related (deterministically) but are known only to the buyer. (The sellers have only probabilistic information about these values.) In a simple two-stage bargaining model, the best strategy of a "soft" buyer is to negotiate first with the low-value seller. (This is because the incentives for the buyer to posture in the first bargain lead to the first seller being more willing to give in.) Chatterjee and Kim also look at the conditions under which simultaneous bargaining might be optimal.

Another issue that becomes important in multi-person bargaining is the role of communication links among the players. For example, when the US began its overtures to Mao's China, it was handicapped by the absence of formal relations. Pakistan, which had communication links to both parties, was able to use its communication power to its own advantage, while facilitating the contacts between the US and China.

Since the early work of Myerson (1977), it has been recognized that the structure of communication among the parties to a negotiation affects the allocation of the gains from bargaining. Bolton and Chatterjee (1996) and

Bolton, Chatterjee and Valley (1999) have conducted and analyzed experiments under varying communication structures, including face-to-face and email interfaces. Their results show that different extensive form bargaining models are needed to explain the data from different communication experiments. For example, if all messages are public, there is an element of competition that is best modelled by postulating simultaneous opportunities to bid against each other. By contrast, settings with less complete communication structures produce results more consistent with sequential offers. Moreover, for most communication structures, the model of Chatterjee et al (1993) generates better explanations of the data than does Myerson's axiomatic framework. Thus extensive forms might be good representations of *actual* bargaining, even though they might not match up feature to feature.

Finally, one way of resolving a longstanding problem in multi-person bargaining is to incorporate well-known ideas of bounded rationality. In the pure bargaining problem of dividing a pie among N bargainers, with proposals and responses being made sequentially, almost any division can be explained as an equilibrium outcome, even those that seem implausible. One way to sharpen predictions in this setting is to model the bounded rationality of the players. Recent work in game theory has sought to model the limitations on an individual decision-maker's ability to implement arbitrarily complex strategies by invoking the concepts of finite automata from computer science. These automata are not supposed to represent the players themselves, but rather the strategies they choose. Natural measures of the size of automata translate into the cost of complexity of strategies. It has been observed recently – by Binmore, Piccione, and Samuelson (1998) for the case of two players and by Chatterjee and Sabourian (2000) for the general case of N players – that modelling complexity in this way helps to eliminate the implausible equilibria and to sharpen the predictions of the model. While this too is at least one step removed from the concerns that motivate practitioners, it provides a rational basis for concentrating on simple strategies, which are easier to implement in practical settings.

2. Criticisms of Game-Theoretic Models

Despite advances in modeling non-cooperative bargaining, criticisms of the game-theoretic approach persist. The critics come from two different camps. One group (many game theorists included) are disturbed by the lack of general results. Negotiation behavior and bargaining outcomes depend intimately on the posited rules and characteristics of the bargaining game. Seemingly small differences in the bargaining setting can mean important differences in equilibrium outcomes. A second group levies the opposite criticism, arguing that game theory models of bargaining are too simple and too abstract – that is, they leave out too much of the richness of actual negotiation. Rather than attribute these complaints to specific individuals, we will consider these kinds of criticisms (whether from practitioners, behavioral researchers, economists, or game-theorists themselves) as a whole. Let's consider the most frequent criticisms.

1. *Common knowledge and rationality assumptions.* Non-cooperative game theory is based on the concept of equilibrium, which assumes that players are optimizing against one another (and moreover that everyone knows this). Moreover, computing an equilibrium requires common knowledge about various components of the game, such as the types of opposing players, where types summarize not only other players' private information, but also their beliefs (and their beliefs about others' beliefs and so on). The critics assert that this is an incredible amount of knowledge to expect from individuals in any real-life situation. This is indisputable and a major thrust of research has been to explain how players learn to play equilibrium without the strong common knowledge assumptions. See, for example, Milgrom and Roberts (1991) and Krishna (1992). The question to ask, however, is whether the assumptions are so untenable that one should abandon the models altogether. Thus, Raiffa (1982) and Sebenius (1992) endorse a sophisticated "decision-analytic" approach, where the focus is on a single decision-maker's optimization problem given his beliefs about his opponents. (The danger, of course, is that the decision-makers expectations may be misleading if they fail to take into account the behavior of the other parties involved.)

In my opinion, this criticism, though theoretically important, may be of limited practical consequence in many situations. Consider a seller and a buyer conferring on the price to be charged for a piece of equipment. It is pretty clear that the engineers on the buyer's side have a good idea, though maybe not an exact value, of what the cost of production would be. Moreover, this is usually based on publicly available information. Similarly, the seller should have a notion of what the buyer will do with the equipment. Common knowledge of cost distributions may not actually hold, but the deviation from it may not be of much significance. Of course, this argument gets harder to make when we are dealing with personal preferences.

Individuals often have very idiosyncratic preferences. However, surely we have a good idea as to the possible range of preferences, especially if we have put in some hours of research into understanding one's opponent.

Are people rational? The more pertinent question might be: Do negotiators think about what agreements will be good for them and how they might seek to obtain them? This may not describe all human behavior, but there clearly is a strong goal-directed component in the behavior of serious negotiators. As a program of research, it is also desirable to investigate the role of systematic deviations from rational behavior in bargaining situations. Thus, the rich array of bargaining experiments to date is essential for identifying where and when actual behavior tracks the theoretical equilibrium benchmark on the one hand, and when it deviates on the other.

2. *Indeterminacy of predicted outcomes.* A second criticism of bargaining theory points out the multiplicity of equilibria that arise in many negotiation settings. Bargaining models under incomplete information, for example, have different equilibria sustained by different assumptions on what an individual in the game would believe if an opponent took an action that he was not supposed to take in equilibrium. The so-called "refinements" literature has tried to find a priori arguments for ruling out as implausible certain kinds of inferences from deviations.

It is true that that game theorists will not be able to make determinate predictions about the outcome of bargaining processes, at least in many cases. This need not, in my opinion, condemn the activity of modeling these processes. There may even be an advantage to being able to explain the range of observed behavior.

3. *Non-robustness and explaining "too much."* Another frequently heard criticism about the extensive-form bargaining models is that the results depend crucially on the procedure of offers and counter-offers and at what stage discounting takes place. In any real-world negotiation, these features are not "etched in stone" but rather evolve endogenously. Consequently, bargaining results dependent on these details may be too specific to be of any use. However, knowing the properties of such procedures, and the contingent nature of the solutions might help understand what kind of procedure players should seek to use. This knowledge could confer a strategic advantage. There

is a literature on what procedures to choose in incomplete information games – for instance, the mechanism design approach pioneered by Myerson (1981) and used in the bargaining paper of Myerson and Satterthwaite (1983). This approach adopts a strictly normative point of view. Players are presumed to choose an efficient mechanism according to some definition of efficiency. But, it is not clear why this should be the case. The actual practice of bargaining also suggests some answers. Casual observation suggests that some procedures are more commonly used than others. Rejecting a proposal usually means either breaking off negotiations or making a counter-offer. Not to make a counter-offer may be construed as not bargaining in good faith. Symmetry is another powerful motivation in the choice of procedure. It does appear that there are only a few generic procedures, such as simultaneous offers or alternating offers, which actually get used in practice. We don't fully understand (theoretically) why this is the case.

A related criticism is the difficulty of empirically testing the game-theoretic models, since there appear to be few results that some extensive form cannot explain. If the content of the theory is measured, as Popper (1959) has proposed, by what it rules out, this would appear to give the theory low empirical content. However, experimental work in game theory is moving ahead, nor should Popperian empirical content be the sole arbiter among theories.

4. *Wrong intuition.* The critics sometimes target specific models as lacking the right intuition, at least in the settings in which they are applied. Thus the search for alternatives and competition is more important in most markets than discounting, yet most applications papers focus on discounting as the determinant of the allocation of surplus. To the extent that this is a call for more and better models to capture important factors in real-world bargaining, I certainly agree.

While all the criticisms listed above have a certain degree of validity, they are not destructive of the general enterprise of game-theoretic modeling of bargaining. In the next section, I shall offer my interpretations of the various models. (This is a personal view and by no means universally shared, as conversations with colleagues have made clear.) The focus will then move to a comparison of the models and the insights of practitioners.

3. Interpreting the Results of Game-Theoretic Models

What is it that the recent models of bargaining have been trying to do? Usually, there are two responses as to what they *should* have been trying to do. Either the models should give advice to individual negotiators, in the same way that the long-time models of operations research prescribed quantitative answers to questions on the basis of mathematical models. Or, they should be "scientific," that is, they should generate quantitative predictions of how individuals will behave in bargaining situations, or at least predictions that are amenable to falsification in the manner espoused by Popper. There is no other way, some would claim; either the model should provide answers or it should provide predictions. We have seen in the last section that there are deficiencies on both scores. Decision analysts criticize bargaining theory for not being prescriptive, while others criticize it for not being physics. The second criticism is also leveled at economics in general, for instance by Rosenberg (1992). I agree that we should seek precise answers and predictions, without necessarily expecting to find them. At the very least, bargaining theory can deliver insights and explanations.

Let us consider the criticism that game theory cannot provide advice to individual players. As pointed out by Luce and Raiffa (1957) years ago, game-theoretic equilibrium concepts are *conditionally normative*. That is, my equilibrium strategy is optimal if my opponent plays his or her equilibrium strategy (both being part of the same equilibrium). So the advice to a player would not necessarily be to play an equilibrium strategy; it depends on whether there is evidence to believe that one's opponent will or will not play his corresponding strategy. This is perfectly consistent with decision analysis. When we have compelling subjective beliefs, we should certainly play best responses given these beliefs. What the equilibrium approach adds is the insight that the subjective beliefs can be wrong. If you claim that the probability of rain tomorrow is .6, it is difficult to say whether you are wrong or right in holding this subjective opinion. But, if you believe your opponent will play tough with probability .6 (and the opponent knows this), the true probability might well turn out to be 1.0 (not .6), and one might conclude that the subjective belief of .6 is actually wrong, given the assumption of rationality about your opponent.

Thus, equilibrium analysis provides the following general advice. Think hard about what your opponent would do if he or she were able to anticipate what you will do or what you think he or she will do and so on. In order to close

the analysis, we may need to follow through to a fixed point of the recursion. Of course, a practical negotiator need not divine the equilibrium all of the time. But, he is likely to go through a number of iterations of interactive thinking. My contention is that thinking about the interactive nature of a negotiation will surely enhance the decision analysis approach. Indeed, the practical negotiator can do better still by examining the evidence on bargaining behavior gleaned from controlled laboratory experiments.

To a great extent, optimization (prescriptive) models are also valued for the kinds of thought processes they engender rather than for their specific results. MBA students (used to) learn linear programming to think in terms of constraints and resources, and decision analysis to be able to separate judgement and preference and to understand the time structure of a decision. Most students find the logic more valuable as an aid to judgment than the "answer" provided by the model.

As for the criticism that game theory and economics are not like physics, there is some ground to believe that physics may not quite be like physics either. For example, Putnam (1974) discusses Newton's law of gravitation and asserts correctly that the law by itself does not imply a single "basic sentence" without auxiliary assumptions, such as: 1. No bodies exist except the sun and the earth. 2. The sun and the earth exist in a hard vacuum. 3. The sun and the earth are subject to no forces except mutually induced gravitational forces. As Putnam claims, from the conjunction of the theory and these auxiliary statements we can deduce predictions such as Kepler's laws. (These assumptions, by the way, sound quite as bad as the assumption as to the common knowledge of rationality.) Putnam's point is that a large part of scientific work seeks understanding and explanations, not solely predictions. In this sense, the game-theoretic models are useful.

To sum up, my view is that game-theoretic models offer us nuanced and contingent explanations, of the type "Do this if that is the situation." These models rightly caution against following negotiation advice that purports to be general and universal ("Never make the first offer"). A few simplistic formulas cannot fit the many differing shades of practical negotiations.

4. Conclusion

I have sought to develop a basic theme. Recently developed game-theoretic models of bargaining suffer from a number of theoretical deficiencies and strong assumptions. The aim of researchers should be to remedy these deficiencies with better models. However, the theory is still valuable, not because it provides sharp quantitative predictions of outcomes in real negotiations, but because it gives us qualitative insights into behavior. By its very nature, the theory is contingent and points out that even seemingly minor differences in a given negotiating situation can dramatically change the nature of the game being played. And the theory also suggests the direction of such a change and what one might do about it. There is nothing that can replace careful study and preparation for a negotiation. Knowledge of the theory provides categories for classifying the material one has about a situation and for analyzing the contents.

* This chapter is a revised and updated version of my earlier paper, "Game Theory and the Practice of Bargaining," which originally appeared in *Group Decision and Negotiation*, 1995. The revision has greatly benefited from detailed comments by William Samuelson. I apologize to the reader for the numerous references to my own work.. Because I have not attempted a general survey of the literature; it is probably inevitable that this discussion is biased toward current topics in which I've been particularly involved and interested.

References

Abreu, D. and F.Gul (2000), "Bargaining and Reputation," *Econometrica*, forthcoming.

Aumann, R. and A. Brandenburger (1995), "Epistemic Conditions for Nash Equilibrium," *Econometrica*, 63, 1161-1180.

Banerji, A. (1999), "Sequencing Strategically: Wage Negotiations under Oligopoly," mimeo, Delhi School of Economics.

Binmore, K., M. Piccione and L. Samuelson (1998), "Evolutionary Stability of Alternating Offers Bargaining Games," *Journal of Economic Theory*, 80, 257-291.

Binmore, K. (1985), "Bargaining and Coalitions," in A.E. Roth (Ed.), *Game-Theoretic Models of Bargaining*, Cambridge University Press, Cambridge, UK.

Binmore, K., A.Rubinstein and A. Wolinsky (1986), "The Nash Bargaining Solution in Economic Modelling," *Rand Journal of Economics*, 17, 176-185.

Binmore, K., A. Shaked, and J. Sutton (1988), "An Outside Option Experiment," *Quarterly Journal of Economics*, 104, 753-770.

Bolton, G.E. and K. Chatterjee (1996), "Coalition Formation, Communication and Coordination: An Exploratory Experiment," in R. Zeckhauser, R. Keeney and J. Sebenius (Eds.), *Wise Choices: Games, Decisions and Negotiations*, Harvard Business School Press, Boston, MA.

Bolton, G., K. Chatterjee and K. Valley (1999), "How Communication Links Affect Coalitional Bargaining." Mimeo, Pennsylvania State University and Harvard University.

Chatterjee, K. and N. Kim (1999), "Strategic choice of bargaining order with one-sided incomplete information," draft manuscript.

Chatterjee, K. and Y.W. Chiu (1999), "When Does Competition Lead to Efficient Investments?" draft manuscript.

Chatterjee, K. and B. Dutta (1998), "Rubinstein Auctions: On Competition for Bargaining Partners," *Games and Economic Behavior*, 23, 119-145.

Chatterjee, K., B. Dutta, D. Ray, and K. Sengupta (1993), "A Non-Cooperative Theory of Coalitional Bargaining," *Review of Economic Studies*, 60, 463-477.

Chatterjee, K. and C.C. Lee (1998), "Bargaining with Incomplete Information about Outside Options," *Games and Economic Behavior*, 22, 203-237.

Chatterjee, K. and H. Sabourian (2000), "Multiperson Bargaining and Strategic Complexity," *Econometrica*, forthcoming.

Chatterjee, K. and L. Samuelson (1987), "Bargaining with Two-Sided Incomplete Information: An Infinite Horizon Model with Alternating Offers," *Review of Economic Studies*, 54, 175-192.

Chatterjee, K. and L. Samuelson (1990), "Perfect Equilibria in Simultaneous Offers Bargaining," *International Journal of Game Theory*, 19, 237-267.

Chatterjee, K. and W. Samuelson (1983), "Bargaining under Incomplete Information," *Operations Research*, 31, 835-851.

Chikte, S.D. and S. D. Deshmukh (1987), "The Role of External Search in Bilateral Bargaining," *Operations Research*, 35, 198-205.

Cramton, P.C. (1992), "Strategic Delay in Bargaining under Two-Sided Uncertainty," *Review of Economic Studies*, 59, 205-225.

Cohen, H. (1989), *You Can Negotiate Anything*, Bantam Books, New York.

van Damme, E., R. Selten, and E. Winter (1990), "Alternating Bid Bargaining with a Smallest Money Unit," *Games and Economic Behavior*, 2, 188-201.

De Meza, D. and B. Lockwood (1998), "Investment, Asset Ownership and Matching the Property Rights Theory of the Theory in Market Equilibrium," Mimeo, Universities of Exeter and Warwick.

Evans, R.A. (1989), "Sequential Bargaining with Correlated Values," *Review of Economic Studies*, 56, 499-510.

Farrell, J. and R. Gibbons (1989), "Cheap Talk Can Matter in Bargaining," *Journal of Economic Theory*, 48, 221-237.

Fudenberg, D., D. Levine and J. Tirole (1985), "Infinite Horizon Models of Bargaining with One-Sided Incomplete Information," in A. E. Roth (Ed.), *Game Theoretic Models of Bargaining*, Cambridge University Press, Cambridge UK.

Grossman, S.J. and O. Hart (1986), "The Costs and Benefits of Ownership: A Theory of Vertical and Lateral Integration," *Journal of Political Economy*, 94, 691-719.

Harsanyi, J.C. (1956), "Approaches to the Bargaining Problem Before and After the Theory of Games: A Critical Discussion of Zeuthen's, Hicks's and Nash's Theories," *Econometrica*, 24, 144-157.

Hart, O. (1995), *Firms, Contracts, and Market Structure*, Oxford University Press, New York.

Hart, O. and J. Moore (1990), "Property Rights and the Nature of the Firm," *Journal of Political Economy*, 98, 1119-1158.

Kreps, D. (1991), *Game Theory and Economic Modelling*, Clarendon Press, Oxford, UK.

Krishna, V. (1991), "Learning in Games with Strategic Complementarities," Mimeo Harvard University. (Now available from Penn State's Department of Economics.)

Lax, D. and J. Sebenius (1986), *The Manager as Negotiator: Bargaining for Cooperation and Competitive Gain*, The Free Press, New York.

Lee, C.C. (1994), "Bargaining and Search with Recall," *Operations Research*, 42, 1100-1109.

Luce, R.D. and H. Raiffa (1957), *Games and Decisions*, John Wiley and Sons, New York.

Marshall, R.C. and A. Merlo (1999), *"Pattern Bargaining,"* mimeo Pennsylvania State University and the University of Minnesota.

Milgrom, P. and J. Roberts (1990): "Rationalizability, Learning and Equilibrium in Games with Strategic Complementarities," *Econometrica*, 58, 1255-1278.

Muthoo, A. (1990): "Bargaining without Commitment," *Games and Economic Behavior*, 2, 291-297.

Myerson, R. (1977), "Graphs and Cooperation in Games," *Mathematics of Operations Research*, 2, 225-229.

Myerson, R., (1981), "Optimal Auction Design," *Mathematics of Operations Research*, 6, 58-73.

Myerson, R. and M. Satterthwaite (1983), "Efficient Mechanisms for Bilateral Trading," *Journal of Economic Theory*, 29, 265-281.

Nash, J.F. (1950), "The Bargaining Problem," *Econometrica*, 18, 155-162.

Nash, J.F. (1953), "Two-Person Cooperative Games," *Econometrica*, 21, 128-140.

Osborne, M. and A. Rubinstein (1990), *Bargaining and Markets*, Academic Press, San Diego, California.

Popper, K. (1959), *The Logic of Scientific Discovery*, Hutchinson.

Putnam, H. (1974): "The 'Corroboration' of Theories," in P.A. Schilpp (Ed.), *The Philosophy of Karl Popper*, Open Court Publishing Co.

Raiffa, H. (1953), "Arbitration Schemes for Generalized Two-Person Games," in H. Kuhn and A.W. Tucker (Eds.), *Contributions to the Theory of Games*, Princeton University Press, Princeton, NJ.

Raiffa, H. (1982), *The Art and Science of Negotiation*, Harvard University Press, Cambridge, MA.

Rosenberg, A. (1992), *Economics – Mathematical Politics or Science of Diminishing Returns*, University of Chicago Press, Chicago.

Rosenschein, J. and G. Zlotkin (1994), *Rules of Encounter*, MIT Press, Cambridge, MA.

Roth, A.E. (1979), *Axiomatic Models of Bargaining*, Springer-Verlag.

Samuelson, W. (1984), "Bargaining under Asymmetric Information," *Econometrica*, 52, 995-1007.

Schelling, T. (1960), *The Strategy of Conflict*, Harvard University Press, Cambridge, MA.

Sebenius, J. (1992), "Negotiation Analysis," *Management Science*, 38, 18-38.

Selten, R. (1975), "Reexamination of the Perfectness Concept for Equilibrium Points in Extensive Games," *International Journal of Game Theory*, 4, 25-55.

Selten, R. (1981), "A Noncooperative Model of Characteristic Function Bargaining" in, *Models of Strategic Rationality (Selected papers of Reinhard Selten)*, Kluwer Academic Publishers, Norwell, Mass, 1990.

Shaked, A. (1986), "A Three-Person Unanimity Game," Speech given at the Los Angeles national meetings of the Institute of Management Sciences and the Operations Research Society of America.

Siegel, S. and L. Fouraker (1960), *Bargaining and Group Decision Making*, McGraw-Hill, New York.

Stahl, I. (1972), *Bargaining Theory*, Stockholm School of Economics, Economic Research Institute, Stockholm.

Vincent, D. (1989), "Bargaining with Common Values," *Journal of Economic Theory*, 48, 47-62.

Winter, E. (1997), "Negotiation in Multi-issue Committees," *Journal of Public Economics*, 65, 323-342.

Notes

[1] The influential early work in bargaining behavior of Siegel and Fouraker (1960) adopted an experimental approach.
[2] The commentary and the interpretations are mine and are not intended to represent Cohen's views, though I'd be happy if they did.
[3] With simultaneous offers this commitment does not exist; incompatible offers are followed by another set of simultaneous offers. See Chatterjee and Samuelson (1990) and Muthoo (1990); the latter paper allows a player to withdraw an offer.
[4] See van Damme, Selten and Winter (1990) for results concerning the smallest money unit of account.
[5] Note that this index is again a shorthand, not an essentialist definition.

10 AUCTIONS IN THEORY AND PRACTICE

William Samuelson

Auctions and competitive bidding institutions are important both for empirical and theoretical reasons. Auctions are among the oldest forms of economic exchange. Today, auctions are used in an increasing range of transactions – from online sales via the internet to the sale of the radio spectrum. In the familiar English auction, bid prices rise until the last and highest bid wins the item. The English auction enjoys a secure place in a wide variety of settings: the sale of art and antiques, rare gems, tobacco and fish, real estate and automobiles, and liquidation sales of all kinds. An alternative institution is the sealed-bid auction where the highest bidder wins the item at the named sealed bid. The sales of public and private companies has been accomplished by sealed bids as have the sales of real estate, best-seller paperback rights, theater bookings of films, U.S. Treasury securities, and offshore oil leases. A third method is the "Dutch" auction, used in the sale of a variety of goods but especially in the sale of flowers in Holland. The auctioneer's initial price is set sufficiently high, and then the price is lowered at intervals. The first buyer who signals a bid obtains the item at the current price. Competitive bidding is also a common means for conducting multiple-source procurements. Here, a single buyer solicits bids from a number of competing suppliers with the objective of obtaining the most attractive terms measured not only in terms of price, but also by product quality, management capability, service performance, and the like. The most common institution for complex procurements is the submission of multiple rounds of sealed bids before the buyer makes a final selection.

Auctions are also of considerable theoretical importance. Auction theory provides rich and flexible models of price formation in the absence of competitive markets. These models lend valuable insight into optimal bidding strategies under different auction institutions. Besides helping to explain current selling institutions, auction theory also provides normative guidelines concerning "market" performance. What types of auction institutions are likely to promote efficiency? Alternatively, what auction methods maximize the seller's expected revenue? Together, auction theory and the corresponding empirical evidence can provide direct answers to these questions.

The aim of this chapter is to examine the use of auctions, paying equal attention to theory and practice. While theory suggests equilibrium bidding as a benchmark, there is considerable empirical evidence (from controlled experiments and field data) that actual bidding behavior only loosely follows this normative prescription. Thus, it is important to consider the design of auction institutions anticipating *actual* bidding behavior. In addition, we will argue for the design of "transparent" auction institutions that enable buyers to formulate bidding strategies reflecting their underlying values.

Any auction institution performs two functions simultaneously. It determines the allocation of the item up for sale and it establishes the sale price. Accordingly, it is natural to consider two aspects of auction performance: value maximization (efficiency) and revenue maximization. Roughly speaking, an auction is efficient if it allocates the item to the highest-value user, maximizing total value in the process (where total value encompasses the sum of seller revenue and buyer profit). Alternatively, if one takes purely the seller's point of view, revenue maximization may be the primary goal. In many instances, the objectives of revenue maximization and value maximization are in harmony. As a general rule:

The most important benefit of an auction is in marshalling competition among the greatest number of potential buyers.

Thus, increasing the number of bidders will increase total value on average and raise the seller's expected revenue. A simple example makes the point.

Example 1. Bidding versus Bargaining. A single piece of artwork is for sale. The seller, whose personal private monetary value (or reservation price) for the work is $400,000, has been approached by a single potential buyer. The sole issue is the sale price, and prior to the negotiations, the seller has given careful thought to the potential price the buyer might be willing to pay. Based on the best available information, the seller believes that the buyer's personal value, denoted by v_i, is in the range of $400,000 to $640,000. To be concrete, suppose the buyer's value is *uniformly* distributed over this range. What price can the seller expect to obtain on average? Clearly, there is ample room for a mutually beneficial agreement. For instance, if the buyer's actual value were $520,000, a negotiated price of $460,000 (halfway between the parties' values) would generate a profit of $60,000 for each side. If the bargainers are equally matched, one would expect the final price to be near this split-the-difference prediction. Moreover, since $520,000 is the buyer's expected value, one-on-one bargaining between equally matched parties could be expected to result in a price of $460,000 on average.

Now suppose the seller puts the artwork up for competitive bid using the usual English auction. The sale has been publicized, and 5 to 10 potential buyers are expected to bid. The seller's best assessment is that *each* buyer's personal value is uniformly distributed between $400,000 and $640,000 and that all values are *independently* distributed. In short, there is considerable dispersion in the different buyers' personal monetary values. If the item is auctioned, what will be the result? First, the high-value buyer can be expected to obtain the item; the auction is efficient. In the English auction (when buyers hold private personal values), each buyer's dominant strategy is to bid up to its reservation value if necessary. Thus, the bidding stops when the current bid price barely edges above the second-highest value among the buyers. The high-value buyer obtains the item at this price by just outbidding the last active bidder. Second, the auction delivers significant revenue to the seller. When the bidding stops, how high a price will the seller claim on average? For 5 bidders, the expected price is $E(P_E)$ = $560,000. For 7 bidders, the expected price is $580,000, and for 9 bidders, the expected price is $592,000. (In each case, we have computed the expected value of the second-highest value among N independently and uniformly distributed values in the interval $400,000 to $640,000.) The key advantage of the auction is that it delivers the "best" price, a considerably higher price than is forthcoming from one-on-one bargaining. Increasing the number of bidders not only increases this price on average, but it also increases the expected total value generated by the auction. (On average, the highest value for the item from among N buyers is increasing in N.)

The analysis in this chapter compares the most common auction institutions across various settings in terms of both efficiency and revenue. Our view is somewhat eclectic. Whether efficiency or revenue is the more important goal depends on the setting. Clearly, obtaining the best price is of primary concern to a private firm conducting an auction sale or (conversely) organizing a competitive procurement. On the other hand, allocative efficiency may be paramount when the government or a public agency is conducting the auction. For instance, the allocation of airport landing rights, the siting of hazardous waste facilities, the sale of the broad-band spectrum, the privatization of state-owned enterprises – all of these can be accomplished by an auction institution where the primary focus is efficiency. Indeed, in the recent spectrum auctions, government regulators chose to put multiple sections of bandwidths up for bid in all regions to promote *ex post* competition among multiple winning bidders. The intent was to foster competition and, thereby, overall market efficiency. Clearly, the government might have significantly increased its own revenues by auctioning exclusive spectrum rights. (Auctioning a monopoly right to the spectrum in a particular region would be more valuable to potential bidders and, thereby, would generate higher bid prices.)

In the present analysis, we evaluate auction performance according to both the revenue-maximizing and value-maximizing criteria. In some instances, we take up a third criteria: auction implementation costs. Between two auction methods that are expected to generate comparable revenues and/or values, one should prefer the one that is simpler or less costly to implement.

In comparing the two most common auction methods, the English and sealed-bid auctions, our analysis delivers a mixed message. First, the English auction can be expected to outperform its sealed-bid counterpart on efficiency grounds. (This conclusion has been emphasized by numerous researchers and is hardly novel.) However, in a number of representative settings and examples, the efficiency advantage of the English auction is relatively small. Furthermore, neither auction method will get high efficiency marks in complex auction settings, for instance, when multiple items are to be allocated and buyer values are non-additive.

Second, contrary to the benchmark prediction of auction theory, the preponderance of practical evidence favors the sealed-bid auction as a revenue generator. Sealed bids can be expected to exceed English prices not only in the well-known case of bidder risk aversion, but also in the less-examined case in which bidders hold asymmetrically drawn values. Moreover, there is considerable experimental and field evidence that actual sealed-bidding behavior is significantly elevated relative to the equilibrium prescription. In short, one implication of our analysis is to redirect attention to the practical virtues of sealed-bid procedures.

Third, the relative performance of various auction institutions in complex environments remains an open question. For instance, there are well-known difficulties in sequentially auctioning multiple items when values are non-additive. Only recently have theoretical and empirical investigations focused on alternative auction institutions. The simultaneous ascending auction has performed well in experiments and in practice. (This method was used in the FCC's multi-billion-dollar spectrum auctions.) Serious attention is also being paid to the sealed-bid combinatorial auction, which allows buyers to bid for individual items and combinations of items. In combination with a Vickrey payment scheme, the combinatorial auction also promises favorable performance in theory and practice.

The goal of this chapter is to provide an introduction to the evolving theory and evidence concerning auctions. The approach is non-technical, so the results should be accessible to the general reader with a basic knowledge of game theory. The chapter reviews many of the seminal findings in auction theory and also considers current topics of interest in the field. The overview is also selective. There is no attempt to be encyclopedic and many interesting research paths are omitted. Rather, the intent is to provide an assessment of what auction theory and evidence seem to tell us. Earlier expositions and surveys of auction theory include: Milgrom (1986), McAfee and McMillan (1987), Wilson (1987), Bulow and Roberts (1989) and Milgrom (1989). More recently, Klemperer (1999) surveys auction theory and Klemper (1999b) provides a comprehensive collection of the classic papers on auctions. Kagel (1995) is the indispensable reference to the experimental research. While there is no broad survey of auctions in the field, Thaler (1988) and Blecherman and Camerer (1996) review research on the "winner's curse." The chapter is organized as follows. Section 1 considers single-object auctions, section 2 examines auctions for multiple objects with interdependent preferences, and section 3 concludes.

1. Single-Object Auctions

To date, there has been a considerable body of analysis examining equilibrium bidding behavior and auction performance when a single item is up for sale. Attention is focused on four auction institutions: the English auction, the sealed-bid auction, the Dutch auction, and the second-price auction. In this last auction, buyers submit sealed bids and the highest bidder obtains the item but pays a price equal to the second-highest bid. It is easy to confirm that the sealed-bid and Dutch auctions are strategically equivalent; buyers should bid exactly the same way in each. (In fact, the Dutch auction is simply a dynamic version of a sealed-bid auction. Instead of waiting for the price to fall, each buyer could simply submit in writing the price at which it would first bid. The highest sealed-bid wins the Dutch auction and the winner pays its written bid.) Thus, one would expect the two auctions to produce the same bidding behavior and the same allocation and revenue results.

Independent Private Values

We start by summarizing bidding and revenue results for a simple environment, the so-called independent private-values model (IPV). Here, as in the previous artwork example, each bidder holds a private value for the item, independent of any other bidder's value. Though no bidder knows any other's value, the probability distributions from which values are drawn are common knowledge. In the symmetric IPV model, bidder values are independent draws from a common distribution, with cumulative distribution function denoted by $F(v)$. The seminal auction investigations of William Vickrey (1961) laid the groundwork for the following general result.

Proposition One. Given symmetric independent private values and risk-neutral bidders, the four auction institutions produce identical allocations for the good and identical expected revenues for the seller.

In each of the auctions, the high-value buyer places the winning bid and obtains the item. Thus, it is hardly surprising that the four auctions are equivalent in terms of allocative efficiency. That the auctions generate equal expected revenues is perhaps more surprising. The intuition, however, is straightforward once one takes a careful look at bidder behavior. Recall that in the English auction, each buyer's dominant strategy is to be willing to bid up to its private value (if necessary). Therefore, the price rises to (or just above) the second highest value: $P_E = v_{2nd}$. Interestingly, the second-price auction also produces this same price. In the second-price procedure, each bidder's dominant strategy is to submit a bid equal to its private value. (Because the high bidder pays the second-highest bid, bidding below one's value can never favorably affect the price paid; it only risks losing the item altogether.[1]) Thus, the seller receives a price set by the second-highest bid, which is again identical to the second-highest value, $P_{2nd} = v_{2nd}$.

Turning to the sealed-bid and Dutch auctions, we already know that they are strategically equivalent. In each, equilibrium bidding behavior is characterized by a common bid function, $b_i = b(v_i)$ such that $b(v_i)$ maximizes the typical bidder's expected profit for each value v_i, against the probability distribution of competing bids by rival bidders $j \neq i$, implied by $F(v)$ and $b(v)$.

The qualitative features of $b(v)$ are straightforward. The bid function is increasing and satisfies $b(v) \leq v$ (with equality only at the lower support of the value distribution). In equilibrium, a typical bidder's expected profit is $E\pi(b_i, v_i) = [v_i - b(v_i)]F(v_i)^{N-1}$. (Given the common bidding function, bidder i is the high bidder among the N competitors if and only if all rival values are smaller than v_i.) More important, it is straightforward to show that the common biding function $b_i = b(v_i)$ is given by:

$$b_i = E[v_{2nd}|v_{2nd} \leq v_i].$$

In words, each buyer sets its bid by assuming it holds the highest value and bids at a level given by the expectation of the next highest value (v_{2nd}) among the population of bidders. Observe that this bidding behavior immediately implies that the sealed-bid and English auctions generate the same expected revenue. In the English auction, bidders are willing to ascend (if necessary) to their true values and the bidding stops at the second-highest value, $P_E = v_{2nd}$. In the sealed-bid auction, the seller receives the high bid b_{Max}, but this bid is shaded below the high value v_{Max} with the net effect that b_{Max} exactly equals $E[v_{2nd}]$. In equilibrium, this shading in the sealed-bid auction exactly compensates for the fact that the bidding stops at the second-highest value in the English auction. The optimal auction approach, pioneered by Myerson (1981) and Riley and Samuelson (1981), provides a very general character-ization of revenue. In the IPV setting, any two auctions that imply the same allocation of the item also generate equal expected revenues. With symmetric buyers, the four auctions imply identical allocations and, therefore, equal expected revenues.

<u>Example 2</u>. Revenue Equivalence. Suppose values are independently and uniformly distributed on the interval [0, 100]. Then, it is easy to confirm that the expectation of the highest value[2] among N independent draws from this distribution is: $E[v_{Max}] = [(N)/(N+1)]100$, while the expectation of the second-highest value is: $E[v_{2nd}] = [(N-1)/(N+1)]100$. The expected revenue of the English auction is given by this latter expression. The common

equilibrium bid strategy is $b_i = E[v_{2nd}|v_{2nd} \leq v_i]. = [(N-1)/N]v_i$ in the sealed-bid auction. Thus, $E[b_{Max}] = [(N-1)/N]E[v_{Max}] = [(N-1)/N][N/(N+1)]100 = [(N-1)/(N+1)]100$. This confirms that under equilibrium bidding behavior, the English and sealed-bid auctions generate the same expected revenue.

Empirical Evidence. The equivalence results of Proposition One constitute an important benchmark. It is natural to ask whether the auctions are equivalent in practice. The most direct empirical evidence comes from the extensive body of controlled auction experiments using student subjects. At the risk of oversimplifying, we can summarize some of the main results from the myriad of auction experiments in the IPV setting. First, actual behavior in the English and second-price auctions closely tracks the game-theoretic prediction; in each case, the vast majority of subjects bid according to their true values. Consequently, these auctions score high marks for efficiency. Surveying a wide range of experiments, Smith (1995) reports an average 95% efficiency rating for the English auction.

Second, bidding strategies in the sealed-bid experiments lie above the risk-neutral equilibrium prediction (Smith, 1995 and Kagel, 1995). One explanation for this finding is risk aversion on the part of subjects. It is well known that risk aversion leads to higher sealed bids (pushing bidders toward lower profit margins but higher "win" probabilities). Indeed, the evidence from many sealed-bid experiments is consistent with risk-averse bidding behavior. At the same time, subjects display significant differences in sealed-bid strategies (perhaps due to differing degrees of risk aversion). Of course, bounded rationality would also explain these strategy differences. It is far from evident how subjects might identify or learn to play a benchmark equilibrium strategy (calling for an optimal amount of bid discounting for each possible reservation price), even in the simplest experimental settings.[3]

Because of elevated bidding strategies, the sealed-bid auction holds a significant revenue advantage over the English auction. For instance, in the well-studied case of independent private values drawn from a common uniform distribution, experimental subjects place average bids that are significantly above the risk-neutral equilibrium prediction, $b_i = [(N-1)/N]v_i$. In fact, many replications of my own classroom experiments[4] point to the linear function $b_i = [N/(N+1)]v_i$ as a good fit of average behavior (at least for experiments with two to six bidders). For instance, for $N = 2$, actual subject bids are in the neighborhood of 60% to 75% of value (not the equilibrium benchmark of 50%). For $N = 5$, bids cluster in the 80% to 90%

range (again above the 80% benchmark). In short, subjects on average do not recognize how much they should shade their bids below values when there are small to moderate numbers of bidders.

Elevated bidding in the sealed-bid auction has a more subtle implication for auction design. With average bidding behavior described by the function $b_i = [N/(N+1)]v_i$, the seller can further enhance its expected revenue by conducting a *hybrid* auction. With N buyers, the seller should conduct an English auction to eliminate all but two final contestants. Then, a sealed-bid contest should be held between these last two. What is the intuition behind this result? As noted above, the gap between actual bidding behavior and equilibrium behavior is most pronounced for $N = 2$. Thus, the seller should take maximum advantage of this overbidding by delaying sealed bids until only two bidders remain. Indeed, the revenue boost from the hybrid auction is significant. Turn again to Example Two, with buyer values independently and uniformly distributed on [0, 100]. For $N = 5$ bidders, the English auction's expected price is $E[v_{2nd}] = 66.7$, the sealed-bid auction's expected price is 69.4, and the hybrid auction's expected price is 72.2. (To calibrate this advantage, note that the seller's maximum potential revenue is $E[v_{Max}] = 83.3$, if it could somehow extract full-value from the winning buyer.) Similarly, for $N = 9$, the expected prices for the English, sealed-bid, and hybrid auctions are 80.0, 81.0, and 83.3 (with $E[v_{Max}] = 90$). In short, under the prescribed bidding behavior, the hybrid (or elimination) auction has a significant revenue advantage.[5] In fact, in many real-world bidding settings (including competitive procurements), the winner is determined by best-and-final sealed bids from a pair of finalists.

As noted above, a practical disadvantage of the sealed-bid institution is the complexity attendant on formulating an optimal bidding strategy. For this reason, bidders will tend to deviate from the equilibrium benchmark in idiosyncratic ways. The upshot of heterogeneous bidding behavior is an increased incidence of inefficient allocations. With differing bidding strategies, the winning bidder need not be the high-value buyer. Auction experiments provide direct evidence on the allocative performance of different auction institutions. The experiments confirm that the English auction earns high efficiency marks; the high-value buyer wins the item about 95% of the time. By comparison, sealed-bid auctions – marked by idiosyncratic bidding behavior – earn lower efficiency ratings. However, the difference between the two auctions is in the eye of the beholder. Stressing the efficiency difference, some commentators, Smith (1995) for example, have noted that across myriad experiments, sealed-bid procedures achieve efficient allocations only about 88% of the time (well below the English performance). Others see only a very modest efficiency advantage for the English auction.

A simple example is useful in illustrating and explaining these opposing points of view. Let's reconsider Example Two in the case of two buyers. The first buyer uses the risk-neutral equilibrium bidding strategy, $b_1 = .5v_1$. The other uses a "severely elevated" bidding strategy, say $b_2 = .8v_2$. Given the dramatic differences in strategies, efficiency suffers. Buyer 2 wins the item whenever $v_2 > .625v_1$, implying that inefficient allocations occur whenever $v_1 > v_2 > .625v_1$. The incidence of inefficient allocations is computed to be 3/16. In other words, the sealed-bid auction achieves efficient allocations only 13/16 or 81.25% of the time. By contrast, the English auction delivers 100% efficiency and generates expected trading gains: $E[v_{Max}] = [(N)/(N+1)]100 = 66.66$. For the sealed-bid auction, it is straight-forward to compute the expected value of the winning bidder, given the bidding strategies b_1 and b_2: $E[v_{Winner}] = 64.32$. Thus, the sealed-bid auction achieves 64.32/66.66 or 96.5% of the maximum possible trading gains. What is the explanation for this surprisingly high efficiency rating? Despite the dramatic difference in bidding strategies, in the predominant number of cases (some 81%), the high value determines the high bid. Even in cases in which the lower-value buyer casts the higher sealed bid, the sacrifice in value ($v_1 - v_2$) is relatively small; on average, the reduction in total trading gains is only 3.5%. By contrast, the revenue difference between the auction methods is significant. The expected revenue of the English auction is $E[v_{2nd}] = (1/3)(100) = 33.33$. The expected revenue of the sealed-bid auction (with the elevated bidder) is $E[P_{Winner}] = 45.21$, some 35.6% greater than the revenue of the English auction.

The message of this example extends to more general settings that allow for affiliated values and asymmetric value distributions (both discussed below). In single-object auctions, even when actual bidding behavior implies significant, first-order, revenue effects, the accompanying efficiency effects are typically less significant and of second order. Thus, the tradeoff between revenue and efficiency is not particularly severe. In these cases, the sealed-bid auction promises enhanced revenues with little sacrifice in efficiency.

We summarize the empirical evidence for the symmetric IPV case in the following statement.

Proposition One'. Given symmetric independent private values, actual bidding behavior approximates the equilibrium prediction for the English auction and exceeds the risk-neutral equilibrium prediction for the sealed-bid auction. The sealed-bid auction generates greater expected revenue with an accompanying modest reduction in efficiency.

Affiliated Values

Now suppose that the value of an item to any buyer depends not only on a private-value component, but also on a common-value element. For instance, the economic value of an oil lease to any buyer depends partly on the common, unknown revenue (possibly zero) it generates over its economic life, and partly on the company's particular and private costs (of drilling and shipping). Clearly, items that are ultimately intended by bidders to be resold on a secondary market will have a significant common-value element. One way to model this situation is to describe each buyer's value by the function: $v_i = U(x_i, x_{-i}, V)$, where x_i is buyer i's private signal, x_{-i} is the vector of signals for the other buyers, and V is a common, unknown random variable (possibly a vector). Again, we focus on the symmetric case in which the buyers share a common value function and where their private signals are drawn from a symmetric probability distribution. Bidding involves **pure common values** if $v_i = V$, for all bidders i – that is, the (unknown) acquisition value is the same for all bidders. Here, bidders have private signals (or value estimates) x_i from which they can draw inferences about the unknown V. (Alternatively, if $v_i = x_i$ for all bidders i, we are back in the IPV setting.) In general, the function U allows for a mixture of private and common values. In addition, it is assumed that the buyers' private signals are affiliated. We can spare the reader a technical definition by simply noting that affiliation is roughly equivalent to positive correlation between the signals.

A key implication of the affiliated-values model is that an optimal bidding strategy must now be based on a buyer's assessment of the item's value, *conditional on winning the auction*. See Wilson (1977) for an early equilibrium bidding model. A naïve bidder who fails to recognize this fact is apt to fall prey to the **winner's curse.** Upon winning the auction, this buyer finds that the price paid far exceeds the actual value of the item.

To understand the winner's curse, consider firms bidding for an off-shore oil lease. Suppose that each bidder's signal is an unbiased estimate of the true value of the lease: $E[x_i|V = v] = v$. Nonetheless, conditional on v, there will be considerable dispersion of these estimates. On average, the highest estimate x_{Max} will be significantly greater than the true value of the tract v. Moreover, the winning bidder is most likely to be the buyer with the highest (most over-optimistic) estimate. If it makes a sealed bid based on its "raw" estimate x_i (i.e. incorporates a small bidding discount from x_i), a naïve bidder will find that its expected value upon winning is still far lower than its bid. By contrast, an astute bidder must first assess $E[V|x_i$ & its bid wins], (a quantity much smaller than $E[V|x_i]$), taking into account the bidding behavior and possible estimates of its rivals, before discounting its sealed bid. One can characterize the buyers' common equilibrium sealed-bid strategy, $b_i = b(x_i)$. The implied bid discounts can be considerable. For estimated statistical models of oil-lease bidding, the equilibrium strategies call for sealed bids in the range of 30% to 40% (depending on the number of bidders) of the firm's unbiased tract estimate (Wilson, 1995).

What about the performance of alternative auction methods when values are affiliated? Under symmetric conditions, all buyers employ a common (non-decreasing) equilibrium bidding strategy (under any of the auction types). The buyer with the highest estimate places the highest bid and wins the item. Therefore, all of the auction institutions (as modeled) imply an efficient allocation of the good. However, it is well-known (Milgrom and Weber, 1982) that the auction methods are not equivalent in terms of expected revenue.

<u>Proposition Two.</u> If values are affiliated and all buyers are risk neutral (and play symmetric roles), the auctions are equivalent in terms of efficiency. However, in terms of expected revenue the auctions are ranked: English, Second-Price, and Sealed-Bid (from best to worst). In addition, releasing public information about the item's value will increase average revenues in all of the auctions.

An intuitive explanation for the revenue differences rests on a comparison of bidding strategies under the different methods. As noted above, a buyer's optimal sealed bid depends only on its estimate x_i. (So too for the Dutch auction, which is strategically equivalent to its sealed-bid counterpart.) Uncertain of the item's underlying value and not knowing the others' estimates, each buyer must bid significantly below its estimate to avoid the winner's curse. In equilibrium, this bidding effect limits the seller's expected revenue from the sealed-bid auction. By comparison, consider the English auction. Each buyer must decide how high to bid (when to drop out), conditional on its own estimate and on the implied estimates of the other bidders (those that have already dropped out and those that remain). In the modeled equilibrium (though not necessarily in actual practice), active bidders can observe the prices at which bidders withdraw and infer precisely departing bidders' estimates.[6] In short, the dynamics of the English auction reveal considerable information about the bidders' estimates, attenuating the risk of the winner's curse.

Thus, relative to the sealed-bid auction, the English auction reveals much more information about bidder estimates, prompting buyers to bid much closer to the item's expected value.[7] (In fact, each bidder's equilibrium strategy is to bid up to the expected value of the item, conditional on the inferred estimates of dropout buyers, and conditional on all active bidder estimates being *equal* to the bidder's own estimate.[8]) For similar reasons, releasing public information to bidders about the item's unknown value increases revenue (by diminishing the risk of the winner's curse).

While the expected-revenue rankings of Proposition Two are grounded in theory, they also have obvious practical implications. First, it is natural to explain the widespread use of the English auction by invoking its relative transparency and potential revenue advantage. Second, several researchers have cited the revenue gains of the English auction when advocating new auction methods to address novel allocation problems – from the sale of treasury securities to the allocation of the radio spectrum. Nonetheless, it is important to measure the revenue differences between the auctions, both in theory and in practice. In recent investigations, Li and Riley (1999) have found the revenue differences in several modeled settings to be quite small.[9] In one canonical example, they consider two bidders with jointly normally distributed values. (The two-bidder case is the most favorable for finding revenue differences. In general, as the numbers of bidders increase, revenues for either auction increase and tend to converge.) Here the correlation coefficient between the values, ρ, is a natural measure of affiliation. If $\rho = 0$, the values are independent and the auctions are revenue equivalent. The results show that the English auction's revenue advantage is very small, peaking at about one-quarter of the standard deviation of values for $\rho = .7$. Thus, even in a setting favoring a revenue difference, the English auction's edge is inconsequential.

Empirical Evidence. The most extensive experimental evidence (see Kagel, 1995, for an indispensable survey) centers on the sealed-bid auction for an item with a common unknown value. Auction experiments with inexperienced subjects, Bazerman and Samuelson (1983) and Kagel and Levin, (1986), show a pervasive winner's curse. Buyers fail to adjust sufficiently their estimates downward (conditional on winning the item); consequently, winning bidders consistently overpay for items. Experienced subjects (returning students who had participated in previous common-value experiments) were less likely to overbid (Kagel and Levin, 1986). Nonetheless, the bidding behavior of these subjects consistently exceeded the risk-neutral equilibrium benchmark.

Particularly striking is the effect wrought by the number of bidders. For small groups (three or four bidders) subjects bid above the equilibrium prediction but earn positive profits, avoiding the winner's curse. When subjects received public information, bids and seller revenue increased, consistent with the prediction of Proposition Two. For large groups (six or seven bidders) the results were exactly reversed. Not only was the winner's curse pervasive, but provision of public information lowered bids and revenue. The contrast between the small numbers and large numbers results is easy to explain. As N increases, it becomes much more likely that the most optimistic estimate will lie far in the right tail of the distribution of estimates, i.e. x_{Max} will greatly exceed the item's true value V. For instance, suppose that estimates are uniformly and independently drawn from the interval $[V - \Delta, V + \Delta]$. Then, this bias is $x_{Max} - V = [(N-1)/(N+1)]\Delta$. To avoid the winner's curse, all buyers must increasingly discount their original estimates, x_i, as N increases. However, even experienced subjects fail to make a large enough adjustment when there are numerous bidders. It is not surprising that releasing public information (by reducing the uncertainty about the item's value) tames some of the bidders' over-optimism and, therefore, reduces winning bids on average.

Levin, Kagel and Richard (1996) also studied subject behavior in English auction experiments. The general finding is that bidders, by and large, avoid the winner's curse. Observing the dropout behavior of competitors conveys information about the true value of the item, therefore, dampening buyer optimism and winning bids. Though winning bidders earn positive profits on average, they continue to bid above the equilibrium benchmark. However, the incidence of overbidding in the English auction is far less than in the sealed-bid auction. Contrary to the prediction of Proposition

Two, the English auction generates less revenue (or equal revenue) than the sealed bid auction in these economic experiments.

While laboratory experiments provide controlled tests of auction theory, field studies provide evidence of actual bidding behavior by experienced professionals under real-life institutions where substantial profits and losses are at stake. Hansen (1986) analyzed timber sales by U.S. government, and found no statistically significant revenue differences between the English and sealed-bid auctions. Among the most studied settings are oil lease auctions in the Gulf of Mexico in the 1950s and 1960s. Pointing to the significant common-value components of these sales, a number of researchers, Capen, Clapp, and Campbell (1971) and Hendricks, Porter, and Boudreau (1987), find evidence of the winner's curse, with winning bidders in sealed-bid auctions earning below market rates of returns on acquired leases. Subsequent studies have found that bidders partially adjust bids according to the degree of tract uncertainty and the number of bidders (or by pooling information), thereby largely avoiding the winner's curse. Evidence of the winner's curse has also been identified in corporate takeovers (Roll, 1986), construction bids (Gaver and Zimmerman, 1977), and bidding for baseball free agents (Blecherman and Camerer, 1996). Across a number of different bidding settings, Brannam, Klein, and Weiss (1987) confirm that increasing the number of bidders (N) improves the winning bid price.

Finally, we mention two studies that mix the experimental and field-study approaches. Dyer, Kagel, and Levin (1989) tested construction executives in a series of common-value auction experiments. They found that, like student subjects, executives, despite their vast experience, succumbed to the winner's curse. Evidently, these practitioners have developed estimation techniques, and bidding rules of thumb to adapt profitably to their own business environments. Indeed, contractor behavior in the construction industry is extensively documented by Dyer and Kagel (1996) in a follow-up study. At the same time, contracting executives seem unable to generalize profitable behavior in the field to the starker challenges of the experimental setting.

A second study is based on my own in-class experiments in which MBA students placed bids for motion pictures that were about to open in theaters nationwide. Each MBA team represented a hypothetical theater, bidding a weekly guarantee for the right to show the film for a four-week run. The team's task was to predict the average gross revenue per week per screen that the film would earn over the run. Needless to say, the gross revenues for film releases are highly uncertain – ranging between $12,000 per screen per week for a blockbuster to $2,000 per screen per week for a "dog." Thus, the setting involves common unknown values. Besides being uncertain, a film's revenue performance (per screen) depends on many factors: the film's genre, its cast, its overall quality and entertainment value (embodied in reviews and word of mouth), the time of year, and the breadth of its release (wide release or city-by-city exclusive engagements). Indeed, student teams have the chance to analyze the revenue track records of past films before formulating their estimates and their bidding strategies for the current films up for bid. Upon award of a film to the high bidder, the winner's profit or loss was reckoned by obtaining the film's actual four-week revenues (as reported by *Variety* magazine). Exercise grades were based on these *ex post* profit or losses. In effect, the exercise places student subjects in a real-world field study.

The table below shows the bid results for eight films auctioned in Fall 1998. Prior to the bidding, each of the six teams reported their best estimates of the films' revenues. The table reports the maximum, second-highest, and average estimates for each film. Next, each team was asked to submit a sealed-bid for each film. Then before the bids were opened, the films were "sold" in order by English auction. Students were instructed that a coin toss would determine whether the high sealed bid or the high English bid would win the film. In effect, the films were auctioned twice under controlled conditions, allowing for a "clean" comparison of the two auction methods.

Films	Bidder Estimates				Winning Bids	
	Max	2nd	Average	Actual	Sealed	English
Stella	6.0	5.5	4.92	5.85	5.1	5.0
Blade	8.3	6.5	5.92	6.23	6.2	6.0
Dead Man	5.0	4.0	3.05	1.95	3.5	2.8
Pi	7.0	5.9	4.83	6.93	6.1	5.0
Wrongfully Accused	5.4	3.6	3.10	1.25	3.2	3.3
Return to Paradise	6.0	4.5	3.77	2.23	5.0	3.4
Dance with Me	5.2	4.3	3.25	2.55	3.4	3.0
Air Bud II	4.5	3.6	3.43	1.40	3.4	4.0
Averages	5.93	4.74	4.03	3.55	4.49	4.06

The table results strongly support our earlier revenue comparisons. In the sealed-bid auctions, the team with the highest estimate was the winning bidder in 7 of 8 cases with the winning bid averaging about 75% of the bidder's estimate. Though students were experienced (they had been introduced to the winner's curse in simpler settings), winning sealed bids were loss making on average. The winning bid for a film exceeded the average weekly estimate for the film (across the six teams) by about $460 on average – $4,490 versus $4,030. Arguably, this is the best measure of over bidding. In addition, the actual weekly revenues (because of three unusually awful movies) averaged still less, $3,550, implying an average bidding loss of about $940 per film per week. By comparison, bidding was more orderly and the winner's curse attenuated in the English auctions. The few active bidders (usually two or three) realized that the other teams did not share their enthusiasm. From the table, we see that the English price (with the exception of one team desperate not to be shut out of the last film) ended below the second highest estimate – that is bidders made some adjustments downward in their original estimates. Thus, the English price for a given film was nearly identical to the average estimate across the six teams – $4,060 versus $4,030 on average. Accordingly, the English auctions generated significantly lower revenues than the sealed-bid auctions, a reduction of $460 per screen per week on average.

We can sum up the empirical results for affiliated-value settings as follows.

Proposition Two´. Given symmetric affiliated values, actual sealed bidding behavior is significantly above the equilibrium benchmark. In risky and competitive settings (high value uncertainty and five or more bidders), winning bidders suffer the winner's curse. In these cases, releasing public information tends to lower seller revenue. In the English auction, bidders adequately adjust their behavior to avoid the winner's curse, though winning bids tend to exceed the equilibrium prediction. On average, English auction revenues fall short of sealed-bid revenues.

Asymmetries

Less attention has been paid to bidding behavior and auction performance when buyer asymmetries are important. For instance, in the IPV case, suppose that buyer values are drawn from different distributions. Maskin and Riley (1999) have derived extensive results for asymmetric environments. The following two-bidder example (modified slightly from their presentation) provides a dramatic illustration.

Example Three. Suppose the "weak" buyer's value is uniformly distributed on the interval [0,60], while the "strong" buyer's value is uniformly distributed on the interval [120, 180]. Both the English and sealed-bid auctions guarantee an efficient outcome; the strong buyer is always the winning bidder. However, the sealed-bid auction generates much more revenue than the English auction. In the latter auction, the weak buyer's value determines where the price stops, implying $E[P_E] = 30$. In the sealed-bid auction, the strong buyer's optimal bid is 60 for any value it holds. (Intuitively, holding high values, the strong buyer's optimal strategy is to make a shut-out bid, rather than risk losing the item.) Thus, the sealed-bid auction delivers twice the expected revenue of the English auction.

The revenue advantage of the sealed-bid auction holds for less extreme distribution asymmetries. Begin with values drawn independently from the same distribution, and then shift one of the distributions so that the so called "strong" bidder tends to have more favorable values.

Example Four. For two bidders, suppose that the weak buyer's value is drawn from the uniform distribution [0, 60] and that the strong buyer's value is drawn from [0,120]. Again, the English auction suffers because the weak buyer tends to cease bidding at a relatively low price. (In fact, expected English revenue is: $E[v_{2nd}] = 25$.) In the sealed-bid auction, the expected price fares better. In equilibrium, the weak buyer bids "relatively close" to full value. Of course, aware of his advantage, the strong buyer increases its bid discount (compared to the case of facing a symmetric rival). Nonetheless, equilibrium revenue in the sealed-bid auction comes to 27.25 – some 9% higher than in the English auction. As a last example, suppose that values are drawn from the uniform distributions [0, 60] and [30, 90] – the strong bidder's distribution is simply shifted to the right. Here, expected English revenue is: $E[v_{2nd}] = 28.75$. In the sealed-bid auction, despite increased discounting by the strong buyer, equilibrium expected revenue is 34.4, about 20% higher than in the English auction.

Maskin and Riley (1999) and Li and Riley (1999) provide general analyses of the revenue differences between the two auctions under value asymmetries. Their main finding is that the sealed-bid auction holds a significant revenue advantage over the English auction when there is a single strong bidder. (The advantage holds even allowing for affiliated values and multiple weak bidders.) On the other hand, revenue differences are inconsequential when there are two or more equally strong bidders or when the seller has an incentive to set a binding reserve (i.e. minimum) price in order to elevate bids.

Reserve Prices and Bidder Preferences. The seller can increase expected revenue by two additional means – setting reserve prices and extending bidding preferences to selected buyers. Let v_0 denote the seller's private value for the item in the event it is not sold at auction. (This can be interpreted as the seller's personal value for the item or as the expected "liquidation" price the item would bring.) The seller benefits from setting a reserve price below which the item will not be sold. In general, the optimal reserve price will be set at a level strictly *greater* than v_0. For instance, in the English auction of Example Three, the seller increases expected revenue from 30 to 120 by setting the reserve at 120 (the strong buyer's lower support). Indeed, frequently, the optimal reserve should be set in the interior of the buyer value distributions, thereby risking the chance that no bid meets the reserve, leaving the item unsold. For instance, in Example Two with $v_0 = 0$, the seller's optimal (i.e. revenue-maximizing) reserve price is $r = 50$, at the midpoint of the [0, 100] uniform distribution, independent of the number (N) of bidders. In the symmetric IPV setting, Myerson (1981) and Riley and Samuelson (1981) prove the following strong result.

> Including a provision for an optimal reserve price, any of the four auction methods of Proposition One constitutes a revenue-maximizing auction *among all possible selling methods.*

Thus, in Example Two, the sealed-bid auction would use the same reserve price, $r = 50$, as the English auction and earn the same expected revenue. Here, instituting the reserve price serves to raise the entire equilibrium sealed-bid strategy, $b(v)$.

As Myerson (1981) first observed, when values are asymmetrically distributed, the seller can also benefit by instituting a system of bidder preferences. Such preferences typically benefit "weak" buyers in order to bolster the competition faced by "stronger" rivals. To understand how preferences work, return to Example Four with buyer values uniformly distributed on [0, 60] and [0, 120]. One way to level the playing field is to give the weak buyer a 50% bidding credit, i.e. the buyer only pays 50% of its bid. Thus, this transforms the weak buyer's value distribution to [0, 120], restoring symmetric values. In either an English or sealed-bid auction, the expected winning bid is: $[(N-1)/(N+1)]120 = (1/3)(120) = 40$. The weak buyer wins half the time, but only pays 50% of its bid. Therefore, the seller's expected revenue is: $(.5)(40) + (.5)(20) = 30$. This sum is considerably greater than the 25 in expected revenue from an English auction without preferences. Indeed, it is straightforward to identify the optimal (i.e. revenue-maximizing) combination of reserve price and bidder preferences.[10] For instance, if $v_0 = 0$ (the seller has no resale value or use value for the item), the seller should grant the weak bidder a 30-unit bidding preference and set the reserve price at $r = 60$. This combination raises the seller's expected revenue to 36.9. (With preferences alone, expected revenue is 32.5. With a reserve price alone, set optimally at $r = 45$, expected revenue is 33.44.)

Empirical Evidence. To my knowledge, there has been little or no published research on bidding behavior under asymmetric conditions. In my own classroom experiments with asymmetric values, the bidding behavior of strong buyers significantly exceeds the risk-neutral equilibrium prediction. For instance, consider a two-bidder example with values uniformly distributed on [0, 120] and [60, 120] for the weak buyer and strong buyer respectively. Here, the equilibrium revenue predictions are 55 for the English auction and 60.25 for the sealed-bid auction. Actual-sealed bidding behavior by student subjects generates expected revenue of 64.8 – some 7.5% greater than the equilibrium prediction (and 18% greater than English expected revenues). Subjects in the role of strong buyers fail to take full advantage of their favorable value distribution. This limited evidence suggests that in asymmetric settings, sellers could exploit the sealed-bid auction framework for revenue gains.

Corns and Schotter (1999) test the efficacy of bidder preferences in laboratory experiments. Employing sealed bids in a procurement setting, their experiments found that granting modest (i.e. 5%) preferences to "weak" bidders increases bidding competition, thereby, inducing more favorable bids from "strong" bidders. The upshot was improved auction performance (i.e. lower procurement costs). In fact, induced sealed bids were so aggressive that actual auction performance surpassed the optimal auction benchmark. Higher (10% or 15%) preferences, however, proved to be counterproductive. Ayers and Cramton (1996) have studied bidder preferences in the 1995 round of the radio spectrum auctions. Bidding preferences (some 40-50% for designated minority- or female-controlled firms) increased government revenues by some 12% according to their estimates. By increasing the degree of competition, the auction generated significantly greater revenues from unsubsidized bidders, more than compensating for the subsidies paid.

These positive findings notwithstanding, there is a significant leap between instituting bidding preferences in theory and in practice. First, there is an obvious conflict between bidder preferences and the usual expectation that auctions should ensure equal treatment of participants. For this reason alone, preferences may be infeasible. Second, to implement optimal preferences, the auctioning party must have extensive (even precise) knowledge of player differences. In actual practice, an auctioning party with limited knowledge can at best hope to implement an approximate preference scheme.

This latter point also applies to setting reserve prices. With limited knowledge of potential buyer values, setting a reserve is more "art than science." Indeed, Ashenfelter (1989) describes auction house reserve policies in exactly this way. Reserve prices have a further limitation. If buyer competition is sufficient, a seller reserve is largely redundant – that is, the revenue impact is insignificant. For instance, in Example Two, once there are five bidders, implementing the optimal reserve price raises revenue by less than one percent.[11] Indeed, under general symmetric conditions (allowing for independent or affiliated values), Bulow and Klemperer (1996) show that the effect of a reserve price is always less than the impact of adding a *single* additional buyer. In short, a reserve price is likely to have a practical impact on seller revenue only if three conditions are met: 1) the seller's own value is high enough that the seller is willing to risk not selling the item, 2) the seller has extensive knowledge of the range of likely buyer values, and 3) the number of bidders is limited.[12]

We can sum up our review of asymmetric auctions in this proposition.

Proposition Three. Given pronounced asymmetries (for instance, one strong buyer) and the absence of a reserve price, the sealed-bid auction will have a revenue advantage over the English auction. Armed with extensive know-ledge of bidder differences, the auctioneer can use bidder preferences to raise its expected revenue.

2. Multi-Object Auctions

Auction institutions and buyer strategies become more complicated when multiple objects are up for sale and values are non-additive. By the term non-additive, we mean that buyer values for collections of objects need not equal the sum of values for individual objects. Subsets of items may be substitutes or complements. For instance, a buyer might wish to buy only one unit of an item for which multiple units are being auctioned. In a less extreme case, the value of additional units might decline with increases in the total quantity acquired. Alternatively, items might be worth considerable more in combination than they are separately. A piece of real estate may be worth much more undivided than when sold separately in pieces. Regional and national combinations of spectrum rights can be expected to be worth considerably more than the separate parts.

In practice, the most frequent means of selling multiple items is by sequential auctions. Ordinary sales of artwork, antiques, jewelry, collectables, and the like by auction houses all adopt sequential auctions, numbering the items and auctioning them in order by lot number. This method presents no difficulties if buyer values are additive – that is, if the value of acquiring items any items A and B together is the sum of the values of acquiring each separately. Additivity may be a good approximation for most buyers in most of these sales, in which case the auctions can be undertaken independently in any order. (Clearly, if buyers seek to achieve or augment particular collections of paintings, the presumption of additivity could be problematic.) Alternatively, if complementarities are easy to identify, the auctioneer can group items in lots accordingly. (Pieces of furniture, place settings of china, or pairs of candlesticks would be sold as respective lots.)

When values are non-additive, however, the choice of auction method is important. Alternative mechanisms for auctioning multiple items include:

1. Sequential English auctions for separate items.
2. Simultaneous English auctions for items.
3. Sequential English auctions for items and also for collections of items.
4. A single sealed-bid auction for items and combination of items.

In applying these auction methods, we consider a pair of polar applications: settings involving substitute values and those with complementary values. At the outset, it is fair to acknowledge that the theory of auction performance under non-additive values is in the early stage of development. Accordingly, we consider only the "private-values" setting for the analysis in the remainder of this section.

Substitute Values

A host of auction settings – from commodity sales to Treasury Bill auctions – involve substitute values. By this, we mean buyer values for additional units decline with increases in the total quantity obtained. Let $v_i(Q)$ denote buyer i's demand schedule, i.e. $v_i(Q)$ measures the buyer's marginal benefit for the Qth unit. Then, $v_i(Q)$ might be a smooth non-increasing function or it might take the form of a downward sloping step function.

Consider the case in which K identical units are to be sold. To begin simply, suppose that each of N buyers has a private value for *at most one unit*, so $v_i(Q) = 0$ for $Q > 1$. Then, analogous to Proposition One, Weber (1983) identifies the following equivalence result concerning the sale of multiple objects.

<u>Proposition Four</u>. Given symmetric independent private values and risk-neutral bidders (each demanding a *single* unit), all auction methods – sequential, simultaneous, or sealed bid – produce identical and efficient allocations of the items and identical expected revenues for the seller.

Like Proposition One, this result is established using the optimal auction approach. To understand the proposition's practical importance, consider briefly how it applies to particular auction methods. Start with a sealed-bid auction, where the K units are sold to the K highest bidders. Three pricing methods are of interest. i) In a **discriminatory** (or first-price) auction, winning buyers pay their bid prices. ii) In a **uniform-price** auction, winning bidders pay a common price set at the highest rejected bid, i.e. the $(K+1)$st highest bid. iii) In a **Vickrey** auction, for items won, each buyer pays a price equal to the reported monetary values foregone by rivals who would have won the items instead.

In the uniform-price auction, each buyer's dominant strategy is to bid its true value (just as in a second-price auction for a single item), so that the uniform price is set equal to the $(K+1)$st highest value among the buyers. Thus, the seller's expected revenue is $KE[v_{K+1}]$. When each buyer's demand is limited to a single unit, the Vickrey auction and the uniform-price auction coincide – that is, they are one and the same. By acquiring one unit, each winning bidder deprives the $(K+1)$st buyer from obtaining it. Therefore, each pays that buyer's foregone monetary value, v_{K+1}. Finally, in the discriminatory auction, buyers shade their sealed bids below their values.

The optimal degree of bid shading also implies expected seller revenue of $K\mathrm{E}[v_{K+1}]$.

The same expected revenue is attained by auctioning the K items simultaneously or sequentially. If simultaneous auctions are held for the items, prices will rise continuously across the board until the common level just exceeds the value of v_{K+1} (where there is no longer any excess demand). Alternatively, consider using successive English auctions to sell the items. In the first auction, the price rises until K active bidders remain. Anticipating the course of the future auctions, none of the remaining K bidders should advance the price any higher; therefore the first item's sale price is v_{K+1}. In the second auction, the price stops at the same level, v_{K+1}, when K-1 active bidders remain (for the K-1 remaining items). By similar reasoning, the price is the same, v_{K+1}, for each of the remaining items. (Of course, this assumes that the buyers can discern the precise number of active bidders at any point in the auction.) As Weber (1983) shows, sequential sealed-bid auctions and sequential second-bid auctions generate the same expected revenue, $K\mathrm{E}[v_{K+1}]$, as well. However, in view of the more complex bidding strategies involved in the discriminatory and sequential auctions, the uniform-price auction provides the simplest practical means of determining what the "market will bear" for the items.

Now consider the more general case where demand is either flat or decreasing and buyers desire *multiple* units. For instance, in Treasury bill auctions, a typical buyer might be willing to pay a relatively high value (i.e. accept a lower implicit interest rate) for a certain volume of bills, but lower values for additional quantities. Consider the customary sealed-bid auction, where buyers are free to place multiple bids at different prices for separate quantities. As Ausubel and Cramton (1998b) show, Proposition Four's equivalence result is decisively rejected. None of the standard auctions can guarantee efficient allocations, and revenue comparisons are problematic. When multiple units are demanded, the typical buyer can potentially profit by understating demand in the standard sealed-bid auctions.

This is even true in the uniform-price auction. Bidding one's true values for units (one's true demand curve) is no longer a dominant strategy (as in the single-unit case). Now, it pays to understate values (particularly for "later" units). While this strategy risks some bids being rejected, it also promises the chance of influencing (i.e. lowering) the uniform price at which all units are purchased. As long as there is the chance that the buyer's bid is pivotal (i.e. determines the final uniform price), there is the incentive to lower bids on "later" units in order to reduce the price paid on "earlier" units. (The intuition is analogous to that lying behind a monopolist's marginal-revenue curve.) Engelbrecht-Wiggans and Kahn (1998) provide a rich characterization of uniform-price equilibria exhibiting demand reduction. They present a dramatic example with two units for sale and two bidders, where each desires two units and each unit's value is drawn independently from a uniform distribution. Here, the unique symmetric equilibrium is for each buyer to bid full value for his first unit and to bid zero for his second. As a result, both units sell for zero prices!

Besides lowering expected revenue, demand reduction in uniform-price auctions is a source of inefficiency. "Large" buyers (those demanding and likely to receive many units) have a greater incentive to reduce demand than do "small" buyers. As a hypothetical example, a large bidder that has value 12 for its 20th unit might bid 6 hoping that this bid is pivotal. By contrast, a small bidder that has value 10 for its 4th unit might bid 8 (i.e. a much smaller discount). Consequently, if P = 7 happens to be established as the uniform price, an inefficient outcome occurs (the larger buyer is denied but the small buyer wins the respective units). As Katzman (1996) and Ausubel and Cramton (1998) emphasize, *differential* demand reduction (not demand reduction per se) is the source of inefficiency. They also show that the inefficiency result extends to the case of correlated values.

Now consider the discriminatory auction in which winning buyers pay their bids. Here, there is a much stronger incentive to shade bids below values (than in the uniform-price auction) since this reduces the purchase price dollar per dollar if the bid wins. Unfortunately, theory alone is insufficient to rank the discriminatory and uniform-price auctions either with respect to revenue or efficiency. For particular examples, one auction can be shown to outperform the other on either criterion. Finally, consider the Vickrey auction allowing for multi-unit demand by buyers. If a buyer wins J items, it pays the amount of the jth-highest rejected bid *other than its own* for the jth item, for j = 1, ...J. By design, the Vickrey auction induces buyers to bid their true marginal values for units demanded. Thus, allocating items according to high bids ensures an efficient outcome. To illustrate the method, suppose that three buyers bid for four items. In descending order, the buyer bids (and, therefore, values) are: **20**, *18*, **18**, *16*, *13*, 11, *10*, 9, 7, **6, 4**, and, 2, where buyer 1's bids are in bold and buyer two's bids are in italics. Buyer 1 wins two units, paying 13 (the highest rejected bid) for the first unit and 11 (the second-highest rejected bid) for the second unit. Buyer 2 pays 11 (not 13 because 13 is buyer 2's own bid) for the first unit and 9 (instead of 10 for the same reason) for the second unit. In this example, the winning buyers pay different prices for their units.[13] Each price reflects the foregone benefit of excluded *rival* buyers. To induce truthful bidding, buyers pay Vickrey prices – never their own bid prices. While the Vickrey auction guarantees efficiency, Maskin and Riley (1989) demonstrate that it need not deliver maximum revenue to the seller. Depending on the particular example constructed, it appears that the three sealed-bid methods – discriminatory, uniform-price or Vickrey auction – can rank in any order with respect to seller revenue.

Empirical Evidence. There is limited experimental evidence concerning bidding behavior and auction performance in multi-item auctions. Cox, Smith and Walker (1984) conducted experimental tests of the discriminatory auction where each buyer bids for a single unit of the available multi-unit supply. In the majority of experimental conditions, subject bidding is consistent with Nash equilibrium (risk-neutral or risk-averse) behavior. In some conditions, however, bids are below the risk-neutral equilibrium prediction. Overall, the revenue data exceed the risk-neutral prediction. Thus, actual bidding behavior favors the discriminatory auction over the uniform-price auction – contrary to Proposition Four's prediction of equivalent revenues. In a related study, Cox, Smith, and Walker (1985) compare the experimental performance of the discriminatory and uniform-price auctions (again with individual demand limited to a single unit) and find no significant revenue differences. More recently, Kagel and Levin (2000) have conducted laboratory experiments where buyers desire multiple units. They compare the uniform-price auction with Ausubel's (1997) ascending bid version of the Vickrey auction. As expected they find greater bid reductions on second units in the uniform-price setting. Nonetheless, the uniform-price auction generates the greater revenue (though it is less efficient than the Vickrey auction). List and Lucking-Reilly (2000) conduct field experiments in the form of sportscard auctions to compare the Vickrey and uniform-price formats. In accord with theoretical predictions, they find that demand reduction is more pronounced in uniform-price auctions than in Vickrey auctions. Indeed, these reduced bids cause frequent changes in the allocation of the goods. In addition, the uniform-price auction generates significantly more zero bids than the Vickrey auction. Overall, however, revenues are not significantly different between the two auctions.

Because of the significant monetary sums at stake, the sale of treasury securities by auction has been the focus of considerable research attention.

In the United States, government debt securities are commonly sold by a discriminatory auction, with winning buyers paying their bids. Starting with Milton Friedman four decades ago, some economists have questioned the performance of the discriminatory auction, backing instead the uniform-price auction under which all winning bidders pay the same market-clearing price. According to these advocates, the uniform-price auction will "level the playing field" for all purchasers, obviate the information advantage of institutional insiders, increase participation by small buyers, and elevate average bids.[14] As we have seen, theory alone does not provide a prediction about the relative revenue merits of the two auctions. The U.S. Treasury has experimented with the uniform-price method, in 1973-1974 and since 1992 for 2-year and 5-year notes. Empirical studies of the alternative auctions' comparative revenues have brought mixed results. For some periods, the uniform-price auction appears to have brought higher government revenues (i.e. lower borrowing costs). In other periods, the revenue results for the two auctions are indistinguishable. Bartolini and Cottarelli (1997) and Malvey, Archibald, and Flynn (1996) are useful references to these studies.

Ausubel and Cramton (1998b) note interesting cases of demand reduction in the spectrum auctions. In these auctions, licenses of the same type in the same market sold at approximately the same prices. Thus, the simultaneous multiple-round auction (to be discussed below) approximated a uniform-price auction for multiple identical items. In one auction, a bidder cut back from bidding on three large licenses to bidding on two large licenses and one small license, apparently to avoid driving the prices up on all the large licenses. A subsequent auction suggests that the winningest bidder also held back its demand. After another bidder defaulted, the large buyer bought 60% of the re-auctioned spectrum at much higher prices than it bid in the original auction. (In the re-auction, this buyer could bid aggressively without affecting the prices for licenses already won in the original auction). Weber (1997) also documents instances of demand reduction in the spectrum auctions.

Complementary Values

When complementarity is the rule, combinations of items are worth more than the sum of the parts. First, consider the common practice of **auctioning items in sequence**. While we have seen that single-object English auctions guarantee efficiency, this result does not generalize to sequential English auctions with non-additive values. Some simple examples suffice to make the point. Consider three Firms, X, Y, and Z, which seek to buy two items F and G, separately or in combination. Each firm knows its own values but has only a probabilistic assessment of its rivals' possible values.

For the value realizations of case 1 below, the efficient allocation calls for Firm X to win both items. But how should Firm X bid? With item F first up for sale, the firm faces a dilemma. It seeks to obtain both items, but the disposition of G is unknown at the time it bids for F. (Clearly, without G, item F is of little value.) Firm X faces an obvious tradeoff under uncertainty. Supposing that the joint distributions of firm values are common knowledge, one can again focus on a Bayesian bidding equilibrium as a benchmark. Suppose that X's equilibrium strategy calls for it to bid up to 5 for F if necessary. Note that X is willing to bid above its single-item value, $X_F = 3$, since winning F enhances X's marginal value for G, $X_{FG} - X_F = 12 - 3 = 9$. Thus, X is willing to risk being stuck with F at a loss, in the

Case 1　　　　　　　　　　　　　Items

	F	G	FG
Firm X	3	3	12
Firm Y	6	1	8
Firm Z	1	4	7

hope of winning G. The first English auction ends with Y winning F at a bid slightly above 5. In turn, Z wins item G at a price slightly above 3. Thus, the sequential auction ends in an inefficient "split": $Y_F + Z_G = 10 < X_{FG} = 12$.

Case 2 (with slight changes in value realizations for Firms Y and Z) shows the opposite case, an inefficient bundled allocation. Again, Firm X's equilibrium strategy calls for bidding as high as 5 and now Firm X wins item F.

Case 2　　　　　　　　　　　　　Items

	F	G	FG
Firm X	3	3	12
Firm Y	4.9	1	8
Firm Z	1	8	10

In turn, X wins item G but is forced to pay a price slightly above 8. Firm X wins the bundle FG (unfortunately and unexpectedly paying a total price of 12.9). The upshot is an inefficient bundled sale: $X_{FG} = 12 < Y_F + Z_G = 12.9$. Finally, it is easy to modify case 2 (by increasing Z's value for G, say to $Z_G = 10$) to show still another type of inefficiency. Now Firm X continues to win item F at a price of 5 but is outbid for item G. The auction inefficiently allocates item F to Firm X (at a bid exceeding X's value for the single item).

In each of these cases, the root cause of the inefficiency is the same. While each buyer's private values are multidimensional (here three dimensional), the sequential auction setup allows for only two-dimensional bidding (i.e. separate bids on each of the two items). Thus, even with the second-bid opportunity, the initial auction cannot guarantee an efficient disposition of the first item. Of the two kinds of inefficiency, it would appear that the first, inefficient split sales, is potentially more serious. Achieving an assembled bundle via sequential auctions (while risking a loss-making partial purchase) would appear to be problematic. The impediment to assembling efficient bundles is commonly known as the exposure problem. In item-by-item bidding, buyers are reluctant to bid for high-valued bundles in fear of ending up with partial bundles or single items, worth far less than the prices paid.

Item-by-item bidding also poses problems for the seller's revenue. As an extreme example, consider case 3 with two firms bidding for two items. Firm X only derives value (say 10) from the bundle and places zero values on separate items. Firm Y only seeks a single item (with either item worth 7). Then the only sequential bidding equilibrium is extreme: Firm Y obtains the first item at a minimum bid, while Firm X tenders no bids at all. (Firm X knows that it must bid at least 7 to win each item and so cannot profit from assembling the bundle at a price of 14.) The upshot is an inefficient allocation (the single item to Y instead of the bundle to X) and minimum revenue for the seller.

To mitigate the exposure problem, many auction researchers have advocated the use of **simultaneous English auctions**. In the simultaneous format, all items are up for bid at the same time, and bidders can keep continuous track of the ascending prices of all items. To date, the FCC spectrum sales have utilized simultaneous English auctions. Among others, McMillan (1994), McAfee and McMillan (1996), Cramton (1995), and Cramton (1997) argue that simultaneous auctions provide bidders much better information about prices, thereby allowing bidders much better opportunities to assemble valuable packages of items. In this setup, bidders can switch among items as prices unfold during the auction. In particular, a bidder can limit its exposure by abandoning items (giving up on the package) if the bidding gets too steep.[15] Or it may decide to pursue an alternative package if the prices (as they unfold) for individual items are cheap enough. Unfortunately, lacking an equilibrium model of simultaneous English auctions, there is no way to confirm the notion that simultaneous auctions promote efficient bundled allocations. While sequential auctions have been extensively used (and are straightforward to implement), simultaneous auctions are yet to be widely used and tested.

An obvious response to the exposure problem posed by auctioning many individual items is to allow **separate bids for bundles** as well. The seller's goal is to sell the items (singly or in combination) so as to maximize total sales revenue. In the examples above, bids would be sought for item F, for item G, and for the bundle FG. If the high bid b_{FG} exceeds the sum of the high single bids, $b_F + b_G$, there is a bundled sale. Otherwise, there are separate sales. Clearly, the advantage of this kind of "complete" auction is that it offers a direct price comparison between the separate and bundled sales alternatives. At the same time, it eliminates the exposure problem. For instance, in cases 1 and 2, Firm X will compete for the FG bundle without being exposed to the risk of overpaying for F and G singly. Similarly, in case 3, Firm X would place no bids for either single item but would obtain the bundle (both items) by outbidding Firm Y at a price slightly greater than 7. The result is an efficient allocation and healthy revenue for the seller.

While bidding for bundles (so-called combinatorial bidding) appears to increase the options for achieving efficient allocations and higher revenues, it is not without certain problems. One problem is *bias*: the bidding process might induce too many (or too few) bundled sales. For instance, the process might lead to a bundled sale when separate sales imply higher total value, or vice versa. To examine the issue of bias, consider the following illustration.

Example Five. Suppose that two goods are up for sale among N bidders of two types. Some bidders only value the separate items (put no added value on the combination). Others only value the bundle (i.e. have no value for a single good). One natural selling procedure is to solicit bids for the separate items in *second-bid sealed auctions*, and then a bid for the bundle also in a second-bid auction. To illustrate, suppose the second-highest sealed bid is 10 for item F, 8 for item G, and 20 for the bundle FG. Because P_{FG} exceeds $P_F + P_G$, there is a bundled sale and the winning bidder pays P_{FG}. As in the single item case (see the discussion following Proposition One) each bidder's dominant strategy is to bid its actual value.[16] Thus, the sale price is determined according to the second-highest value (either for separate sales or the bundle). For this selling procedure to achieve efficient allocations all the time, it must be that $v_{maxFG} \geq v_{maxF} + v_{maxG}$ whenever $P_{FG} > P_F + P_G$, where v_{max} denotes the maximum buyer value, whether for an item or for the bundle.

Now consider the most interesting case where $E(v_{maxFG}) = E(v_{maxF}) + E(v_{maxG})$ – on average, the maximum values from separate or bundled sales are equal. The question of bias turns on a comparison of $E(P_{FG})$ and $E(P_F) + E(P_G)$. There is a bias toward selling the bundle if the second-highest bundled value tends to be higher on average than the sum of the second-highest separate values. It is easy to construct examples in which the bias can go either way. For instance, suppose each bidder's values, v_F and v_G, are drawn identically and independently from all others' and $v_{FG} = v_F + v_G + \Delta$, where Δ (> 0) is deterministic and such that $E(v_{maxFG}) = E(v_{maxF}) + E(v_{maxG})$. Then it is easy to show that the bias is toward bundled sales, $E(P_{FG}) > E(P_F) + E(P_G)$. (However, if each bidder's Δ is an independently drawn random variable with large enough variance, the bias shifts toward separate sales.)

In sequential auctions, a second obvious source of bias stems from the order of sales. There is a favorable bias toward the sale that occurs last. Consider sequential English auctions. If the bundled auction is last, it is favored. Its winning bid has the last chance to beat the sum of the highest standing bids for the separate items. If the bundled auction occurs first, the later separate auctions have the last chance to beat the bundled price.[17]

Finally, a third source of bias stems from the "free-rider" problem faced by bidders for the separate items. Case 4 provides a simple illustration.

Case 4	Items		
	F	G	FG
Firm X	2	3	12
Firm Y	8	1	8
Firm Z	1	6	11

Suppose the bundle is auctioned first and generates a standing price of 11. Item F is next and brings a standing price of 4 (Y raises the bidding to 4 even after X drops out). In the final auction for G, Z bids 6 (its full value). The upshot is a bundled sale ($11 > 4 + 6$), despite the fact that separate sales are efficient. Even if the auctions are simultaneous, the free-rider problem persists. Each buyer for an individual item has an incentive to restrain its own bid, hoping that the other buyer will be the one to bid higher and top the bundled bid.

The relative efficiency of the English auctions (sequential of simultaneous) versus the combinatorial auction (allowing for bundled bidding) would seem to depend upon the severity of the "exposure" problem for the former and the "free rider" problem of the latter. Furthermore, with non-additive values, the revenue performance of the alternative auction methods is an open question. (To date, we lack models of equilibrium bidding behavior for any of these auction methods.) It is natural to hypothesize that revenue is likely to increase with the degree of competition afforded by the auction. Thus, one could conjecture that the expected revenue rankings would be: the combinatorial auction, simultaneous English auctions, and sequential English auctions in that order.

While permitting bundled bidding would seem to allow the seller extra chances to find higher bids, this is not always the case. Consider the following peculiar case.

Case 5

	Items		
	F	G	FG
Firm X	10	6	26
Firm Y	4	8	22

Here, each buyer knows the other's precise values. First, consider sequential English auctions for F and G. In the unique sequential equilibrium, X will win both items, paying 8 for G in the second auction and 16 for F in the first auction. (Firm Y can afford to bid up to $16 - \varepsilon$ for F; if Y wins F, it can then bid $6 + \varepsilon$ and beat X for item G.) The seller claims 24 in total revenue, and the sale of FG to X is efficient. Instead, suppose that the seller includes a final bundled auction for FG, hoping to glean greater revenue. Now the buyers can coordinate their bidding. Suppose Y allows X to win F at a price of 4 let's say. Then, X has an incentive to allow Y to win G at a profitable price (say 5). Now, neither side has an incentive to bid seriously in the bundled auction. (If the bundled price were to pass the sum of the separate prices (4 + 5 = 9), the bidding would rise to 22, leaving each with lower profits than generated by separate sales. Paradoxically, allowing bundled bidding serves to deter bidding on the separate items. In this perverse case, the result is an inefficient split allocation and minimal seller revenue.

A final choice of selling procedure is to employ a **Vickrey auction** in which buyers submit sealed bids for all items and combination of items. Based on the bids, the seller implements the value-maximizing allocation of the items. The important feature of the Vickrey auction is that a winning bidder pays a price set according to the "cost" (in terms of reduced value) that his award imposes on the *other* buyers. As is well-known, each buyer's dominant strategy is to bid his actual values for all items and combination of items. Thus, the Vickrey auction's allocation is guaranteed to be efficient.

To grasp the pricing implications of Vickrey auctions, consider our earlier combinatorial examples. In case 1, the value-maximizing assignment is the bundle FG to X. The Vickrey price is $P_{FG} = 6 + 4 = 10$. (This is simply the opportunity cost of the next best assignment, item F to Y and item G to Z.) There are two points to note. First, if the second-best bundled value had been 11 instead of 8, that would have been the bundled price. In general, whenever a bundle is sold, its price is the greater of the next-best bundled price and the sum of the best individual prices. Second, a buyer never is put in a position of "bidding against himself." To illustrate, suppose we modify case 1 by switching the values of buyers X and Z for item G. Now X holds the highest value for item G as well as for bundle FG. Buyer X continues to win the bundle but now pays $P_{FG} = 6 + 3 = 9$. He pays the next-best value for item G among *the other buyers*. In a Vickrey auction, the key to truthful revelation of values is that the price a buyer pays is always independent of its bid. (By contrast, in sequential English auctions, buyer X would be uncertain whether to pursue the single item G or the bundle. In effect, he may end up bidding against himself.)

Now consider case 4 (case 2 is similar). Here, efficiency dictates separate sales – item F to Y and item G to Z. Buyer Y's payment is $P_F = 12 - 6 = 6$. (Buyer Y's presence shifts the allocation of the *other buyers* from $X_{FG} = 12$ to $Z_G = 6$.) Analogously, Buyer Z's payment is $P_G = 12 - 8 = 4$. Once again we see that Vickrey payments are not only item-specific, but also buyer-specific. By guaranteeing efficiency, the Vickrey auction solves the above-mentioned free-rider problem when it comes to separate sales. Unlike a sequential auction, a buyer of an individual item has no incentive to free ride on the higher bids of others. Instead, the buyer pays only the marginal cost that its purchase imposes on others. (This price can be quite low. For example, if Z_G were 9, instead of 6, P_F would have been $12 - 9 = 3$.)

By construction, the Vickrey payment for an individual item is at least as great as the second-best value for that item. However, the sum of the Vickrey payments will frequently fall short of the highest (or even the second-highest) bundled value. Indeed, this is true in case 4.

Empirical Evidence. Prior to the launch of the FCC spectrum auctions, several research teams tested a variety of auction methods for multiple items in controlled experiments. Plott (1997) compared the performance of two candidate procedures: 1) simultaneous English auctions for individual items (with a release provision) and 2) sequential English auctions preceded by an initial sealed-bid auction for the complete bundle of items. Overall, the results showed that simultaneous auctions achieved slightly higher efficiency marks (95%) than sequential auctions (92.5%). Buyers were moderately successful in assembling efficient bundles via the simultaneous individual auctions. The alternative procedure – holding a sealed-bid auction for the bundle – created a significant bias toward bundled sales, even when such sales were inefficient. In short, the exposure problem proved to be less serious than the free-rider problem.

Ledyard, Porter, and Rangel (1997) tested these same two methods as well as a combinatorial auction under a variety of demand conditions. The combinatorial auction did not use Vickrey payments. Instead, it allowed subjects to place ascending bids for single items or for any combinations of items they chose. At each stage the highest standing bids were announced. In addition, there was provision for a "standby" list of bids that were not large enough to displace a current winning bid (but which might become winning if raised in unison). This provided information to bidders to help them coordinate and overcome the free-rider problem. These experiments confirmed the efficiency advantage of simultaneous auctions versus sequential ones. In addition, in "hard" environments (where an efficient allocation depended upon a complex "meshing" of bidders' non-additive preferences), the combinatorial auction significantly outperformed the simultaneous auction. Allowing bids for packages led to increased efficiency and revenue and eliminated the exposure problem (thereby preventing bidder losses). Earlier experiments reported by Bykowsky, Cull, and Ledyard (1995) document these same performance comparisons.

Finally, the experiments of Isaac and James (1998) make a strong case for the feasibility of combinatorial auctions using Vickrey payments. In a series of controlled experiments involving two goods (either sold separately or bundled), the Vickrey mechanism achieved an average 96% efficiency rating. Interestingly, subjects submitted bids "close" to true values only 50% of the time. Nonetheless, deviations from their dominant strategies led to relatively small efficiency losses. (In a comparison set of experiments that excluded bundled bidding, efficiency fell to 91%.) The authors conclude that despite its added complexity, combinatorial bidding is crucial to promote efficiency.

Early assessments of the spectrum auctions indicate that conducting simultaneous auctions for individual licenses has worked well. Cramton (1997) and Ausubel et al (1997) document that bidders in the early spectrum auctions had local synergies for sets of licenses and frequently achieved desired combinations. Cramton notes that similar licenses sold for similar prices. The fact that the license auctions were open simultaneously (i.e. did not close sequentially) allowed buyers to modify the focus of their bids as current prices changed. However, the auctions were time consuming, lasting from as short as a week to as long as three months to complete. Despite the complex environment and large number of licenses, the process of license-by-license bidding went smoothly (though slowly). However, it is still an open question whether provisions for combinatorial bidding would further enhance efficiency and government revenues.

3. Conclusion

This chapter has reviewed many of the main findings concerning auctions in theory and practice. Nonetheless, a number of important topics have been omitted. Here, the focus has been exclusively on one-sided auctions, either for sales or procurements. Double auctions (where buyers and sellers both place bids) are of obvious practical importance and, beginning with the seminal work of Wilson (1985), have received increasing theoretical attention.

The possibility of bidder collusion is an important theoretical and practical concern. (The present review, like much of the literature, has held fast to the presumption of non-cooperative behavior by bidders.) The following chapter by Marshall and Maurer looks closely at the theory and evidence concerning bidder collusion. In particular, it demonstrates that collusion is much more likely to be a concern in the English auction than in the sealed-bid auction. Robinson (1985) and McAfee and McMillan (1992) are other useful references.

Transaction costs are a potentially important aspect of auctions (as they are for any other means of exchange). Potential buyers frequently incur real costs of entering the auction or of placing multiple bids. Bid preparation costs are a significant factor in competitive procurements. Similarly, in a tender-offer battle, would-be acquirers face significant costs – in assessing the target company's value and in implementing bids. Because many of the spectrum auctions spanned months, bidding costs were also significant. Bidding costs have a number of implications. First, bid costs provide a rationale for preemptive bids and jump bids. In an English auction, a high-value buyer makes a jump bid to signal its willingness to bid higher and, therefore, deter rivals from bidding. Avery (1998) demonstrates equilibria involving jump bidding.

Second, including bid costs alters the usual measures of auction performance. Samuelson (1985) extends the symmetric IPV model by including a common bid-preparation cost for all buyers. After observing its private value, each buyer submits a bid if and only if its expected profit net of bid cost is non-negative. Two main results emerge. First, any of the standard auctions are value maximizing as long as the seller sets its reserve price at its private value, $r = v_0$. With a "truthful" reserve price, equilibrium entry ensures social optimality. Low-value buyers are naturally screened out of the auction. (As always, revenue maximization dictates setting a reserve, $r > v_0$, at the expense of efficiency.) Second, with bidding costs, increasing the number of potential buyers (N) no longer guarantees increased value or increased seller revenues. When bidding is costly and buyers make independent bid decisions, there is an obvious coordination problem. In particular circumstances, too many or too few buyers may enter the bidding. In addition, increasing the number of potential buyers deters individual entry. If buyer value differences are small ($F(v)$ has a small variance), an increase in N can significantly increase coordination costs. In this instance, increased competition can reduce expected value (and expected revenue). Thus, a surprising result emerges. Under certain conditions, the seller can increase its expected value by reducing the allowable number of competitors. French and McCormick (1984) and Levin and Smith (1994) examine related models of costly entry. Wang (1993) compares the performance of auctions versus posted prices when auctioning is costly.

Many important auction settings involve multi-dimensional preferences, therefore, prompting the possibility of multi-dimensional bids. For instance, in a competitive procurement, contractor selection typically depends on a number of factors: bid price, quality measures, timetable of delivery and so on. Samuelson (1983) proposed that the auctioneer should announce a scoring rule (a function of the multi-dimensional bids) to select the winning bid. Recently, Zheng (2000) and Branco (1997) have examined optimal scoring rules in multi-dimensional settings. A related feature of procurement settings is that the winning bidder typically operates under an incentive contract, whereby the bidder's compensation depends on its *ex post* performance. As Riley (1988) first emphasized, when the winner's private value or cost (or a noisy estimate of such) can be observed *ex post*, the auctioning party can improve its payoff (and reduce the winning bidder's informational rent) by basing compensation upon this observed information. In the procurement setting, this implies interesting contract tradeoffs. For instance, moving in the direction of cost-plus contracts has the twin positive effects of reducing bidder informational rents and bid premiums (if bidders are risk averse). But, it has the twin negative effects of making it more difficult to identify the most efficient bidder (since bids are less linked to private values) and raises the specter of moral hazard (by blunting bidders' efforts to improve values or lower their costs). Englebrecht-Wiggins, Shubik, and Stark (1983), Laffont and Tirole (1987), and Samuelson (1988) provide useful analyses of these procurement issues.

With the advent of electronic commerce, the sweep of auctions is growing ever wider. The explosive growth of EBay, the internet auction company that brings thousands of buyers and sellers together each day, is one example. The planned implementation of initial public offerings of stock via the internet is another. While the traditional auction forms – the English and sealed-bid auctions – have been the focus of much attention and analysis, computerized exchange allows a new and wider realm of auction methods.

References

Ashenfelter, O. (1989), "How Auctions Work for Wine and Art," *Journal of Economic Perspectives* 3, 23-36.

Ausubel, L.M. (1997), "An Efficient Ascending Bid Auction for Multiple Items, University of Maryland, Working Paper 97-06.

Ausubel, L.M. and P.C. Cramton (1998), "Auctioning Securities, University of Maryland, mimeo.

Ausubel, L.M. and P.C. Cramton (1998b), "Demand Reduction and Inefficiency in Multi-Unit Auctions, University of Maryland, mimeo.

Ausubel, L.M. et al (1997), "Synergies in Wireless Telephony: Evidence from the Broadband PCS Auctions, *Journal of Economics and Management Strategy* 6, 497-527.

Avery, C. (1998), "Strategic Jump Bidding in English Auctions," *Review of Economic Studies* 65, 185-210.

Ayers, I. And P. Cramton (1996), "Deficit Reduction Through Diversity: How Affirmative Action at the FCC Increased Auction Competition," *Stanford Law Review* 48, 761-815.

Bartolini, L. and C. Cottarelli (1997), "Designing Effective Auctions for Treasury Securities," *Current Issues in Economics and Finance* 3, Federal Reserve Bank of New York.

Bazerman, M.H. and W.F Samuelson (1983), "I Won the Auction but I Don't Want the Prize," *Journal of Conflict Resolution* 27, 618-634.

Blecherman, B. and C.F. Camerer (1996), "Is There a Winner's Curse in the Market for Baseball Players? Evidence from the Field," California Institute of Technology, Working paper 966.

Branco, F. (1997), "The Design of Multidimensional Auctions," *Rand Journal of Economics* 28, 63-81.

Brannam, L., J.D. Klein, and L.W. Weiss (1987), "The Price Effects of Increased Competition in Auction Markets," *Review of Economics and Statistics* 69, 24-32.

Bulow, J. and P. Klemperer (1996), "Auctions Versus Negotiations," *American Economic Review* 86, 180-194.

Bulow, J. and J. Roberts (1989), "The Simple Economics of Optimal Auctions," *Journal of Political Economy* 97, 1060-1090.

Bykowsky, M.M., R.J. Cull, and J.O. Ledyard (1995), "Mutually Destructive Bidding: The FCC Auction Design Problem," California Institute of Technology, Working paper 916.

Capen, E.C., R.V. Clapp, and W.M. Campbell (1971), "Competitive Bidding in High-Risk Situations," *Journal of Petroleum Technology* 23, 641-53.

Cassady, R. (1967), *Auctions and Auctioneering*, University of California Press, Los Angeles.

Corns, A. and A. Schotter (1999), "Can Affirmative Action be Cost Effective? An Experimental Examination of Price-Preference Auctions," *American Economic Review* 91, 291-304.

Cox, J.C., V.L. Smith, and J.M. Walker (1984), "Theory and Behavior of Multiple Unit Discriminative Auctions," *Journal of Finance* 39, 983-1010.

Cox, J.C., V.L. Smith, and J.M. Walker (1985), "Expected Revenue in Discriminative and Uniform Sealed-bid Auctions," in V.L. Smith (Ed.) *Research in Experimental Economics* Volume 3.

Cramton, P.C. (1997), "The FCC Spectrum Auctions: An Early Assessment," *Journal of Economics and Management Strategy* 6, 431-496.

Cramton, P.C. (1995), "Money Out of Thin Air: The Nationwide Narrowband PCS Auction, *Journal of Economics and Management Strategy* 4, 267-343.

Dyer, D. and J.H. Kagel (1996), "Bidding in Common Value Auctions: How the Commercial Construction Industry Corrects for the Winner's Curse," *Management Science* 42, 1463-1475.

Dyer, D. and J.H. Kagel and D. Levin(1989), "A Comparison of Naïve and Experienced Bidders in Common Value Offer Auctions: A Laboratory Analysis," *Economic Journal* 99, 108-115.

Englebrecht-Wiggans, R. and C. Kahn (1998), "Multi-unit Auctions with Uniform Prices," *Economic Theory* 12, 227-258.

Englebrecht-Wiggans, R., M. Shubik, and R.M. Stark (1983), *Auctions, Bidding, and Contracting: Uses and Theory*, New York University Press, New York.

French, K.R. and R.E. McCormick (1984), "Sealed Bids, Sunk Costs, and the Process of Competition," *Journal of Business* 57, 417-41.

Gaver, K.M. and J.L. Zimmerman (1977), "An Analysis of Competitive Bidding on BART Contracts," *Journal of Business* 50, 279-295.

Hansen, R.G. (1986), "Sealed-Bid versus Open Auctions: The Empirical Evidence," *Economic Inquiry* 24, 125-142.

Hendricks, K., R.H. Porter, and B. Boudreau (1987), "Information, Returns, and Bidding Behavior in OCS Auctions: 1954-1969," *Journal of Industrial Economics* 35, 517-542.

Isaac, R.M. and D. James (1998), "Robustness of the Incentive Compatible Combinatorial Auction," University of Arizona, mimeo.

Kagel, J.H. (1995), "Auctions: A Survey of Experimental Research," in J.H. Kagel and A.E. Roth (eds), *Handbook of Experimental Economics*, Princeton University Press, Princeton, NJ, 501-586.

Kagel, J.H. and D. Levin (2000), "Behavior in Multi-Unit Demand Auctions: Experiments with Uniform-Price and Dynamic Auctions," *Econometrica,* forthcoming.

Kagel, J.H. and D. Levin (1986), "The Winner's Curse and Public Information in Common Value Auctions," *American Economic Review* 76, 894-920.

Katzman, B. (1996), "Multi-Unit Auctions with Incomplete Information," Working Paper, University of Miami.

Klemperer, P.D. (1999), "Auction Theory: A Guide to the Literature," *Journal of Economic Surveys*, 13.

Klemperer, P.D. (1999b), *The Economic Theory of Auctions*, Edward Elgar, Cheltenham UK.

Laffont, J-J. and J. Tirole (1987), "Auctioning Incentive Contracts," *Journal of Political Economy* 95, 921-937.

Ledyard, J., D. Porter, and A. Rangel (1997), "Experiments Testing Multiobject Allocation Mechanisms," *Journal of Economics and Management Strategy* 6, 639-675.

Levin, D., J.H. Kagel, and J-F. Richard (1996), "Revenue Effects and Information Processing in English Common Value Auctions," *American Economic Review* 86, 442-460.

Levin, D. and J.L. Smith (1994), "Equilibrium in Auctions with Entry," *American Economic Review*, 84 585-599.

Li, H. and J.G. Riley (1999), "Auction Choice," University of California, Los Angeles, mimeo.

List, J.A. and D. Lucking-Reiley (2000), "Demand Reduction in Multiunit Auctions: Evidence from a Sportscard Field Experiment," *American Economic Review* 90, 961-972.

Lucking-Reiley, D. (1997), "Magic on the Internet: Evidence from Field Experiments on Reserve Prices in Auctions:," Vanderbilt University, mimeo.

Lucking-Reiley, D. (1999), "Using Field Experiments to Test Equivalence between Auction Formats: Magic on the Internet," *American Economic Review* 89, 1063-1080.

Malvey, P.F., C.M. Archibald, and S.T. Flynn (1996), "Uniform-Price Auctions: Evaluation of the Treasury's Experience," Working Paper, U.S. Treasury.

Maskin, E. and J. Riley (1999), "Asymmetric Auctions," *Review of Economic Studies* 67, 413-438.

Maskin, E. and J. Riley (1989), "Optimal Multi-Unit Auctions," in F. Hahn (ed), *The Economics of Missing Markets, Information, and Games*, Oxford University Press, Oxford, 312- 335.

McAfee, R.P. and J. McMillan (1996), "Analyzing the Airwaves Auction," *Journal of Economic Perspectives* 10, 159-175.

McAfee, R.P. and J. McMillan (1987), "Auctions and Bidding," *Journal of Economic Literature* 25, 699-738.

McAfee, R.P. and J. McMillan (1992), "Bidding Rings," *American Economic Review* 82, 579-599.

McMillan, J. (1994), "Selling Spectrum Rights," *Journal of Economic Perspectives* 8, 145-162.

Milgrom, P.R. (1986), "Auction Theory," in: T. Bewley, *Advances in Economic Theory*, Cambridge University Press, Cambridge.

Milgrom, P.R. (1989), "Auctions and Bidding: a Primer," *Journal of Economic Perspectives* 3, 3-22.

Milgrom, P.R. and R.J. Weber, (1982), "A Theory of Auctions and Competitive Bidding," *Econometrica* 50, 1089-1122.

Myerson, R.B. (1981), "Optimal Auction Design," *Mathematics of Operations Research* 6, 58-73.

Plott, C.R. (1997), "Laboratory Experimental Testbeds: Application to the PCS Auction," *Journal of Economics and Management Strategy* 6, 605-638..

Riley, J.G. (1988), "Ex Post Information in Auctions," *Review of Economic Studies* 55, 409-430.

Riley, J.G. and W.F. Samuelson (1981), "Optimal Auctions," *American Economic Review* 71, 381-92.

Robinson, M.S. (1985), "Collusion and the Choice of Auction," *RAND Journal of Economics* 16, 141-145.

Roll, R. (1986), "The Hubris Hypothesis of Corporate Takeovers," *Journal of Business* 59, 197-216.

Samuelson, W.F. (1988), "Bidding for Contracts," *Management Science* 71, 381-392.

Samuelson, W.F. (1985), "Competitive Bidding with Entry Costs," *Economic Letters* 71, 381-92.

Samuelson, W.F. (1983), "Competitive Bidding for Defense Contracts," in R. Englebrecht-Wiggans et al, *Auctions, Bidding, and Contracting: Uses and Theory*, New York University Press, New York.

Smith, V.L. (1995), "Auctions," in J. Eatwell, M. Milgate, and P. Newman (Eds.), *Allocation, Information, and Markets, The New Palgrave*, W.W. Norton & Co., New York.

Thaler, R.H. (1988), "The Winner's Curse," *Journal of Economic Perspectives* 2, 191-202.

Vickrey, W (1961), "Counterspeculation, Auctions, and Competitive Sealed Tenders," *Journal of Finance* 16, 8-37.

Wang, R. (1993), "Auctions Versus Posted Posted-Price Selling," *American Economic Review* 83, 838-851.

Weber, R.J. (1983), "Multiple-Object Auctions," in R. Englebrecht-Wiggans et al, *Auctions, Bidding, and Contracting: Uses and Theory*, New York University Press, New York.

Weber, R.J. (1997), "Making More from Less: Strategic Demand Reduction in the FCC Spectrum Auctions," *Journal of Economics and Management Strategy* 6, 529-548.

Wilson, R. (1977), "A Bidding Model of Perfect Competition," *Review of Economic Studies* 44, 511-518.

Wilson, R. (1985), "Incentive Efficiency of Double Auctions," *Econometrica* 53, 1101-15.

Wilson, R. (1987), "Auction Theory," in J. Eatwell, M. Milgate, and P. Newman, *The New Pelgrave: A Dictionary of Economic Theory*, MacMillan, London.

Wilson, R. (1995), "Bidding," in J. Eatwell, M. Milgate, and P. Newman (Eds.), *Allocation, Information, and Markets, The New Palgrave*, W.W. Norton & Co., New York.

Zheng, C.Z. (2000), "Optimal Auctions in a Multidimensional World," Northwestern University, mimeo.

Notes

[1] Indeed, the major auction houses increase participation by allowing buyers who cannot be present at the auction to submit phone or mail bids in advance. The submission authorizes the house to bid on the buyer's behalf. A buyer submitting a $1,000 bid might win the public auction when the house makes the last bid of $800 on its behalf. For the mail bidder, the institution is effectively a second-price auction. The mail bidder's dominant strategy is to place a bid equal to its true reservation price. Indeed, if all bidders agreed to submit their bids in advance, a virtual English auction could be implemented simply by opening the bids and awarding the item to the highest bidder at the second-highest bid.

[2] More generally, the expected value of the kth order statistic of N independent draws from a uniform distribution on the interval [0, 1] is given by $(N+1-k)/(N+1)$.

[3] The empirical evidence concerning the strategic equivalence of the sealed-bid and Dutch auctions is mixed. A series of laboratory experiments summarized in Kagel (1995) find higher bids in the sealed-bid format than in the Dutch format (where the latter bids approximate the Nash equilibrium prediction). By contrast, Lucking-Reiley's (1999) field experiments involving internet auctions of some $2,000 worth of trading cards found higher Dutch revenues (due in part to greater numbers of bidders). In a separate series of auctions, Lucking-Reiley found equivalent revenues for the English and second-price auctions.

[4] Unlike monetary experiments, performance in these classroom exercises contributes to a portion of the student's final grade. As a consequence, MBA students take them very seriously and have strong performance incentives. Classroom experiments can also generate considerably more data than their monetary counterparts. In the auction experiments, each student submits a complete bidding strategy. From these strategies, it is straightforward to derive the class's empirical bid distribution. Thus, the student's payoff is determined by pitting the submitted strategy against this distribution (and integrating over the range of the student's possible reservation prices). As an added advantage, the task of maximizing this overall average payoff induces risk-neutral preferences on the part of subjects.

[5] The case for the hybrid auction depends on the precise pattern of elevated bidding. For instance, if buyers are risk averse (with coefficient of relative risk aversion r), then the common equilibrium bidding strategy is of the form: $b = [(N-1)/(N-r)]v$, when values are independently uniformly distributed. In this case, a single sealed-bid auction among the N bidders – rather than a hybrid auction – is revenue maximizing.

[6] To prove precise theoretical results, researchers have invoked an "ideal" version of the English auction described by Cassady (1967). To begin all buyers are active. Then, the auctioneer continuously raises the price (by an electronic price "clock") with bidders signaling withdrawal by releasing a button. Withdrawn bidders cannot reenter the bidding. The winning bid and bidder are determined when only one buyer remains active. In this electronic English auction, buyers have perfect information at all times about dropout prices and the number of remaining bidders. Obviously, informal English auctions provide buyers only partial information about these elements. Accordingly, the English bidding behavior implicit in Proposition Two should be understood to hold only approximately for actual English auctions.

[7] The revenue comparisons can also be understood by invoking the "linkage principle" (Milgrom and Weber, 1982). As in the IPV setting, the greater the extent of private information a buyer holds, then the greater is its expected profit. (As an extreme case, if all bidders held the same information, they would all place equilibrium bids equal to the expected value of the item, and all would earn zero expected profit.) In the sealed-bid auction, the winning bid depends only on the bidder's own estimate. Thus, relative to the other auctions, the player holds the most private information and earns the greatest expected profit. By contrast, in the English auction, a buyer's bid depends not only on its own estimate but also on the estimates of the other buyers who drop out. Because bidder estimates are positively correlated, the winning bid increases more steeply with the buyer's own estimate, reducing its expected profit. Consequently, the seller obtains greater expected revenue from the English auction than the sealed-bid auction.

[8] Note the intuition behind assuming that all other active buyers have the *same value* as the bidder's own estimate. This establishes this limit price as a break-even bid. The buyer earns a zero profit if it wins the item and is forced to pay exactly this bid. Dropping out at a lower price means possibly forgoing a

positive profit. Dropping out at a higher price is also detrimental. Waiting to drop out at a higher price makes a difference only when it means winning the item (instead of being outbid) in which case the buyer ends up overpaying. Equilibrium behavior in the second –price auction is similar (except that there is no dropout information). Each buyer's bid is set at the expected value of the item, conditional on the highest of the other buyers' estimates being equal to the bidder's own estimate.

[9] Perhaps due to the absence of closed-form solutions, there have been few research efforts to date to measure these revenue differences. This should change with the development of numerical solutions for bidding strategies as in Li and Riley (1999).

[10] Define $J_i(v_i) \equiv v_i - (1-F_i(v_i))/f_i(v_i)$, where J_i is assumed to be increasing, and F_i and f_i are the cumulative distribution function and density function of buyer i. As Myerson (1981) shows, an optimal system of bidder preferences awards the item to buyer i if and only if $J_i(v_i) > Max(J_k(v_k), v_0)$ for all $k \neq i$. In words, the winning bidder must beat all other buyers' J's and the seller's reservation value v_0. If the buyer distributions are identical, all buyers are treated equally, and the winning bid must exceed the seller's reserve price given by: $r = J^{-1}(v_0)$. With asymmetric distributions, some bidders receive preferences. In the present example, $J_W = 2v_W - 60$ for the weak buyer, $J_S = 2v_S - 120$ for the strong buyer, and $v_0 = 0$. Thus, the strong buyer wins the item if and only if $v_S > Max(v_W + 30, r = 60)$, implying a 30 unit preference for the weak buyer. Intuitively, the optimal preference bolsters the effective competition facing the strong buyer but does not fully level the playing field.

[11] Lucking-Reiley (1997) conducts a fascinating field experiment by auctioning real goods (Magic cards) using different reserve prices via the Internet. The empirical findings accord with the predictions of auction theory. Imposing reserve prices in sealed-bid auctions reduces the number of bidders and increases the frequency of goods going unsold. A more subtle result is that buyers raised their entire bidding schedules in response to increased reserve prices. Reserve prices increased seller revenue on units that were sold. Over all goods (those sold and unsold), expected revenue peaked at moderate levels of reserve prices.

[12] McMillan (1994) discusses the disastrous revenue experience of New Zealand's spectrum auction, caused by the government's failure to set appropriate reserve prices in the face of few bidders.

[13] However, if 13 had been bid by buyer 3, the winning bidders would have paid the same prices, 13 and 11.

[14] Indeed, many proponents of the uniform-price auction, including Nobel Prize winning economists Milton Friedman and Merton Miller, have claimed incorrectly that it guarantees truthful bidding. It is true that a small buyer – one that has a negligible chance of being pivotal, i.e. setting the common price -- should place bids very close to its true values in the uniform-price auction. However, there is considerable evidence of high buyer concentration in treasury auctions. As few as five large primary dealers frequently account for 50% of the winning bids.

[15] Relative to sequential auctions, the simultaneous auction has two other advantages. First, continuous price revelation should mitigate the winner's curse and lead to elevated seller revenues (Proposition Two). Second, the simultaneous auction ensures that identical items sell for identical prices. By contrast, a famous sequential auction – the 1981 sale of seven identical licenses for the use of RCA's communication satellite -- generated bids ranging from $14.4 million (in the first auction) to $10.7 million (in the sixth auction). Because of the widely differing prices, the FCC later nullified the auction as "unjustly discriminatory."

[16] If bidders value both individual items and bundles, truthful bidding no longer constitutes a dominant strategy. Suppose a bidder anticipates winning the bundle at a small profit by bidding truthfully. He might instead lower his bundle bid in order instead to win a single good (for which he is high bidder) at a greater profit.

[17] Note that a bidder no longer has a dominant strategy (bidding up to one's value if necessary) in sequential English auctions. The high bidder at an early auction might seek to continue to raise the bidding (even after the last competing bidder has dropped out) in order to better compete against the (uncertain) best bids forthcoming at the subsequent auctions.

11 THE ECONOMICS OF AUCTIONS AND BIDDER COLLUSION

Robert C. Marshall and Michael J. Meurer

1. INTRODUCTION

Auctions and procurements are pervasive mechanisms of exchange.[1] Most government acquisitions are competitvely procured (see Kelman 1990). In the private sector, a large number of commodities are sold by auctions such as antiques, art, rugs, and used machinery. The assets of bankrupt businesses are typically liquidated by means of auction. The federal government is the biggest auctioneer in the country. Offshore oil leases as well as timber from national forests are sold by means of auction. But the largest of all auctions are those for government securities. The Treasury sells over $2.5 trillion of bills, notes, and bonds by means of auction every year to refinance debt and finance the deficit.

Auctions and procurements are popular despite their vulnerability to bidder collusion. Criminal and civil enforcement of the antitrust laws has deterred price-fixing in some market settings – but not bidder collusion. A spate of cases in the 1980's and more recent high-profile cases serve as a reminder that the success of anti-collusive policies is limited in auction and procurement markets.[2]

In a posted price market when sellers join together to fix prices, society suffers an efficiency loss from reduced output in the market. Output falls since the colluding firms raise the market price in a quest for monopoly profits. However, when bidders collude in an auction market, the efficiency effect is not so clear. One reason is that private information is significant in typical auction markets but not posted price markets. In posted price markets transactions usually involve products with salient features that are easy to discern. In many bidding markets the transaction involves an informationally complex product that requires significant training and expertise to properly assess. Examples include offshore oil tracts and weapons contracts. Costly and dispersed information changes the nature of competition as compared to posted price markets. Horizontal agreements between bidders affect the distribution of information which in turn affects the expected winning bid, the profits to the parties, and the incentives of the parties to gather information in advance of an auction or procurement. It seems reasonable that informational issues should play a prominent role in assessing the illegality of bidder collusion.

The ability of the auctioneer to combat collusion is a second factor that distinguishes posted price from auction markets. In posted price markets, the victims of collusion are small and powerless to stop it. In contrast, an auctioneer often has market power and can react strategically to bidder collusion. There is a possibility that the countervailing power of the auctioneer may mitigate the adverse effects of collusion. A parallel argument is used to justify labor unions – when employers have monopsony power the consequent deadweight loss can be reduced by allowing workers to organize and bargain as a unit.

In order to analyze efficiency issues we discuss basic theoretical structures through which one can think about bidder collusion. We show that bidding rings at oral auctions are more resistant to cheating by cartel members than price-fixing agreements in other markets. Conventional wisdom holds that the greatest obstacle to collusion (besides illegality) is the incentive of colluders to cheat by cutting their prices below the cartel price.[3] Although this wisdom holds at a sealed bid auction it fails to hold at an oral ascending bid auction. At a sealed bid auction the designated winning bidder in the ring, the member with the highest valuation, must shade his bid below his valuation. Since he pays what he bids, shading is the source of the collusive gain. This action leaves the designated winner vulnerable to a cheater who can submit a bid slightly above the collusive bid and win the item. To deter cheating, potential cheaters must each get a large share of the collusive gain which makes collusion problematic. At an oral auction the designated ring bidder follows the same strategy that he would as a non-cooperative bidder. Since there is no shading he is not vulnerable to cheating. The gain to collusive bidding comes from the strategic behavior of co-conspirators who would have lost by acting as non-collusive bidders. These bidders can potentially affect the price paid by suppressing their bids at the auction.

Besides cheating, cartels must contend with the problem of entry. Monopoly profits are likely to attract entrants who erode the profits and destabilize the cartel. OPEC's experience illustrates the normal situation. New suppliers have entered the petroleum market attracted by high collusive prices.[4] Entrants have captured some of OPEC's customers at prices near the collusive price. Over time OPEC's share of the market and ability to maintain high prices have declined. Entry into an oral auction in which a bidding ring is active is no more profitable than entry to an oral auction that has no collusion. A potential entrant knows that the ring will respond to entry. The representative of the ring at the auction will remain active against an outsider up to the highest valuation of any ring member. This is exactly the same as non-cooperative bidding. Consequently, an entrant can earn no profit in excess of what he could earn if all bidders acted non-cooperatively.[5] The only hope for an entrant to share the gains from collusion is to force his way into the ring.

Auction and procurement markets differ from posted price markets with regard to limited collusion. Cartels in posted price markets strive to be inclusive. Since posted price markets usually feature relatively homogeneous goods an excluded firm destabilizes a cartel by undercutting the cartel's price and capturing a large market share. While an excluded bidder can harm a ring by reducing its collusive gain at an auction the threat is less severe for specific kinds of auctions. A ring may exclude certain bidders because they do not add much to the collusive gain. The stability of less than all inclusive collusion at certain kinds of auctions is one more reason why auctions are particularly vulnerable to collusion. This difference from posted price markets is also relevant to the issue of antitrust injury.

In the following section we present an analysis of bidder behavior, both non-cooperative and collusive, for different auction schemes. We then discuss the efficiency consequences of bidder collusion.

2. NON-COOPERATIVE BIDDING AND BIDDER COLLUSION

An auction is a mechanism of exchange whereby a seller, following a simple set of procedural rules, evaluates the simultaneous offers of potential buyers to determine a winner and payments for bidders to make. A procurement is the flip side of an auction – a buyer evaluates the simultaneous offers of potential sellers.[6] For simplicity, the following discussion will be presented in terms of auctions. Also, for simplicity, we initially assume the auctioneer does not take strategic actions, such as setting a reserve price.[7]

For the sale of a single item, there are four standard auction schemes. At the **English, or oral ascending bid** auction bidders appear before the auctioneer and, through open outcry, raise the bid price until no bidder remains who is willing to pay a higher price. Then the item is sold to the highest bidder at the amount of their last bid. This scheme is used for liquidation auctions, the sale of timber by the Forest Service, the sale of art, antiques, rugs, industrial machinery and many other items. A variant has been used by the FCC for the sale of frequency spectra.

At a **Dutch, or oral descending bid auction** bidders appear before the auctioneer who starts by asking a very high price for the item. When no one "takes" the item she progressively drops the price. The first bidder to stop the price descent by open outcry wins the item for the price at which he stopped the bidding. This is a common mechanism used in western Europe for the sale of vegetables, flowers, and other foodstuff.

At a **first price sealed bid** auction bidders submit sealed bids to the auctioneer. The highest bidder wins the auction and pays the amount of his bid to the auctioneer. First price mechanisms are used for most procurements. The Forest Service and Bureau of Land Management use them on occasion for the sale of timber (most of the sales are by English auction). The Mineral Management Service uses them for the sale of offshore oil lease tracts. The Dutch and first price auctions are strategically equivalent in certain modelling environments.[8] In subsequent analysis we focus only on the first price.

At a **second price auction** bidders submit sealed bids to the auctioneer. The highest bidder wins and pays the amount of the second highest bid. This mechanism is rarely used in practice. However, it is enormously important as an analytic device. In many modelling environments it is equivalent to the English auction. In both auctions, under reasonable assumptions, the price paid is exactly equal to the second highest valuation among all bidders.

Multiple objects can be by schemes that are extensions of the single object first price and second price auctions. At a **discriminatory** auction bidders submit sealed bids that specify how many units of the commodity they are willing to buy at a specific price. These bids are aggregated by the auctioneer. The auctioneer determines the highest bid price at which the last item available will be sold (we call this price the "market clearing price"). Then all bidders bidding that amount or more are allocated items. Like a first price auction, they each pay the auctioneer the amount of their bids.[9]

At a **uniform price** auction bidders submit sealed bids that specify how many units of the commodity they are willing to buy at a specific price. These bids are aggregated by the auctioneer. She determines the highest bid price at which the last item available will be sold. Then all bidders bidding that amount or more are allocated items. They pay the auctioneer the amount of the market clearing price. The Treasury uses both discriminatory and uniform price auctions for the sale of bills, notes, and bonds.

Most economic analysis of auctions is presented in models with incomplete information.[10] This means that bidders possess private information – at least one bidder knows something that another bidder does not know. Often the private information concerns a bidder's personal valuation of an item or, in the case of procurements, a bidder's performance cost.

A benchmark informational framework is called the **independent private values** (IPV) model. As the name implies bidders obtain individual-specific values from some common underlying probability distribution. Knowledge of one's own valuation provides no useful information about what another bidder holds as a valuation (this is the independence). Each bidder's valuation is known only to them (this is the privacy).

Clearly this formulation does not encompass some fairly common auction settings. Consider the sale of offshore oil lease tracts. If the pool of reserves, the quality of the oil, and the cost of extraction were known to all bidders then each bidder would have the same valuation for the tract. In practice, this common underlying value is not known, but each individual bidder obtains private information about the salient unknown characteristics. The modelling framework in which bidders obtain conditionally independent signals about the true underlying value of the item being sold is called the **common value** (CV) model. A variant of the CV model allows for the possibility that some bidders have better information about the true value than other bidders.

There are probably no pure IPV or CV settings, but the models are extremely helpful in analyzing bidding behavior. The IPV model is used, for example, in procurement settings in which seller costs have a random and independent component. It is also used to analyze auctions for collectibles such as art or antiques. The CV model is used in procurement settings in which the common cost of performance is uncertain. It is also used to analyze auctions for many kinds of natural resources.

In the subsequent analysis we assume that bidders have complete information (no bidder holds private information) whenever we can. Although this is less realistic it makes the analysis much easier to follow. Some issues are essentially about private information. When we analyze incomplete information models the reader should be mindful of the auction format and informational environment.

2.1 BIDDER COLLUSION AT A SINGLE OBJECT AUCTION

We begin with the most basic modeling framework from the theory of auctions. A single non-divisible object is to be sold to one of several risk-neutral bidders. First we explain non-cooperative behavior at English and first price auctions. Second, we address the susceptibility of each auction scheme to bidder collusion. Third, we discuss the anti-collusive effect of entry by new bidders.

2.1.1 Non-Cooperative Behavior. A remarkable result established independently by a number of authors in the early 1980s is called the Revenue Equivalence Theorem.[11] It holds that in the basic auction model the first price and English auctions generate, on average, exactly the same revenue for the seller. This result is surprising because the strategic behavior of bidders is so different at the two auctions. First consider an English auction. When should a bidder withdraw from the bidding? A bidder should withdraw when the price reaches his valuation (presuming arbitrarily small bid increments). Any alternative behavior is less profitable. Bidding in excess of his valuation is foolish because he may win at a price above his valuation. On the other hand, withdrawing below his valuation offers no benefit and may cause a bidder to lose an item he otherwise would have won. This logic is independent of the strategic behavior of any other bidder. In the language of game theory, it is a dominant strategy for bidders to remain active up to their true valuations. The auctioneer's revenue from this auction is therefore equal to the magnitude of the second highest valuation.

What does strategic behavior look like for a first price auction? A specific example is helpful. Suppose there are five bidders who have valuations of 5, 4, 3, 2, and 1. Bidders know the valuation held by others as well as their own. Clearly, no bidder will bid his valuation for the item. If such a bid wins it will leave the winner with no surplus from the auction since the winner pays the amount of his bid. So each bidder will have an incentive to bid some increment below his valuation. In deciding how much to shade his bid below his valuation a bidder optimally trades off the reduction in probability of winning from reducing his bid with the increase in surplus that comes from winning with a lower bid. In equilibrium, all bidders shade their bids below their valuations in a way that is mutually consistent – namely, no one wants to change their strategic behavior in light of how all others are behaving. In our example, bidder 5 wins the first price auction with a bid of 4.[12] Thus the revenue is the same as at the English auction.

Although the two auctions produce the same expected revenue for the seller the strategic behavior of the bidders is very different. At the first price auction each bidder shades his bid. The surplus of the winner is strictly determined by the difference between his valuation and the amount of his bid. At the English auction each bidder remains active up to his valuation. The surplus of the winner is determined by the difference between his valuation and the valuation of the runner-up.

At the risk of seeming redundant we need to emphasize the difference in the source of the winner's surplus for the two auctions. For the first price auction, the magnitude of the winner's surplus depends strictly on the actions and characteristics of the winner. For the English auction, the magnitude of the winner's surplus depends on the characteristics of the winner, and actions of another bidder, the runner-up. This is a critically important distinction for understanding the susceptibility of each auction scheme to collusion.

2.1.2   Collusion. Bid-rigging cases catalogue varied and subtle methods of collusion. Colluding bidders specify rules of communication within the ring, bidding behavior, the ultimate allocation of commodities obtained by the ring amongst ring members, and any payments between members.[13] In contrast, we examine a rather austere model of collusion. We prefer an austere model because it focuses one's attention on factors crucial to ring stability.[14]

The stability of a bidding ring depends on at least three factors. Participation must be individually rational. In other words, bidders must voluntarily opt to be members of the ring (individual rationality constraint). Next, bidders cannot have an incentive to cheat. Specifically, the ring must be designed so that bidders do not pretend to participate in the ring and then disregard the directions of the ring, for example, by using surrogate bidders at the main auction. We call this the "no-cheat" constraint. Finally, bidders must be truthful in reporting private information to one another (incentive compatibility constraint). We discuss the first two constraints but not the third. Our initial modeling environments are ones of complete information so the question of how information revelation occurs within rings does not arise. When we turn to incomplete information we suppose that truthful revelation occurs. When appropriate we will comment on the effect of the incentive compatibility constraint on collusion.

How might bidders at an English auction organize themselves to bid collusively? We continue with our five bidder example. Our first question is what grouping of bidders can effectively collude. Consider collusion between any number of bidders but where the ring does not contain both bidders 4 and 5. In this circumstance the ring cannot realize a gain to collusion. We refer to such rings as "ineffective". A ring that contains bidder 5 but not bidder 4 will pay 4 for the item. A coalition that does not contain bidder 5 will never prevail as a winner.

Next consider the minimally effective ring of two bidders – bidders 5 and 4. Collusion can operate as follows. Bidder 5 bids until he wins the item as long as the latest bid is less than or equal to 5.[15] Bidder 4 is silent. In exchange, bidder 4 receives a sidepayment from bidder 5. The individual rationality constraints are satisfied because bidders 4 and 5 both get a higher profit in the ring than from non-cooperative bidding. Bidder 4 profits by the amount of the sidepayment. Bidder 5 profits by obtaining the item for a lower price, 3 instead of 4. But what about cheating? Does bidder 4 have any incentive to deviate? Remarkably, the answer is no! The logic stems from the strategic actions of bidder 5. Bidder 5 does nothing differently when colluding than when acting non-cooperatively. Bidder 4 cannot win the item at a price less than 5, and so cannot profitably cheat. This logic applies to rings of any size for the English auction.

Continuing with our example we ask how collusion might be organized at a first price auction.[16] Again, a ring that does not consist of bidders 5 and 4 cannot possibly be profitable. Therefore, we consider the minimally effective coalition of bidders 4 and 5. Clearly, there is a potential collusive gain. Bidder 4 could bid strictly less than 3 and then bidder 5 could win the item for a bid of 3.[17] This produces a collusive gain of one. But what compensation does bidder 4 require to submit a bid below 3? Suppose bidder 5 offers to pay bidder 4 an amount of .5 for bidder 4 to bid an amount less than 3. Then bidder 4 will face the following tradeoff – accept .5 and not win the object or act as if the sidepayment is acceptable but then bid slightly above 3 at the main auction. By beating bidder 5 with such a bid bidder 4 would obtain a payoff close to one, instead of just .5. So, what is the smallest sidepayment that will dissuade bidder 4 from cheating on the collusive agreement? Bidder 4 must be paid at least what he could obtain by cheating – a payment of 1. But such a payment means that bidder 5 does not benefit at all from the collusion – he pays 3 for the object, makes a sidepayment of 1 to bidder 4, and earns a surplus of 1. This is exactly what he could have achieved from non-cooperative behavior. If organizing collusion is at all costly, or if there is any potential penalty associated with collusion, then bidder 5 will prefer non-cooperative behavior to collusion.[18]

Surprisingly, there are still no gains to collusion for bidder 5 at a first price auction if the ring expands in size. In fact, the problem of deterring cheating becomes worse. Consider the ring with bidders 5, 4, and 3. It could potentially reduce the price paid for the item to 2. However, when bidder 5 bids 2 (or just above 2) bidders 4 and 3 will see the possibility of a profitable unilateral deviation from the collusive agreement. To dissuade bidder 3 from bidding in excess of 2 he must receive a payment of 1. Bidder 4 must receive a payment of 2. Note that bidder 5 strictly prefers non-cooperative behavior to making these sidepayments. The reason is that the marginal cost associated with securing a reduction in the price paid from 4 to 3 is still 1 but the marginal cost of securing a reduction in the price paid from 3 to 2 is not 1 but 2. Both bidders 3 and 4 must be compensated in order not to deviate when the bid is suppressed from 3 to 2.

Collusion is not sustainable at a first price auction because it is too costly for the highest valuation bidder to create incentives that stop pivotal ring bidders from cheating.[19] This is not a problem at an English auction.[20] Why is there such a difference between the two schemes? In order to secure a collusive gain at a first price auction the high valued bidder must decrease his bid relative to what he was bidding non-cooperatively. As his bid falls the door swings open for deviant behavior by complementary bidders. To secure a collusive gain at an English auction the high valued bidder acts exactly as he would non-cooperatively. The gain comes from the suppression of bids by complementary bidders. The fact that the high valued bidder remains active up to his valuation implies that there is no opportunity for profitable deviant behavior by his conspirators.[21]

2.1.3 Entry. To study entry and collusion at English auctions consider a model in which there is a ring with $n - 1$ members and one potential entrant. Suppose that all bidders have the identical valuation of V for the single item to be auctioned. Further suppose that the cost of entry is K, and that the auctioneer does not fix a minimal acceptable bid (known as a reserve price). We fix the parameter values so that $nK > V > K$. If there is no threat of entry then ring members share equally in the collusive gain of V. A potential entrant sees the possibility of capturing V/n if admitted to the ring. Since this profit is less than the cost of entry, the outside firm would not enter the market to join the ring. But what about entering the market to bid non-cooperatively? This cannot be profitable either. At an English auction, the ring will bid up to its valuation, V, and keep the entrant from getting any profit at all.[22]

Now consider the same situation with a first price auction. Our previous analysis showed that collusion would not be profitable at a one-shot first price auction. Thus the impact of entry on collusive profits is a moot point. However, for the purpose of comparison, we assume that a ring of $n-1$ bidders is stable and can win the item for a price of zero. As was true of the English auction, the potential entrant cannot profitably enter to join the cartel because the entry costs are too high compared to an equal share of collusive profits. Unlike the English auction, the new firm can make a profit gross of entry cost from noncooperative behavior. If the ring always submits the fully collusive sealed bid of zero, then the entrant can bid slightly above zero and always take the item at a profit of $V - K$. To guard against this possibility, the ring will sometimes submit a positive bid ranging all the way up to $V - K$. This strategy is optimal because it discourages (but does not completely deter) entry. The ring balances the increased purchase price against the increased probability of winning the item given the threat of entry. The shift of auction format from English to first price reduces the ring's expected profit from V to K.[23]

We note a feature of this story. With an English auction, potential entrants see large positive profits being earned by incumbents but realize that they cannot profitably enter. This stands in sharp contrast to the usual notion in microeconomics that positive profits will attract entry until there are no more profits for the market participants.[24]

2.2 MULTIPLE UNITS FOR SALE

We now consider the sale of multiple objects. We study a simple model in which two objects will be sold, simultaneously, by means of auction. We continue to assume that there are five bidders with valuations 5, 4, 3, 2, or 1 for a single object. None of the bidders places any value on additional objects. We compare the discriminatory and uniform price auctions.

2.2.1 Non-cooperative Behavior.
We begin with non-cooperative behavior. At a uniform price auction the two winners each pay the highest losing bid. All noncooperative bidders will submit bids exactly equal to their valuations. Relative to truthful bidding, bidding in excess of one's valuation is unprofitable. The only difference it could produce is winning an object at a price in excess of its value. Bidding below one's

valuation would not change the price paid upon winning and may result in losing an item that would have yielded positive surplus. Consequently, each bidder reports their valuation truthfully. Just like the English auction we have a dominant strategy equilibrium. In our example, the winning bids are 5 and 4 and each of the winners pays 3.

Like the first price auction, bidders shade their bids below their valuations at a discriminatory auction. The winning bids at the discriminatory auction are both 3. These bids are submitted by the bidders with valuations of 5 and 4.[25] They both pay what they bid. The revenue equivalence theorem still holds in this setting. The revenue from non-cooperative bidding is 6 at either type of auction.

2.2.2 Collusion.
Collusion is relatively easy to achieve at the uniform auction (like the English auction). In the five bidder uniform price example the ring $\{5,4,3,2,1\}$ wins two items at a price of zero to gain a surplus of 9. Bidders 5 and 4 each bid their valuations. The other ring bidders all bid zero in exchange for sidepayments from bidders 5 and 4. The no cheat constraint is satisfied for any positive side-payments, because cheating cannot bring a positive profit to bidders 1, 2, or 3. Just as with the English auction, the high valuation ring members are protected from cheating by others because they do not shade their bids.

Rings that are not all-inclusive are less profitable, but limited membership is common in bidding rings. In practice, membership is not all-inclusive for a number of reasons (i.e. the ring wants to decrease the probability of detection, some bidders have no interest in committing a felony, and ring members do not want to share information with a large group). The following subsets of bidders constitute profitable and sustainable rings at the uniform price auction -- $\{5,4,3\}$, $\{5,4,3,2\}$, $\{5,4,2\}$, $\{5,4,2,1\}$, $\{5,3\}$, $\{5,3,2\}$, $\{5,3,2,1\}$, $\{4,3\}$, $\{4,3,2\}$, and $\{4,3,2,1\}$. The subsets $\{5,4,3,1\}$, $\{5,3,1\}$, and $\{4,3,1\}$ are stable, but they are excluded from the list because bidder 1 could not contribute anything. Subsets that include bidder 3 achieve a collusive gain by depressing the amount paid below the non-cooperative price of 3. Obtaining bidder 3's voluntary participation in the coalition potentially comes at a small cost since bidder 3 would have earned zero surplus by bidding non-cooperatively. Since bidders 4 and 5 will always bid their true valuations, bidder 3 will never be in a position to win an object by cheating on the collusive agreement. The same is true for bidders 2 and 1. Any sidepayment to suppress their bids would be acceptable and there would be no chance for profitable deviation.

Subsets that do not contain bidder 3 are sometimes not profitable.[26] Exceptions are the subsets $\{5,4,2\}$ and $\{5,4,2,1\}$. These rings are profitable if bidder 4 bids zero. If bidders 5 and 4 both try to win an item they each pay 3. If bidders 4 and 2 (and possibly bidder 1) depress their bids, then bidder 5 can win an item for a price of one (zero). The reduction in purchase price more than compensates for the loss of the item by bidder 4. Bidder 4 can be induced to bid zero with an adequate sidepayment from bidder 5.

In contrast to the uniform price auction, at the discriminatory auction the all-inclusive ring cannot achieve winning bids of zero and still satisfy both the no-cheat and individual rationality constraints. Bidders 3, 2, and 1 could all cheat by submitting bids just above the zero bids submitted by 5 and 4. The total profit from cheating is 6; so the smallest aggregate sidepayment sufficient to stop cheating is 6. This payment is so large that bidders 5 and 4 would oppose a winning bid of zero by the ring. They would favor more limited collusion that required smaller side-payments and created less of a problem with cheating.

One profitable ring at the discriminatory auction is $\{5, 4, 3\}$. Bidders 5 and 4 submit bids of 2. Bidder 3 does not bid and receives an aggregate payment from 5 and 4 of at least 1. The no cheat constraint is satisfied for bidder 3 who could earn at most 1 by cheating with a bid just above the bid of 5 or 4. The individual rationality constraints are satisfied for bidders 5 and 4 because their total gain over the non-cooperative outcome is 2 compared to a payment of 1.

If the ring is enlarged to $\{5, 4, 3, 2\}$ no additional gain is possible. The winning bids cannot be pushed below 2 without violating one of the constraints. For example, if the ring attempted to achieve winning bids of 1.5 for bidders 5 and 4, then bidder 3 could get 1.5 from cheating and bidder 2 could get .5 from cheating. The total sidepayments made by bidders 5 and 4 would have to be increased from 1 to 2. This increased side-payment cost of 1 exactly equals the total collusive gain of 1 from the reduced purchase price. If adding bidder 2 is costly to bidders 5 and 4 (*e.g.*, from increased probability of detection), then they would oppose expansion of the ring. Whenever the ring attempts to depress the winning bid into the range from 1 to 2, the cost in deterring cheating exactly matches the gain in reduced purchase price.[27]

Two conclusions emerge from this analysis. First, a large number of coalitions are profitable and sustainable at the uniform price auction whereas only a single limited coalition is profitable and sustainable at the discriminatory auction. Second and relatedly, the robustness of the one-shot, single item, first price auction to collusion does not fully extend to the discriminatory auction.

2.3 PRIVATE INFORMATION

Until now we have assumed that bidders know each others' valuations. We now tackle the question of whether the existence of incomplete information creates an incentive for bidders to collude. We show that certain common value information environments stimulate collusion. There is no commentary on the linkage between private information and collusion in the antitrust literature on price-fixing. The reason is that private information is often relatively unimportant in posted price markets and the posted price model guides thinking about price-fixing.

2.3.1 Common Values Auctions and the Winner's Curse.

We use the auction market for off-shore oil tract leases to explain the effect of private information about a common value on bidding and the incentive to collude (see Hendricks & Porter (1988)). There are two types of offshore oil tract leases let via auction by the federal government: wildcat tracts and drainage tracts. A wildcat tract is remote from other tracts so there is little evidence on oil deposits besides seismic analysis. A drainage tract neighbors a tract that is being successfully tapped. The rate of production on the neighboring tract provides information that is useful in estimating the magnitude of reserves under the drainage tract. We will discuss the influence of information on optimal bidding in each of these types of auctions to explain the "winner's curse" and incentives for collusion. We will start with wildcat tract auctions where bidders are likely to be symmetrically informed.

The term "winner's curse" was coined to explain the poor performance of oil companies in the early days of offshore oil tract auctions. A dominant factor in determining a bid submitted by an oil company is the estimate provided by the company's geologists about the expected amount of oil that can be removed from the tract. There is substantial variability associated with the geologic estimates. Naturally, the non-cooperative equilibrium bids increase as the geologic estimate of oil reserves increases. In the early days, the oil companies' bids were too aggressive, and winners tended to regret their acquisition.[28] This was because the bidders did not account for the fact that the event of winning the auction was informative. The fact that bidder A won the auction was probably attributable to the fact that A's geologists provided the most favorable or most optimistic estimate of oil reserves. The most optimistic estimate from a large number of estimates is almost surely not the most accurate estimate. Thus the winner was cursed with a tract that was less valuable than they estimated.[29] Eventually, the bidders learned to shade their bids down to adjust for the winner's curse, and now they make a normal expected rate of return on wildcat tracts.

2.3.2   Collusion to Preserve an Informational Advantage.
We shift our attention to auctions involving drainage tracts. At wildcat auctions bidders are likely to hold similar information about the value of the oil reserves. At drainage tract auctions there is likely to be informational asymmetries. Firms that hold leases at neighboring tracts are likely to be better informed about the reserves available under a drainage tract (see Reese (1978)). For modelling purposes we assume that neighboring bidders are completely informed about the magnitude of the reserves. We consider a model with two bidders – one informed and one who is uncertain about the tract's value.[30] Instead of analyzing a first price auction we will analyze a second price auction. We make this choice for analytic convenience. Assume that the object is worth one of three values $V_1 < V_2 < V_3$. Bidder 1 knows the true value, but bidder 2 only knows the probabilities of the different possible realizations. This is all common knowledge amongst the bidders.

In equilibrium bidder 1 simply bids the true value of the item, and bidder 2 bids V_1.[31] Bidder 1 wins for a price of V_1 when the true value is either V_2 or V_3, while if the value is V_1 then one of the bidders is arbitrarily chosen to win at a price of V_1. Bidder 2 makes zero profit, but the informational advantage of bidder 1 leads to positive expected profit. This result reflects a general phenomenon. A bidder with strictly worse information than some other bidder cannot make positive expected returns (all else equal), and a bidder with information unavailable to any other bidder can make positive expected returns.[32]

When bidders are asymmetrically informed as in the case of drainage tract auctions, some interesting effects arise from collusion. In our example bidder 1 has no interest in colluding with bidder 2 given a reserve set by the seller of V_1. Furthermore, the addition of a third bidder with the same information as bidder 2 would not affect bidding in the noncooperative setting, and would provide no incentive for any of the parties to collude.[33] In contrast, the addition of a third bidder with the same information as bidder 1 would lead to an equilibrium in which all three bidders always get zero profit. Bidders 1 and 3 would both bid the true value, and bidder 2 would bid V_1. Bidders 1 and 3 could recover their informational rents by colluding. Bidder 1 would bid the true value and bidder 3 would bid V_1 and receive a sidepayment from bidder 1. In response bidder 2 would still bid V_1. Thus, collusion restores the informational rents for bidders 1 and 3 that non-cooperative behavior totally dissipates.[34]

2.3.3 Collusion to Overcome an Informational Disadvantage.
Our next model shows that asymmetric information can also provide an incentive for collusion among the less informed bidders. Suppose the object for sale consists of two distinct components, but it is sold as one unit at a second price auction. Each component has a value of zero with probability π and a value V with probability $1 - \pi$. These value realizations are independent, so the item is worth either 0 or V or $2V$. There are three bidders. Bidder 1 observes the value of both components and, thus, bidder 1 is completely informed. Bidder 2 observes only the value of component A while bidder 3 sees only the value of component B.

Non-cooperative bidding in this context results in bidder 1 bidding the value of the object. Bidder 2 bids the value he observes for component A while bidder 3 bids the value he observes for component B.[35] Bidder 1 always wins. When the item is worth nothing bidder 1 wins for a price of zero. When the object is worth V or $2V$ bidder 1 wins for a price of V. Bidder 1 gets positive expected profit while bidders 2 and 3 get zero profits.[36]

Now suppose bidders 2 and 3 collude. They share information and consequently know with certainty the underlying common value.[37] Both the coalition and bidder 1 will bid the common value. All bidders get zero profit.[38] This example is of critical importance. Collusion increases the revenue to the auctioneer! If the supply is elastic then output rises and collusion is socially beneficial. Collusion is socially beneficial here because the less informed bidders share information which eliminates the bid shading that would otherwise occur to prevent winner's curse. Even though there are effectively fewer bidders (two rather than three), the average bids are higher when the curse is removed.[39]

2.3.4 Disclosure of Information by the Auctioneer.
Besides the use of information by bidders we also want to comment on the provision of information by the auctioneer or procurer. In the offshore oil lease auctions, the federal government provides seismic information to prospective bidders. In contrast, at timber auctions in the Pacific Northwest the federal government withholds information about the quality of timber on neighboring tracts (see Baldwin, Marshall, and Richard (1997)). The government should provide all relevant information to bidders in a CV setting because it eases the effect of the winner's curse and thereby allows bidders to bid more aggressively. The end result is a higher expected winning bid in auctions, and lower expected winning bid in procurements. Furthermore, the government release of information may be particularly helpful in cases in which there are asymmetri-

cally informed bidders. Auction revenues may increase dramatically if the government puts the less informed bidders on the same footing as better informed bidders. Also, the government provision of good seismic information, for example, can help avoid some of the costs of duplicative seismic studies that individual bidders would conduct. In an IPV setting the issue does not arise since the auctioneer would not have any private information to disclose.

3. THE EFFICIENCY EFFECTS OF BIDDER COLLUSION

In all of antitrust law the per se rule is most entrenched in the area of horizontal price fixing. Application of the per se doctrine signals a consensus that horizontal price fixing almost always restricts output and causes social harm.

One factor that contributes to cartel stability is the ease of detecting deviations from agreed prices. Detection is easier in markets with homogeneous goods. The manufacturing of sanitary pottery often yielded defective, but merchantable products, called seconds. The cartel required manufacturers to destroy all seconds. The motivation for this policy was that the sale of seconds would offer manufacturers the chance to offer price discounts, that were larger than the reduction in value due to the defect. This is a means of chiseling on the price set by the cartel. Since the extent of the defect, and implied reduction in value, could vary considerably, the problem of detecting "excessive" discounts would be enormous. Thus, for the sake of cartel stability, the cartel destroyed merchantable output. Essentially the same argument applies to collusion by buyers to post prices below the competitive price. In both cases the social loss is attributable to the decline in the quantity transacted. The social harm caused by monopsony and collusion among buyers is output restriction, just as in the case of monopoly or collusion among sellers. In an auction context, the traditional view holds that bidder collusion depresses seller revenue. In turn, marginal sellers see the depressed revenue and choose not to bring their items to the market. This quantity restriction implies a deadweight loss that the antitrust laws are supposed to deter and correct.[40]

The traditional analysis is too crude for reasonable application to most auction markets. It certainly applies to a fresh fish auction or the sealed bid procurement of sewer pipe. Collusion in these markets affects output by reducing the returns to fishing and ultimately the supply of fresh fish, or by increasing the cost of sewer pipe and possibly jeopardizing governmental demand for new water treatment programs. In these markets there are many auctioneers and procurers. The products are homogeneous and there are no subtle informational problems to thwart competitive forces. But in most auction and procurement markets the market power of auctioneers or procurers, and the scarcity of information, may dampen the effects of competition. Competition by bidders does not always lead to an efficient outcome. Specifically, colluding bidders may bring countervailing power to bear on an auctioneer or procurer who also has market power. Further, collusion may be the only means of protecting the rents that flow from investments that raise the value of a transaction in an auction or procurement market. Thus we have come to question whether the per se rule is appropriate for all bid-rigging cases.[41] We provide standards to determine whether bid-rigging should be characterized as price-fixing and per se illegal or as a horizontal restraint that is subject to rule of reason analysis.

3.1 COLLUSION AS COUNTERVAILING POWER

The term countervailing power was first used by Galbraith to describe his vision of the typical market in a modern economy (see Galbraith (1952)). He intended to highlight the departure from the competitive model that could be seen in many markets. Instead of a large number of price-taking buyers and sellers, there were a small number of powerful buyers and sellers. He claimed that efficiency losses associated with monopoly power would be diminished over time as buyers organized and gained countervailing monopsony power. Galbraith's views have languished for many years, but his notion of countervailing power seems quite apt in many auction and procurement markets.[42]

There are two key concepts in our countervailing power story – market power and bargaining power. If there is a single seller of a commodity, or a small number of sellers, or if most of the commodity is provided by a very few suppliers then there is significant market power on the supply side. If there is a single buyer, or very few buyers, or if buyers have cartelized then there is significant market power on the demand side. Bargaining power is a different concept. If sellers (buyers) can credibly commit to a pricing institution, for example by declaring a take-it or leave-it price, then sellers (buyers) have bargaining power. To illustrate, suppose a monopoly seller has a value of 0 for the single unit they have available for sale while the sole buyer in the market has a value of 1 for the unit. Suppose these values are common knowledge. If the seller has all the bargaining power then the buyer will pay 1 for the unit. If the buyer has all the bargaining power then the seller will receive 0 for the unit. If the bargaining power is shared equally then the item will be sold for a price of 1/2.

Reaching the efficient output in a market depends on who has market and bargaining power and how they use it. If all power rests in the hands of a monopoly seller, that seller will normally use the power to restrict output and achieve monopoly profit.[43] Likewise, a monopsonist with all market and bargaining power will restrict output inefficiently. Although market power and bargaining power are separate concepts, it is intuitive to think that market power engenders bargaining power. For example, if a monopolist faces many small buyers then it would seem unreasonable to think of these buyers calling out a take-it or leave-it offer to the monopolist. It is most natural to think of the monopolist as credibly committing to a price. However, microeconomists do not have a theoretical construct which describes how bargaining power endogenously evolves from market power. Of necessity therefore, our comments here are heuristic.

As a starting point, we compare unionization by workers in a monopsonistic labor market with an auction market in which colluding bidders face a monopolistic seller. Facing individual workers a monopsonist will call out a profit maximizing wage below the competitive wage. Compared to a competitive outcome, too few workers will be employed. When workers unionize it is reasonable to think that they can call out a minimally acceptable wage, or at least bargain to a wage above the monopsony wage where potentially more workers will be employed and deadweight loss will be reduced. But this presumes that bargaining power has shifted – if the monopsonist retained all bargaining power after unionization then the monopsonist would simply call out the same wage that he called out to the non-unionized workers.

There is widespread acceptance of the notion that unionization raises employment and improves efficiency in monopsonistic labor markets. This attitude has never been transplanted to the field of bidder collusion. But the analogy is close. Most auctioneers have some degree of market power. Items sold at auction are often highly differentiated. In certain auction markets, fine art for example, there are few sellers. In certain procurement markets, automobiles for example, there are few buyers. It is rare to have a pure monopolist auctioneer, but it is also rare to have a company town with a pure monopsonist employer. After all, workers are mobile and can retrain themselves for alternative occupations.

We do see two significant distinctions between bidder collusion and worker unionization. First, unions are legal, bid rings are not. Intuitively, a monopsonistic employer would forfeit much less bargaining power to an illegal cartel of workers than she would to a legal one (especially in a regime that banned permanent replacement workers). A bid ring is constrained in its operations by fear of detection. Its bargaining position must be compromised to limit the disclosure of information that may provide enforcement authorities with verifiable information regarding the existence of the ring. The seller decides how many units to bring to the market. The seller decides upon a minimally acceptable price for units sold. An all-inclusive ring can test the commitment power of an auctioneer by withholding all bids and waiting to see if the auctioneer will offer the items again at a lower reserve price. But a ring cannot typically enter negotiations about the quantity or quality of items for sale at the auction.

Second, the mechanism of exchange differs between the labor market and auctions. Non-unionized, blue collar and clerical workers participate in a posted price labor market. The wage rate and benefit package is offered on a take-it or leave-it basis. Unionization changes the wage-setting mechanism into a bilateral negotiation. Non-cooperative bidders participate in an auction. If bidder collusion were legalized, the auction would also be likely be transformed into a negotiation. The difference in the starting points reflects the informational differences between the two settings. A seller chooses an auction rather than posting a price because she is not well informed about the likely equilibrium price. The informational disadvantage of auctioneers would adversely affect their bargaining power in a bilateral negotiation.

The upshot from these observations is that bidding rings probably are less powerful than unions since they must lurk in the shadows to avoid antitrust prosecution. Rings, like unions, may have a desirable effect on efficiency. As bargaining and market power shift to the union or the ring, the quantity brought to market may increase. The following example illustrates this effect.

Consider a market in which the auctioneer can choose to bring either one or two items to the market. Suppose that there are three bidders who desire a single item and have valuations of 5, 3, and 1. The bidders know each others' valuations, and the auctioneer knows these three valuations are present, but not which bidder has which valuation. The method of auction is not important here, but to be concrete we suppose that a uniform price (highest rejected bid) sealed bid auction is used. If two items are sold, then the two highest bidders win an item and they each pay the third highest price. If one item is sold, then the highest bidder wins and pays the second highest bid. To start we assume that the bidders behave non-cooperatively. Then if one item is sold, the winner bids 5 and pays 3. If two items are sold, the winners bid 5 and 3, and they each pay 1 for the items. The auctioneer will of course choose to sell only one item (even if the second item has no value to her) because the revenue is higher. Now consider the case in which the two high value bidders collude, and this collusion is known to the auctioneer. If one item is sold, the highest valuation bidder bids 5, the second highest valuation bidder suppresses his bid to 1 or less, and the third bidder who is not in the ring bids 1. The ring takes the item at a price of 1. If two items are sold, then the two high valuation bidders take the items at a price of 1. Thus, the auctioneer will offer two items for sale if her valuation of a retained item is less than 1. Comparison of the two cases shows that collusion can increase output.[44] Although output rises,[45] revenue to the auctioneer falls, thus she has reason to complain about the collusion, but the gains to the colluding bidders outweigh the losses to the auctioneer.[46]

Besides adjusting quantity, there are a variety of other bargaining tactics that an auctioneer can use to combat bidder collusion.

i. Entry fees. These are relatively rare. Perhaps the purchase of a booklet which describes the items to be sold could be viewed as an entry fee.

ii. Reserve Prices. These are very common at both government and private sales and procurements.

iii. Quantity restrictions. The Mineral Management Service does not sell all feasible Gulf drilling tracts at one time. This would not be revenue maximizing.

iv. Ex ante denial of joint venture status. With very rare exception joint ventures are not approved for Forest Service Timber Sales but, on the other hand, are frequently approved for offshore oil lease bidding.

A secret reserve may remove the possibility of tacit collusion in which ring members bid the reserve. See McAfee & McMillan, (1992). If the auctioneer cannot prove bidder collusion in the courts, she may resort to self-help remedies that disrupt a suspected ring. For example, the auctioneer might retain an item or award it to a non-ring member even though the ring would be willing to pay more. Either of these tactics creates an ex post inefficiency assuming that some ring member had the highest valuation and resale is costly. In contrast to the quantity adjustment example, in the following examples collusion leads to less efficient outcomes.

In the private sector, auctioneers may attempt to combat bidder rings at English auctions by using a "quick knock." When a quick knock is used, the auctioneer ignores the attempts of the ring to raise the current high bid, and awards (or knocks) the item to a non-ring member. This strategy is only effective when the auctioneer knows who the ring members are. Further, the auctioneer must expect that the bidders in the ring will attend future auctions. The quick knock is only worthwhile if it disrupts the ring, and the short run loss is outweighed by the long run gain from more competitive bidding in future auctions. Although the quick knock may be profitable to the auctioneer, it is inefficient because the highest value bidder might be in the ring. If resale is costly, then the award to a non-ring member is socially costly.

An alternative to the quick knock is provided by a protecting bidder. When using this tactic, the auctioneer instructs the protecting bidder to raise the prevailing coalition bid (perhaps above what the protecting bidder would pay of his own accord) in an attempt to elicit a higher counterbid from the coalition. Sometimes, the coalition withdraws from the bidding leaving the protecting bidder with the item. The protecting bidder and auctioneer will typically have agreed upon some discounted price for items awarded to the protecting bidder in this way. As was true with the quick knock, the use of a protecting bidder leads to inefficiency when resale is costly, and the protecting bidder wins an item but some ring member has a higher valuation.

Frequently, the auctioneer acts as her own protecting bidder. She does this by announcing a reserve price, which means that the auctioneer retains the item if no bid exceeds the reserve.[47] Reserve policies are often used in procurements as well, in which case the reserve price sets the maximum acceptable bid. If an auctioneer suspects, but cannot prove collusion, it may be optimal for her to raise the reserve price above what it would be if all bidders acted non-cooperatively.[48] The reserve price compensates to some extent for the lack of competition between the bidders. The increase in the reserve is inefficient because a higher reserve implies a higher probability of retention by the auctioneer.[49]

The preceding discussion shows that bidder collusion may cause a pro-competitive increase in output. The gist of the argument is that market power held by the auctioneer is countered with market power in the hands of the colluding bidders. The countervailing power argument has a lot of intuitive appeal, but we have shown that when the auctioneer retains some bargaining power collusion may exacerbate inefficiencies. It seems sensible to consider relaxing the per se rule against bidder collusion. If the auctioneer (or procurer) has significant market power then the rule of reason would allow colludng bidders the opportunity to demonstrate whether the effect of collusion is likely to be an increase in expected output and efficiency.

3.2 PRE-AUCTION INVESTMENTS AND BIDDER COLLUSION

A second theory that justifies a more lenient attitude toward bid-rigging is based on investment incentives. Cooperative behavior by bidders may be socially desirable because it is effective in stimulating socially productive investments that would not be profitable in the absence of collusion. Competitive bidding diminishes investment incentives for two types of ex ante investments by bidders – (i) investments that produce socially wasteful information about the common value of the item at auction and (ii) investments that directly raise the private and social value of the item at auction.

Ex ante investment in information[50] is our main concern. Noncooperative bidders have a weak incentive to acquire information before an auction. When the item for sale has a common value component, a single bidder with superior information can earn an informational rent. But that rent disappears if any other bidder acquires the same information. In markets like antiques and used machinery, dealers invest in learning the market value of items, and the preferences of specific retail customers. With offshore oil exploration and timber, firms make investments specific to a particular tract that improves their information about that tract. Such investments are greatly affected by the prospect of using the information to profit at the auction. If firms anticipate that the best informed bidders will collude at the auction, then there is a strong incentive to make ex ante investments to become well informed.[51] If the firms anticipate noncooperative bidding, or collusion by less informed bidders, then this incentive is muted. The problem with the noncooperative outcome is that buyers do not have any property right at the time of their investment. There is still some incentive as each firm hopes that it is the only one whose investment successfully yields relevant information, or that it is the only firm that makes the investment necessary to gain certain information.

Whether the extra investment in information production induced by collusion is socially desirable is unclear. Just because a bidder is willing to make a costly investment in information gathering does not mean that the information is socially valuable. Bidders may be eager to obtain information that does nothing more than improve the precision of their estimate of the value of an item at auction. Such information has no social value. Eventually, the value of the item will be revealed to the winner regardless of whether they made an investment. But this information eases the winner's curse on a bidder and allows them to gain expected profits at the auction.[52] This scenario matches certain aspects of an oil tract lease auction. Oil companies each invest in geologic reports regarding the potential size of the oil pool and the cost of extraction.[53]

Many types of informational investments have positive private and social value. Bidders at natural resource auctions gather information that allows for more efficient extraction, harvesting, or processing. For example, by understanding the species and quality composition of timber in a given section of a forest a mill may be able to customize its production process to reduce the cost of converting logs into wood products. If more than one bidder makes such an investment then non-cooperative bidding dissipates the associated rents. Socially valuable investments would be dissuaded, unless the bidders were to collude and preserve the rent.

Dealers gather socially valuable information that facilitates their intermediary role. For example, dealers of used merchandise, such as used industrial machinery dealers, typically make significant ex ante investments in the development of an expertise. Some machinery dealers specialize in presses while others primarily handle specific kinds of saws. This expertise allows them to more quickly reallocate machinery from low valued users to high valued users. An implication of the expertise is that they know who the high valued users are for a given type of machine tool. When buying machine tools at auction two bidders with expertise in presses will bid away all rents to their expertise if they act non-cooperatively. Collusion averts this rent dissipation.

By now the reader may have some enthusiasm for the efficiency promoting aspects of bidder collusion. That enthusiasm should be tempered by two considerations. First, the social value of the extra investment must outweigh the social cost of redundant investment by different bidders, as well as the effect of the price distortion and other social costs caused by collusion. Second, there may be other alternatives for encouraging the ex ante information investment. Auctioneers who benefit from the investment can encourage it by permitting joint ventures or joint bidding. Joint bidding occurs at certain DoD procurements, offshore oil sales, and the current spectrum sales by the Federal Communications Commission. The fact that an auctioneer does not permit joint bidding should not be dispositive, though. It is possible that a socially valuable investment is of no particular value to an auctioneer. This is easiest to see when the investment has lasting value over a sequence of auctions, an auctioneer who runs one sale would not want to promote an investment that would mostly benefit the bidders and other auctioneers.

4. CONCLUDING REMARKS

The fundamental message of this article is that standard supply and demand analysis of cartel behavior is often deficient when applied to collusion by bidders at an auction or procurement. Supply and demand analysis is static, presumes perfect information is held by all participants, presumes there are many sellers and many buyers in the market, and presumes the commodity in question is homogeneous. In markets where auctions are used as allocation mechanisms there is often significant market power held by the sellers, the item being sold is highly heterogeneous (even within a given sale), and significant resources must be expended to understand the item being offered in order to formulate a bid.

Using a game theoretic approach we showed that collusion is more stable at oral ascending bid and uniform price auctions than at other auction formats. It is rather remarkable that the Forest Service continues to use oral ascending bid sales in spite of the suspicion of collusive bidding.[54] Perhaps the explanation is that the agency has been captured by the industry.[55] Our finding also makes us concerned about possible collusion at uniform price Treasury auctions and the ascending bid auctions run by the FCC for spectrum licenses.

Finally, we defined circumstances in which bid-rigging might not be inefficient. If the auctioneer has market power and uses that market power to restrict output then bidder collusion may produce countervailing market power that is socially beneficial. In addition, collusion may stimulate socially valuable ex ante investments in information gathering.

References

Amorosi, Ginsburg, and Gold, *Antitrust Violations*, 31 **American Criminal Law Journal**, 423 (1994).

J. Anton & D. Yao, *Split Award, Procurement, and Innovation* 20 **Rand Journal of Economics**. 538 (1989)

L. Baldwin, R. Marshall, and J-F. Richard, "Bidder Collusion at Forest Service Timber Sales", **Journal of Political Economy**, August 1997

S. Bikhchandani & C. Huang, *The Economics of Treasury Securities Markets*, 7, **Journal of Economic Perspectives** 117, 118 (1993).

Blair & Harrison, *Cooperative Buying, Monopsony Power, and Antitrust Policy*, 86 **Northwestern Law Review**. 331, 333-36 (1992)

E. Capen, R. Clapp, & W. Campbell, *Competitive Bidding in High-Risk Situations*, 23 **Journal of Petroleum Engineering**, 641 (1971)

D. Carlton amd J. Perloff, **Modern Industrial Organization,** Addison-Wesley, 217 (1990).

R. Cassady, **Auctions and Auctioneering,** U.C. Berkeley Press, 16-18 (1967).

V. Chari and R. Weber, *How the U.S. Treasury Should Auction Its Debt*, **Federal Reserve Bank of Minneapolis Quarterly Review** 3 Fall (1992).

R. Engelbrecht-Wiggans, P. Milgrom & R. Weber, *Competitive Bidding and Proprietary Information*, 11 **Journal of Mathematical Economics,** 161 (1983)

V. Fehl & W. Guth, *Internal and External Stability of Bidder Cartels in Auctions and Public Tenders*, 5 **International Journal of Industrial Organization,** 303 (1987)

L. Froeb, *Auctions and Antitrust*,U.S. Department of Justice, mimeo (1989).

J. K. Galbraith, **American Capitalism: The Concept of Countervailing Power**, M.E. Sharpe, (1952)

D. Gambetta **The Sicilian Mafia** 214-20 (1993)

GAO Report, *Changes in Antitrust Enforcement Policies and Activities of the Justice Department*, 4 vol. 59, no. 1495, December 7, 1990.

D. Graham & R. Marshall, *Collusive Bidder Behavior at Single Object Second Price and English Auctions*, 95 **Journal of Political Economy,** 1217 (1987)

D. Graham, R. Marshall, & J-F. Richard, *Differential Payments within a Bidder Coalition and the Shapley Value*, 80 **American Economic Review,** 493 (1990)

W. Guth & B. Peleg, *On Ring Formation in Auctions*, September 1993, mimeo.

Hendricks & Porter, *An Empirical Study of an Auction ith Asymmetric Information*, 78 **American Economic Review,** 865 (1988).

K. Hendricks & R. Porter, *Collusion in Auctions*, **Annales D'Economie et de Statistique** 217, 218 (1989)

J. Hirshleifer and J. Riley, **The Analytics of Information and Uncertainty**, Cambridge University Press, 373-75 (1993).

S. Kelman, **Procurement and Public Management,** American Enterprise Institute Press, 15-16 (1990).

B. LeBrun, **Asymmetry in Auctions**, Ph.D. Dissertation, Catholic University of Louvain (1991).

G. Mailath, & P. Zemsky, *Collusion in Second Price Auctions with Heterogeneous Bidders*, 3 **Games & Economic Behavior,** 467 (1991)

R. Marshall, M. Meurer, & J-F. Richard, *Litigation Settlement and Collusion*, 104 **Quarterly Journal of Economics,** 211 (1994)

R. Marshall, M. Meurer, J-F. Richard, & W. Stromquist, *Numerical Analysis of Asymmetric First Price Auctions*, 7 **Games & Economic Behavior,** 193 (1994)

E. Maskin & J. Riley, *Asymmetric Auctions* unpublished manuscript (1991).

P. McAfee & J. McMillan, *Auction Theory,* **Journal of Economic Literature,** . (1987).

P. McAfee & J. McMillan, *Bidding Rings*, 82 **American Economic Review,** 579 (1992)

P. Milgrom, *Auction Theory*, T. Bewley ed. **Advances in Economic Theory,** Cambridge University Press, (1987)

P. Milgrom & R. Weber, *A Theory of Auctions and Competitive Bidding*, 50 **Econometrica** 1089 (1982).

R. Myerson, *Optimal Auction Design*, 6 **Mathematics of Operations Research,** 58 (1981)

R. Porter & D. Zona, *Detection of Bid Rigging in Procurement Auctions* 101 **Journal of Political Economy,** 578 (1993)

R. Posner, *The Social Costs of Monopoly and Regulation*, 83 **Journal of Political Economy**, 807 (1975).

P. Reese, *Competitive Bidding for Offshore Petroleum Leases*, 9 **Bell Journal of Economics.** 369, 381 (1978)

J. Riley & W. Samuelson, *Optimal Auctions*, 71 **American Economic Review,** 381 (1981)

M. Robinson, *Collusion and Choice of Auction*, 16 **Rand Journal of Economics,** 141 (1985)

F. M. Scherer and D. Ross, **Industrial Market Structure and Economic Performance,** Houghton Mifflin, 331 (1990)

G. Stigler, *A Theory of Oligopoly*, 72 **Journal of Political Economy,** 44 (1964).

R. Sultan, **Pricing in the Electrical Oligopoly**, vol. 1, Harvard Business School Publishing, 38-39 (1974)

J. Tirole, *Collusion and the Theory of Organizations*, Institut D'Economie Industrielle, Working Paper (1992)

W. Vickrey, *Counterspeculation, Auctions, and Competitive Sealed Tenders*, 16 **Journal of Finance** 8 (1961).

Notes

1. The content of this work draws significantly from two earlier papers by the authors: *Bidder Collusion: A Basic Analysis of Some Fundamental Issues*, and, *Should Bid-Rigging Always be an Antitrust Violation?*.

2. See, Froeb (1989) – 81 percent of criminal cases under Sherman Section One from 1979 to 1988 were in auction markets. During that time period there were 245 bid-rigging or price fixing cases involving road construction, and 43 cases involving government procurement. See GAO Report, *(1990)*.

3. The susceptibility of cartels to secret price-cuts by members has played a prominent role in the analysis of cartel stability. See Stigler (1964).

4. See Carlton and Perloff at 246 (1990). (In June 1985 "non-OPEC production is 33% higher than in 1979, undercutting OPEC's prices.")

5. This argument applies to a single object oral auction. It sometimes does not apply to multi-object oral auctions. A coalition might find it optimal to let an entrant win the first item brought up for sale so as to eliminate this source of competition on later items. Note that in this scenario entry is successful only because the coalition allows it.

6. Procurements are typically far more difficult to analyze than auctions. Except when buying a homogeneous commodity, sellers will not only specify a price in their bid but will also specify the product they plan to provide. The products might differ significantly between firms. Then the buyer will need to score each firm's bid in order to rank firms by the surplus they are offering.

7. Of course, in practice, auctioneers use reserve prices, entrance fees, phantom bidding techniques and supply restrictions to raise the price paid for the item they are selling. To include the auctioneer as a player in the description of the games involves stating additional contingencies in the transactions that add little to one's understanding of the central issues regarding bidder behavior. Later, we explicitly discuss the strategic measures that an auctioneer might take to fight collusion.

8. See Vickrey (1961) as well as Milgrom & Weber (1982).

9. To complete the formal specification of these auction games we should specify what happens in the case of a tie. For a single object, a random allocation device (like a coin flip) determines the winner. For multiple objects, a "tie" occurs when the demand at the market clearing price exceeds the quantity available. Bidders at the market clearing price then receive items pro rata.

10. See Milgrom and Weber (1982), Riley and Samuelson (1981), Myerson (1981), McAfee and McMillan (1984).

11. See Myerson (1981), Riley and Samuelson (1981), and Milgrom and Weber (1982).

12. The reader may wonder whether bidder 4 can take the item with a bid of 4 and whether bidder 5 should bid slightly above 4. This possibility is just a technical nuisance that we could deal with in several different ways. One way is to invoke the notion of trembling hand perfect equilibrium which is a refinement of Nash equilibrium. A bid of 4 by bidder 4 always yields a profit of zero whether or not bidder 4 wins the item. A more sensible strategy is for bidder 4 to select a random bid slightly below 4 in the hope that bidder 5 mistakenly submits a bid (*i.e.* trembles) below 4. For the details of this argument see Hirshleifer & Riley *supra* note 16 at 374. The IPV counterpart to this example would have five bidders independently draw private valuations from the uniform distribution on the interval zero to six. Then each bidder would optimally submit a bid equal to (4/5) times their individual valuation realization. Of course, there is no guarantee that valuation realizations will be so accommodating as to produce 4 as the actual revenue. However, on average, the revenue will be 4.

13. A practical example would be a bid rotation scheme such as the electrical contractors conspiracy. (See Sultan (1974)). In that case bidders decided ex ante who would submit the winning bid at a given auction. There were no side payments. Another example is provided by collusion among antiques dealers. At the main auction no coalition member bids against another coalition member. If a member of the ring wins the item then it is the property of the coalition – ultimate ownership is determined in a secondary auction conducted by the coalition after the main auction. The difference between the winning bid at the secondary auction and the main auction is divided as sidepayments in some predetermined manner amongst members of the ring. See United States v. Ronald Pook, 1988 U.S. Dist. Lexis 3398 (E.D.Pa. 1988).

14. Microeconomic theory offers various models of mechanisms for bidder collusion. Mechanisms are discussed in: Graham & Marshall, (1987); Graham, Marshall, & Richard, (1990); Guth & Peleg, (1993); Mailath, & Zemsky, (1991); Marshall, Meurer, & Richard, (1994); Marshall, Meurer, Richard, & Stromquist, (1994); McAfee & McMillan, (1992); Tirole, (1992).

15. Bidder 5 would drop out of the auction if the bidding ever went above 5, but this cannot happen in equilibrium.

16. In our example the bidders know the valuations of other bidders. If they do not, then we cannot identify a collusive mechanism that produces a payoff in excess of non-cooperative behavior. Specifically, there is no such mechanism identified in the literature. However, a bounding argument has been numerically constructed which shows that the best possible payoff for a coalition at a first price auction is never larger than the payoff to a coalition at an English auction. See Marshall, Meurer, Richard and Stromquist (1994).

17. Actually bidder 5 would bid slightly above 3 to beat bidder 3. When this technical issue arises later in the Article we will treat it the same way.

18. The situation is no different if bidder 5 chooses another bid such as 3.5. Bidder 4 could earn .5 by cheating, and a sidepayment of .5 to prevent cheating would wipe out 5's profit from collusion.

19. Factors outside the scope of our discussion can promote collusion at a one-shot first-price auction. See, e.g., Marshall, Meurer, & Richard, (1994 – litigation settlement between colluding bidders); Porter & Zona, (1993 – labor union enforced collusion); Anton & Yao, (1989 – split award procurement).

20. The greater susceptibility of English auctions to collusion has been discussed by Robinson, (1985); and Fehl & Guth, (1987).

21. We have not yet commented on cheating by bidder 5. He has no interest in cheating in terms of his bid, but he has a strong incentive to cheat the ring by reneging on his promised sidepayments. This problem affects first price and English auctions equally. In practice, rings can overcome this problem when bidder 5 has some long-term stake in the transaction. We can think of four factors that induce bidders to make (unenforceable) sidepayments. First, many rings participate in a sequence of auctions, or bid sequentially on many items at a single auction. Sidepayments are made so that bidders are allowed continued participation in the ring. Second, colluding bidders often have other business relations with each other. Cheating in the ring may sour these relations. Third, a general concern about his reputation (honor among white-collar thieves) might induce bidder 5 to make the sidepayment. And fourth, the sidepayments might actually be enforceable through the coercive power of organized crime. Gambetta (1993) reports that an important function of the Sicilian Mafia is facilitating bid-rigging.

22. There is an intriguing normative issue here. Posner (1975) has argued that the social loss from collusion is not just the usual deadweight loss but the entirety of monopoly profit as well. The latter is competed away in rent seeking activities by members of the coalition. However, this effect is dampened for auction schemes where the coalition's mechanism is inherently stable and, in addition, entry is difficult.

23. There is no pure strategy Nash equilibrium for the first price auction. The ring will submit a single sealed bid x from the interval x∈[0, V-K]. The bid is chosen according to the cumulative distribution function F(x) = K/[V-x]. The mixed strategy Nash equilibrium has a mass point at x = 0 such that the probability of x = 0 is K/V, or F(0) = K/V. The entrant will choose not to enter, thereby avoiding the cost K, with the probability K/V. When the potential bidder does enter, then they submit a bid y∈(0, V-K]. The cumulative distribution function conditional on entry for the entrant's bid is $G(y) = \frac{K}{V-K} \frac{y}{V-y}$. There are no mass points in this distribution. The equilibrium profit to the potential entrant is zero, because this is what the firm gets in the case of no entry. Given that entry actually does occur, the entrant gets an expected profit of K from the auction, but also sinks the entry cost of K, implying again a net profit of zero for the entrant. The ring gets an expected profit of K from its bidding strategy.

24. There are alternative means of protecting collusive profits from entrants. For example, a union allegedly played a role in enforcing highway construction bid-rigging in Long Island. *See* Porter & Zona, (1993).

25. Just like the first price auction, this is a trembling hand perfect equilibrium.

26. It is worth noting that surplus maximizing bidding behavior may frequently entail inefficient outcomes. Suppose the five values were instead $\{5, 4-\epsilon, 3, 2, 1\}$. Consider the coalition of $\{5,4-\epsilon\}$. Recall that $\{5,4\}$ was not profitable but now $\{5,4-\epsilon\}$ is profitable. Bidder 4 simply bids below 2. The coalition earns a surplus of 3 whereas non-cooperative behavior would have yielded 3-ϵ. Bidder 4 would get a side payment slightly in excess of 1-ϵ while bidder 5 would get a net surplus slightly below 1+ϵ. The coalition gains by intentionally not winning an object that would have yielded a positive surplus had the coalition bid truthfully for the two highest values.

27. The rings $\{5,4,3,2\}$, $\{5,4,3,1\}$, and $\{5,4,3,2,1\}$ are feasible at the discriminatory auction. These rings could follow the same behavior as the $\{5,4,3\}$ ring and ignore bidders 2 and 1. In a one-shot setting these larger rings will not be formed, because they offer no advantage and may create disadvantages such as an increased risk of detection.

28. See Capen, Clapp, & Campbell, (1971). Of course, this was an "out of equilibrum" phenomenon. In equilibrium the winner's curse refers to the shading that occurs in bids to reflect the fact that winning is bad news in terms of the informational content of one's private signal.

29. The winner's curse grows more severe as the number of bidders or uncertainty grows. See, Bikhchandani & Huang, 1991.

30. The argument was first developed by Hendricks and Porter, (1988).

31. If bidder 2 were to bid any amount they would always earn a zero profit in light of bidder 1's strategy. Why then would bidder 2 bid V_1? Suppose, regardless of the value realization, that bidder 1's value was slightly in excess of the realization V_i, say $V_i + \epsilon$. This might be case because knowing the value allows bidder 1 to make some minor value enhancing investment that bidder 2 cannot make. Then bidder 2 would never want to win with a bid in excess of V_1 since they would always earn a negative surplus. We rule out equilibria that would emerge with $\epsilon = 0$ that do not exist when ϵ is arbitrarily small.

32. See Engelbrecht-Wiggans, Milgrom & Weber, (1983). An empirical study of bidding for off-shore oil tracts shows that non-neighbors earned zero expected profits from bidding on drainage tracts, but neighbors got positive expected profits. *See* Hendricks & Porter, (1988).

33. Hendricks and Porter (1988) observe this pattern in oil tract auctions. They note that collusion apparently was limited to neighbors of the tract up for auction.

34. With regard to off-shore oil drainage tracts, neighboring firms apparently were able to collude to retain the informational rents that derive from their superior information about the value of a drainage tract (Hendricks and Porter (1988)). There were 74 tracts with multiple neighbors, but only at 17 of these tracts did more than 1 neighbor bid. Furthermore, the profits to a winning neighbor were not affected by the presence of multiple neighbors. Finally, increasing the number of neighbors to a particular drainage tract, decreased the probability that a particular neighbor would bid. *See also,* United States v. Champion *Int'l* Corp., 557 F.2d 1270, 1272 (1977) (better informed bidders at timber auctions would exchange information prior to bidding).

35. As in the previous subsection, we are constructing this equilibrium from the premise that a bidder who knows the value of an item with certainty can make it worth ϵ more than its "true" value through some form of ex ante investment. This small value premium is bid by all such bidders. We then let ϵ tend toward zero.

36. Bidder 1's expected profit is $(1-\pi)^2 V$. The probability that both components have a high value times 2V minus V.

37. This may be difficult to accomplish. Bidders have an incentive to distort the information they provide to other ring members. Here and throughout this article we have suppressed this issue. The problem does not exist if the information is easily verified when it is shared. The problem can be overcome in a repeated auction setting.

38. Notice that if there is some cost to collusion then bidders 2 and 3 would rather not collude. It is not difficult to introduce some heterogeneity into the model would give bidders 2 and 3 a positive incentive to collude. For example, if collusion between bidders 2 and 3 allows for better risk sharing or some productive synergy, then they could get a positive profit out of collusion and the auctioneer would still benefit from more aggressive bidding.

39. The CV setting yields stronger incentives to collude than the IPV setting. There is no counterpart to the winner's curse in the IPV setting – the fact of winning cannot convey disappointing information about the value of the object for sale. An IPV auction winner learns only that others do not share his passion for a particular item. He does not learn that he has bad taste. Furthermore, the existence of asymmetric information is not much of a problem at IPV auctions. If individual preferences account for the differences in valuations, then no one can be better informed than anyone else.

40. The second inefficiency created by collusion is peculiar to markets involving the government as a buyer or seller. In these markets, collusion leads to increased government expenditures at procurements, and decreased revenues at auctions. We disregarded such wealth transfers above, stating that they are a distributional issue. The difference in the case of the government is that raising governments funds through distortionary taxes creates inefficiency. The increased revenue spent in procurements because of collusion is not simply a wealth transfer. If the revenue lost by the government as an auctioneer when facing colluding bidders is replaced by distorting taxes, then, once again, there is an efficiency loss.

41. In recent years, the Supreme Court has reexamined various horizontal agreements that in the past would have been quickly condemned as price fixing and per se illegal. There is a trend away from the per se rule toward the rule of reason in horizontal cases. The Court applied a rule of reason standard to practices that impinged on price setting in Professional Engineers,National Society of Professional Engineers v. U.S., 435 U.S. 679 (1978). NCAA v. Oklahoma,National Collegiate Athelethic *Ass'n v.* Board of Regents of the Univ. of Okla., 468 U.S. 85 (1984) and Broadcast Music.Broadcast Music, Inc. v. Columbia Broadcasting, Inc., 441 U.S. 1 (1979) The Court emphasized the possible pro-competitive effects of horizontal agreements in these markets, and in Broadcast Music, the Court permitted price fixing because it was ancillary to a legitimate pro-competitive purpose.

42. The British Restrictive Practices Court is sympathetic to the countervailing power argument applied to price-fixing. See Scherer and Ross, (1990). The Capper-Volstead Act (1922) exempts agricultural cooperatives from antitrust law in order to promote marketing efficiency and counterbalance the market power of suppliers and customers.

43. One must be careful though, a monopolist who has complete knowledge about buyers preferences, will not choose to inefficiently restrict output. Instead she will practice perfect price discrimination and offer the competitive output. We focus on the realistic case in which buyers and sellers have private information.

44. Less than all inclusive collusion is common in auction and procurement markets. But we should note that it is crucial for our result. The output effect disappears if all three bidders collude.

45. The output expanding effect of collusion is lost if the valuations of the buyers are changed to 5, 3, and 2. When the two high value bidders collude and two items are offered their strategy changes. If they do win two items at a price of 2, then there combined profit is 4. If instead the second highest bidder suppresses his bid to 0, then the ring wins one item and gets a profit of 5. The auctioneer's best response to collusion is to sell only one item. In contrast, with noncooperative bidding, the auctioneer's optimal choice is to sell two items. Thus countervailing power depresses equilibrium output.

46. In the noncooperative setting the auctioneer earns a profit of 3 and the winning bidder gets a profit of 2. In the collusive setting the auctioneer gets a profit of 2 and the ring gets a profit of 5 + 3 - 2 or 6. Under collusion the total profit is 8 compared to 5 in the noncooperative case.

47. United States v. Seville Indus. Mach. Corp., 696 F.Supp. 986, 991 (D.N.J. 1988) (even if the reserve is binding and no bid is accepted, there is still a Sherman Act violation).

48. For example, if two bidders independently draw their valuations from the uniform distribution on the interval [0, 1], then the optimal reserve is .5. If these bidders collude, then the optimal reserve is approximately .58.

49. A phantom bidder may be used to implement a reserve policy. The auctioneer tries to force the ring's winning bid up by pretending to receive competing bids from some bidder in the back of the room or over the telephone. Like the reserve, phantom bidding leads to inefficient retention of the item by the auctioneer.

50. We do not discuss ex post investment because usually collusion does not affect investment decisions that are made once an item has been auctioned or a contract let. Besides the acquisition of information, bidders make ex ante investments in physical assets. The incentive to make these investments may also be too low with non-cooperative bidding.

51. Hendricks & Porter, (1988 – at oil lease auctions joint ventures sometimes formed after seismic surveys, but more stable ventures formed before surveys).

52. Chari and Weber (1992) argue that the information gathered by bidders in Treasury auctions has no social value.

53. It is unlikely that these investments would be wholly redundant (i.e. produce exactly the same information). Nevertheless, there is the potential for substantial overlap in the information obtained by individual bidders.

54. Many authors have noted that governments could take many steps to protect themselves against collusion.

55. There is a vast amount of analysis and evidence that shows how certain regulatory agencies have been captured by industry so that the regulations work to provide rents for industry members.

12 ACTIVITY RULES FOR AN ITERATED DOUBLE AUCTION

Robert Wilson

This chapter reports an application of game theory to market design. Like most practical work, it uses a few key principles derived from theoretical studies, rather than any particular model or explicit mathematical analysis.

The purpose of market design is to increase the efficiency of the market outcome by suppressing strategic behavior or rendering it ineffective. One part of this task is to eliminate loopholes in the procedural rules that might be exploited by a wily trader, but the more fundamental part is to devise rules that promote efficiency. In the case of an iterated multi-market auction, the key requirement is reliable price discovery. That is, the rules should encourage suppliers to reveal their costs and demanders their values, steadily throughout the bidding process. This is necessary because any one supplier typically relies on the pattern of prices across the markets to devise its optimal bidding strategy, taking account of its variable and fixed costs of operation. Similarly, each demander relies on the pattern of prices to construct its optimal plan of purchases, taking account of complementarities and substitution among the products offered in the several markets. Efficiency of the final outcome therefore depends on early, and as the auction proceeds, progressively more accurate revelation by all traders.

The present application is to the design of a wholesale market for forward trades of electrical power among suppliers (generators) and demanders (large customers and power marketers). Such a market is typically conducted a day ahead of delivery, and consists of 24 separate markets for delivery during the 24 hours of the next day. Each of the hourly markets clears independently of the others, and all trades are settled at the clearing prices established at the close of the final iteration. This application was developed within the particular institutional features of the California Power Exchange (PX), which started operations on April 1, 1998 (see the website www.calpx.com for reports of transactions and prices). The PX is a public-benefit corporation that competes with other wholesale markets conducted by private parties.

Each of the PX's day-ahead hourly markets is a double auction. Each supplier submits an offered supply schedule indicating the quantity it is willing to provide at each price. Similarly, each demander's bid is a demand schedule indicating the quantity it wants to purchase at each price. The supply schedules are aggregated by computing the total supply offered at each price; similarly, the demand schedules are aggregated by summing to find the total demand at each price. The market is then "cleared" by finding the (least) price at which aggregate supply equals aggregate demand. In the static version of the double auction this closes the market: each supplier and demander is assigned the quantity it offered or bid at that price, and all transactions are made at the clearing price. The design task, however, called for an iterated double auction. In this dynamic version the entire process is repeated several times, allowing suppliers and demanders to alter their submissions in response to the prices and quantities resulting from previous iterations. The market closes only when no submissions are revised, or when a convergence criterion has been met.

The motive for an iterated auction is the important role of price discovery. As described in Section 1, a supplier needs to anticipate the pattern of clearing prices across the entire 24 hourly markets in order to make well-informed decisions about which generating units to start. In particular, the duration of a unit's consecutive hours of operation is a major determinant of whether the costs of start-up can be recovered. The efficiency of the market outcome is partly dependent on a reliable and informative process of price discovery as the iterations proceed.

The basic design problem can be stated simply. The "gaming" behavior that could undermine price discovery, and thereby efficiency, is the strategy called "hiding in the grass." This refers to the tactic of deferring serious bidding until the close of the auction. If the rules allow such a strategy then each trader prefers to wait until the final iteration, when it can see the pattern of hourly prices revealed by others' bids, and then devise its own optimal bids accordingly; moreover, by waiting it avoids affecting interim prices via its own bids. But if many traders do this then price discovery is impaired, and the efficiency gains from an iterative process are lost, since only the final iteration reflects sufficient serious bids to establish the pattern of prices. Thus, the underlying difficulty is a free-rider problem in which each trader prefers that others provide the bids that reveal the pattern of hourly prices.

One solution, albeit a partial one, is to impose "activity rules" of the kind used in the FCC spectrum auctions. The role of such rules is to encourage serious bidding right from the start. The key idea is to confront traders with irreversible decisions throughout the iterative process. At each stage a trader faces a "use it or lose it" decision regarding the bidding options available in later iterations. Activity rules must be designed carefully to minimize adverse effects on efficiency from restricting traders' bidding strategies, but if designed well then they benefit each trader by encouraging others to bid seriously in each iteration. The resulting progressive revelation of the pattern of prices across the markets enables each trader to take advantage of this information in constructing its own bids.

The activity rules described here are based on the principle of revealed preference: a bidder's refusal to improve a previous clearing price is presumptive evidence that it cannot do so profitably. This principle is represented by an "Exclusion Rule" that prevents later improvements in an offer that fails to improve the previous clearing price at the first opportunity. When other routine procedural rules are included, the resulting activity rules perform well in experimental tests, as described by Plott (1997).

The motivation for the activity rules is described in Section 1. The specific rules proposed for the PX are described in Section 2 and elaborated in Section 3. The full set of activity rules is summarized in the Appendix.

1. The Role of Activity Rules

Self-scheduling is a principal feature of the PX auction. Bids and offers are for delivered energy only – transmission losses are absorbed by demanders; all traders incur usage charges for transmission across congested inter-zonal lines; and fixed cost components such as start-up and no-load hourly running costs are absorbed by suppliers, who offer energy from their portfolios of generation assets. The iterative character of the PX is motivated primarily by suppliers' need to recover the fixed costs of daily operations: as the pattern of hourly prices is revealed during the iterations, suppliers are better able to schedule the plants in their portfolios to meet the energy commitments in their accepted offers.

There are several other market designs that provide some assurance that fixed costs are covered. One type allows offers on a full-cost basis; this type includes bilateral bid-ask markets and auctions that allow combination tenders for multiple hours. A second type is represented by the PX auction protocol, in which an iterative auction process enables a supplier to select its operating regime, withdrawing from hours with prices insufficient to cover its total costs. If price discovery is early and reliable then self-scheduling is feasible and there is no need for the system operator to optimize operating schedules.

The role of withdrawals in the PX is due to an interaction between the tender format and the pricing rule. The tender format requires separate offers for each hour. The uniform-price rule stems from the legislated requirements that in each hourly market all energy is traded at the market clearing price, exclusive of transmission usage charges, and that the PX takes no net position. Uniform pricing can be implemented without withdrawals, as in the uniform-price double auction studied by McCabe, Rassenti, and Smith (1993). Alternatively, one can forego the uniform pricing rule by using a dynamic bid-ask market. In such a market, each trader can post bids or offers, or accept any posted bid or offer; each transaction is a binding bilateral contract immediately upon acceptance. Dynamic markets with continual transactions preclude a uniform price but they have the advantage that they ensure impatience to trade. This impatience is borne of fear that profitable opportunities will be missed: when a demander posts a good bid, each supplier is eager to accept it before a competing supplier grabs it first. In such markets the volume of trade rises fairly steadily as the dispatch time approaches, and the accuracy of traders' predictions about the best bid and ask prices that will prevail at the close improves correspondingly.

Impatience to trade is one way to solve the fundamental problem of reliable price discovery. Any dynamic or iterative process provides a sequence of price signals to traders. If these interim prices are good predictors of the final prices that will prevail at the close, then they enable suppliers to make accurate judgments about which plants to operate and in which hours. In turn, early resolution about which plants to operate in each hour ensures stable convergence, since later iterations focus on the simpler task of finding the clearing prices for energy.

Price discovery is more problematic in the PX because no transactions occur until the close of the final iteration. Activity rules are needed to ensure that price discovery is reliable. Without activity rules, and with uniform pricing, no trader has any positive incentive to make serious bids or offers until the final iteration; and without serious bids and offers, the tentative clearing prices in early iterations are unreliable predictors of the final clearing prices. Indeed, any large trader has the opposite incentive: it withholds information about its own final offers in the early iterations, preferring instead to rely on others to provide such information contributing to price discovery. So in the absence of impatience of trade, activity rules are imposed in order to force all traders to reveal early some credible signals about the bids and offers they will tender in the final iteration.

In designing activity rules, the guiding principle is that they should be the least restrictive rules sufficient to assure reliable price discovery. Ideally, they impose no limit on the efficiency attainable at the close of the market. In particular, they should impose no significant restrictions or disadvantages on suppliers who elect to offer their actual costs. The only effect of the activity rules is to suppress gaming, or render it ineffective, by imposing constraints on revisions of offers during the iterative process. These constraints create increasingly strong incentives for cost-based offers. If the activity rules are successful, as the experimental evidence indicates they are, then suppliers learn that there is little to be gained by strategic bidding – it may delay convergence somewhat, but the final outcome is largely determined by cost-based offers in the closing iterations.

To preserve self-scheduling, the activity rules cannot be invasive; e.g., they cannot rely on any additional solicitation of reports about traders' private information. On the other hand, activity rules can be designed using the principle of "revealed preference." By interpreting previous offers as reliable indicators of what is feasible and profitable for the supplier, constraints can be imposed on subsequent offers. As the auction progresses, these constraints narrow the supplier's allowed strategies, until in the final iteration there is little room for offers that differ significantly from actual costs. Realistically, costs must be interpreted as opportunity costs rather than actual running costs, since each supplier also has opportunities to trade in other markets. In addition, opportunity costs must be interpreted in relation to market power. Activity rules cannot prevent a supplier from realizing the profit obtained when it offers the higher cost of the next plant along the aggregate supply function.

As a practical matter activity rules must be easily understood by traders, and simple to implement. The activity rules should be applied automatically: the portion of any submitted tender that violates the rules is discarded without any "negotiation" with the trader.

Activity rules are generally of two kinds. One kind pertains to the opening and closing of the auction, and the other pertains to the ways in which tenders can be revised or withdrawn from one iteration to the next. The rules treat demanders and suppliers symmetrically: the rules for demanders differ only by interpreting price decrements as price increments. To avoid confusion from separate phrasing regarding demanders and suppliers, I refer here only to the rules for suppliers.

I first describe the activity rules for the general case. This formulation is then developed in more detail for a practical implementation.

2. General Statement of the Activity Rules

The activity rules can be derived from a single formulation that is quite general in its application. To express this formulation succinctly, it is useful to interpret the tendered supply function as a bundle of contingent offers: each offer consists of a price for a particular increment of supplied energy. For example, one point on a tendered supply function might offer a price of \$23 per MWh for the 87th MWh delivered in the hour from 10 to 11 AM. Thus, I interpret a point (p,q,t) on the tender as offering the price p for the q-th increment of energy supplied in hour t.

The rule has three parts. In each iteration after the first, for each quantity increment included in the supply tender submitted in the first iteration:

1. The price cannot be increased.

2. The price can be decreased only if the new price is less than the clearing price in the previous iteration by at least a specified price decrement

(e.g., $1.00 or $0.10/MWh). We say in this case that the new price improves the previous clearing price.

3. The price cannot improve any previous clearing price not improved at the first opportunity.

Part 1 is a fundamental requirement for a competitive auction. Part 2's requirement that a price change improves the clearing price eliminates extraneous revisions. A minimum decrement avoids stalling the auction.

Part 3 is the key provision. To make it precise requires the following clarification: the "first opportunity" is the first iteration following an iteration in which the offered price exceeds the clearing price. For instance, if a supplier offers a price of $25 in iteration 1, in which the clearing price is $23, then iteration 2 is the first opportunity to improve this clearing price. If the supplier offers a price less than $23 in iteration 1 then for present purposes it has no obligation or "opportunity" in iteration 2 to improve the $23 clearing price obtained in iteration 1. Therefore, Part 3 imposes no restriction on suppliers who offer prices below the clearing price; in particular, these suppliers are not disadvantaged by refusing to improve the clearing price in the next iteration. However, among those suppliers who offer exactly the $23 clearing price there may be some whose offers are rejected according to the Rationing Rule. For these suppliers, iteration 2 is indeed the first opportunity to improve the previous clearing price.

With this clarification, Part 3 says the following, expressed via the example. Suppose the specified price decrement is $0.50. If in iteration 2 a supplier who offered $25 in iteration 1 does not improve iteration 1's clearing price of $23 then this is taken as *de facto* evidence that its cost increment for this quantity increment exceeds $22.50. Consequently, this supplier is precluded from offering a price equal to or less than $22.50 in any subsequent iteration. However, if the clearing price later rises above $23, say to $24 in iteration 5, then the supplier can in the next iteration 6 improve this clearing price by offering any price between $22.50 and $23.50. But if it fails to do so then, thereafter it cannot offer any price equal to or less than $23.50. Similarly, a supplier who offers exactly the clearing price of $23 in iteration 1 and is rationed, and then declines to improve its offer to a price at or below $22.50 in iteration 2, cannot offer a price in this range later.

The effect of Part 3 is to "freeze" any part of a supplier's tendered supply function for which there is presumptive evidence that its cost exceeds a previous clearing price. It is only frozen, not rejected irrevocably, because there remains the possibility that it is "thawed" if the clearing price rises sufficiently in some later iteration. Part 3 prevents a supplier from profiting by withholding supply until the final iteration.

This general form of the activity rule is not sufficient by itself. The reason is that it allows suppliers to offer very high prices in the first iteration. If demanders similarly offer very low prices in the first iteration then the auction gets off to a slow start due to the resulting gap between supply and demand. This is an inherent problem in all auctions; the usual way of correcting this deficiency is an Opening Rule that governs the first iteration.

The Competitive Process

Activity rules of this form produce a characteristic process of competition among suppliers. After each iteration the supply offers are divided into those that are infra-marginal, because their offered prices are less than the clearing price, and those that are extra-marginal, because their offered prices are more than the clearing price (or they are rationed). In the next iteration, each extra-marginal offer must improve the previous clearing price or forego all subsequent opportunities to offer lower prices – because it is frozen, perhaps permanently if later clearing prices remain below the previous clearing price. Thus, if the previous clearing price exceeds the supplier's cost then the incentive to revise the offered price is quite strong, since this is the supplier's last opportunity. However, when the offer is revised its position in the merit order (the offers ordered in terms of increasing cost to form the aggregate supply function) improves. This improvement relegates some previously infra-marginal offer to a later position in the merit order. The previously infra-marginal offer becomes extra-marginal, and the supplier who submitted it now faces a similar problem. The resulting process resembles a tug-of-war among the marginal suppliers to determine which offers will be accepted at the clearing price. This battle is resolved when the clearing price is driven down to the cost of some contenders, who then prefer to let their offers be frozen. The characteristic pattern is that in each iteration there are many bids and offers near the previous clearing price; but if one side of the market must be rationed, say the suppliers, then those whose offers are excluded and their costs are less, find it advantageous to reduce their prices.

3. An Implementation for the PX

This section describes a fairly complete set of procedural rules for the PX auction. These rules implement the main ideas elaborated in Section 2.

The Auction Process and the Bid Format

The auction can operate in a discrete or continuous mode. In each case there are 24 forward markets for delivery in the hours of the next day, and a clearing price is computed separately for each hourly market. In the version with discrete iterations, the auction operates in batch mode: all clearing prices are updated after each iteration. In the version with continuous market clearing, the arrival of each revised bid or offer prompts a revision of the clearing price in that market, which is then broadcast to all traders. These designs are associated with different formats for tenders. In the continuous version it suffices that each tender specifies a single price and a single quantity or interval for each hourly market. In the discrete version a tender is an entire demand or supply schedule for each hour, presumably in the form of a piecewise-linear function or a step function. In the following I do not address the continuous version, and focus instead on the discrete version.

In the discrete version, after each iteration the current tenders are used to calculate the clearing price for each hourly market independently. Each tender is specific to a particular hourly market, and consists of a piecewise-linear or step function that states the supply offered at each price. This function is interpreted as a bundle of contingent offers: each point (p,q,t) on the tender is an offer to deliver the quantity q in hour t at any price not less than p. Similarly, a step on the schedule offers a price p for any quantity within a corresponding min-max interval [m, M].

The activity rules apply separately to each price-quantity pair (p,q) on the tender for a specific hour t. Thus, when checking the activity rules, no distinction is necessary regarding the exact form in which the tender is submitted: the same rules apply to tenders that are points, intervals, piecewise-linear, or step functions. For simplicity in the exposition, however, I assume that schedules are step functions.

Each tender is a binding bid or offer that remains in force until it is revised or ultimately rejected by the PX. A revised tender replaces all previous tenders for the same portfolio and hour. Except for those withdrawn or replaced, all tenders continue in force for the next iteration. At the close of the auction, those supply tenders with prices above the clearing price are rejected, with ties at the clearing price resolved by a Rationing Rule. The remaining offers are accepted, and each becomes automatically a binding contract, with the PX as the counter-party, for the offered quantity at the final clearing price.

The Opening Rule

The first part of the Opening Rule is simple:

Opening Rule (1): A new tender can be submitted only in the first iteration.

In particular, in each later iteration the only tenders allowed are revisions of ones submitted in the first iteration. This rule ensures that the maximum supply in each hourly market is revealed in the first iteration. This rule is essential for effective price discovery, else a trader could wait until the final iteration to submit its first tenders.

The second part of the Opening Rule is intended to get the auction off to a quick start.

Opening Rule (2): At its option, the PX can specify a seed price for the first iteration.

A seed price is an initial prediction of the final clearing price, which plays the role of the previous clearing price in applying the Exclusion and Revision Rules described below. After the first iteration that part of a supply tender that exceeds the seed price is frozen with the seed price as its Activation Price. The seed price can be based on expert judgment, or it could simply be the final clearing price in that hourly market the previous day or week.

The Exclusion and Revision Rules

I first describe these rules along the lines of Section 2 and then motivate them. All tenders that were not withdrawn after previous iterations are automatically carried over to the current iteration. Based on the history of the auction, the steps on these tenders are divided into those that are frozen and those that are active: active steps can be revised, whereas frozen steps cannot.

Activity Rules for an Iterated Double Auction

All steps are active in the first iteration. In each iteration after the first:

Exclusion Rule: A previously active step on a supply tender becomes frozen after the current iteration if its offered price was not revised to improve the previous clearing price, and in the previous iteration its offered price was above this clearing price – called its <u>Activation Price</u>. A frozen step cannot be revised. A frozen step becomes active again after an iteration in which the clearing price is higher than its Activation Price.

The Exclusion Rule operates as follows. If a tender's offered price for a particular step was less than the clearing price in the previous iteration then the supplier has no obligation to revise the offered price, but is not excluded from doing so. However, if its offered price exceeds the previous clearing price (or equal and the step is rationed), then its offered price must be revised to less than the previous clearing price, else it is frozen until the clearing price regains the previous level. For example, if the previous clearing price was $23 and the supplier now declines to offer a revised price less than $23 then this step cannot be revised again until after the clearing price rises above $23. As described in Section 2, the Exclusion Rule is based on the inference that refusal to improve the previous clearing price signals that the revised price would be insufficient to recover the supplier's cost.

The restriction that frozen steps cannot be revised is essential to reliable price discovery. Otherwise, a supplier could wait until the last iteration to revise, and in the meantime other traders would be getting no information about lower prices the supplier might be willing to offer. Thus, each tendered supply price that is above the clearing price in one iteration must be revised in the next iteration lest it thereafter be excluded from revisions until the clearing price rises again to comparable levels.

Revision Rule: An active step can be divided into two active steps with the same offered price. An active step can be revised only by offering a lower price that improves the previous clearing price. That is, the revised step must offer a new price for the same quantity interval that is less than the previously offered price, and less than the previous clearing price by at least the specified price decrement.

This particular phrasing of the Revision Rule is peculiar to the present supposition that each tender is represented as a step function. In this case, an active step corresponding to an offered price for an interval [m, M] of quantities can be revised by breaking it into two steps with intervals [m, k] and [k, M]. Then, one step is revised to offer a new price that improves the previous clearing price, and the second step is frozen. For the frozen step, the offered price is unchanged and its Activation Price is the previous clearing price.

The clearing price is computed using all steps on the current tenders, both frozen and active. This reflects the fact that even frozen steps remain binding offers to the PX. However, those steps that offer a higher price for a smaller quantity than another step are excluded from the merit order used for the computation, so they have no effect on the clearing price obtained.

It is important to realize that the price decrement (and a comparable price increment for demanders) is an important design parameter that can substantially affect the rate of convergence of the iterative process. In a worst-case scenario the clearing price moves by no more than the price decrement from one iteration to the next. The appropriate magnitude cannot be determined a priori; rather, it must be based on judgment, experience, and predictions about current supply and demand conditions, especially the price elasticities and variances of supply and demand. A practical procedure might start in iteration 2 with a large value, say $1.00/MWh, and then decrease it steadily in later iterations to a final value, say $0.20/MWh. However, experimental evidence indicates that a small decrement need not produce clearing prices closer to the theoretical clearing price. A large decrement has the advantage that it produces stronger pressure on suppliers to tender initial offers closer to actual costs. With a large decrement, a price slightly above actual cost cannot be revised profitably, so a supplier must contend with the risk that a profitable opportunity will be missed.

Another important ingredient is the Rationing Rule. In a typical iteration there can be many offers at the clearing price, and if demand at that price is less than supply, then some of the supply steps must be rationed. The experimental evidence indicates that it is best to reject entire steps rather than allocate the marginal demand *pro rata* among the supply steps at the margin. This avoids a proliferation of subdivided steps and accelerates convergence.

Activity Rules for an Iterated Double Auction

The Withdrawal Rule

The following formulation assumes that after withdrawals the clearing prices are re-computed before the next iteration. Re-computing the clearing prices is desirable to ensure that other traders can take account of this information when revising their tenders for the next iteration.

Withdrawal Rule: After each iteration except the last, each supplier has the option to withdraw a tender entirely and irrevocably from any hourly market. The clearing prices are re-calculated after the withdrawal round. For the purposes of the Exclusion and Revision Rules and setting Activation Prices, these become the clearing prices for this iteration.

Withdrawals are allowed to enable a supplier to exit one or more markets when prices are insufficient to recover fixed costs, but after the final iteration an accepted tender cannot be withdrawn and the supplier is financially liable for delivery. It is clear that withdrawals cannot be revoked easily, else a supplier could withdraw until it re-enters in the final iteration. It might be argued that efficiency could be enhanced by allowing revocation of withdrawals if prices rise later. I have studied this problem but find revocation rules vulnerable to gaming. Within the strictures of the PX protocol, my solution is the Revision Rule, which is constructed explicitly to enable a supplier to offer tenders that cover its average costs. Consequently, my conclusion is that there is no need, and no easy prospect, to allow revocation of withdrawals. Withdrawals might be excluded (to prevent price manipulations followed by unpenalized withdrawals) but this would interfere with self-scheduling.

The Closing Rule

Closing Rule. All the hourly markets close simultaneously. They close automatically after any iteration in which no tender is revised, or a convergence criterion is satisfied.

Both theory and experiments show that the markets converge naturally, but the number of iterations required can exceed the time allowed. However, experiments show that there is little efficiency loss if the markets are closed after progress has slowed sufficiently. The primary criterion is a small ratio of active extra-marginal offers to those infra-marginal ones that would be displaced by another iteration, which signals that the current clearing price is close to the theoretical clearing price. Because quantities typically converge faster than prices, the efficiency loss from using a convergence criterion is likely small.

4. Conclusion

The purpose of activity rules is to encourage convergence to an efficient outcome by suppressing gaming. The rules proposed here are based on the principle of "revealed preference." Essentially, a supplier's refusal to improve a previous clearing price is taken as evidence that such a lower price would not recover its cost, and that therefore it can be prohibited from offering this price later. The resulting process forces suppliers at the margin to compete: each extra-marginal bidder improving the previous clearing price ejects some infra-marginal bidder who is thereby forced to reduce its offered price or forego any profit it might obtain. Each refusal freezes a step of the tender, until possibly the clearing price rises that high again later.

These rules are complemented by procedures for opening and closing the auction, and allowance for withdrawals. All tenders must be submitted at the opening to preclude a strategy of waiting until the final iteration that would impair price discovery. Withdrawals must be irrevocable and in any case withdrawals after the final iteration must be excluded.

The small-scale experimental tests conducted by Charles Plott (1997) indicate that, absent market power, these activity rules suppress gaming and drive the iterative process to nearly efficient prices and quantities in a moderate number of iterations.

Appendix: A Standard Set of Activity Rules

The following "standard" version of the activity rules was used for the experimental tests. This version is stated for supply tenders; symmetric rules apply to demand tenders. The tenders are assumed to be offered supply schedules that are step functions.

Tenders: Each step of each tender is a binding offer to trade at any price not less than the offered price. Each tender remains in force until it is withdrawn

Activity Rules for an Iterated Double Auction

or validly revised by the trader, or rejected by the PX. A revised tender replaces the previous tender for the same portfolio. At the close of the auction, those steps with prices above the final clearing price are rejected; ties at the clearing price are resolved via the Rationing Rule: "first come, first served" based on the time stamp of each new or revised tender. The remaining steps are accepted, and each becomes automatically a binding contract, with the PX as the counter-party, for the tendered or rationed quantity at the final clearing price – except a step at the margin, for which only a portion of the offered quantity might be accepted.

Opening Rule: (1) A new tender can be submitted only in the first iteration. After the first iteration, the only valid tenders are those submitted in the first iteration and revised later. (2) The PX can specify a seed price to start the auction.

Exclusion Rule: An active step on a supply tender becomes frozen after the current iteration if its offered price is not validly revised to improve the previous clearing price, and in the previous iteration its offered price was above this clearing price – called its Activation Price. A frozen step cannot be revised. A frozen step becomes active again after an iteration in which the clearing price is higher than its Activation Price.

Revision Rule: An active step can be divided into two active steps with the same offered price. An active step can be revised only by offering a lower price that improves the previous clearing price. That is, the revised step must offer a new price for the same quantity interval that is less than the previously offered price, and also less than the previous clearing price by at least the specified price decrement.

Withdrawal Rule: After each iteration except the last, each supplier has the option to withdraw a tender entirely and irrevocably from any hourly market. If the clearing prices are re-calculated after the withdrawal round then for the purposes of the Exclusion and Revision Rules these become the clearing prices for this iteration.

Closing Rule: All hourly markets close simultaneously. They close automatically after an iteration in which no tender is revised, or a specified convergence criterion is met, or when the available time expires. The results of the final iteration become binding transactions with the PX at the final clearing price.

* Portions of the applied work reported here were funded by the California Trust for Power Industry Restructuring, and the basic theoretical work was conducted under a grant from the National Science Foundation.

References

McCabe, K.S., S. Rassenti, and V. Smith (1993), "Designing a Uniform-Price Double Auction: An Experimental Evaluation," Chapter 11 in D. Friedman and J. Rust (Eds.), *The Double Auction Market*, Addison Wesley.

Plott, C. (1997), "Experimental Tests of the Power Exchange Mechanism," *Report to the California Trust for Power Industry Restructuring*, (www.energyonline.company/wepex/reports/reports2).

INDEX

Allen, F. and Morris, S.,
   Higher order beliefs, 33-38
Baiman, S. and Demski, J.,
   Monitoring performance, 53-57
Banker, R. and Datar, S., Performance
   measures with multiple signals, 58-61
Bidder Collusion and
   Countervailing Power, 355-359
   Efficiency, 354-355
   English auctions, 346-347
   Entry, 347-348
   First-price auctions, 345-346
   Multiple items, 348-350
   Pre-auction investments, 136-137
   Private Information, 351-354
Bockus, K. and Gigler, F.,
   Auditor resignation and audit pricing, 85-89
Brandenburger, A. and Stuart, H.,
   Monopoly Game, 201-204
Cohen, H., *You Can Negotiate Anything*, 274-276
Double Auctions and
   Activity rules, 375-380
   California Power Exchange, 372-375
   Exclusion and revision rules, 380-383
Fama, E., Testing market efficiency, 19-20
Farrell, J. and Saloner, G.,
   Competition in software, 121-125
Farrell, J. and Weizsacker, G., Amnesty Dilemma:
   Product quality choice, 227-230
   Managerial Slack, 230-232
   Lending to a sovereign power, 232-236
Financial signals, 23-27
Franchising and
   Agent effort, 143-144
   Asset Specificity, 170-171
   Contract Terms, 162-163
   Costly monitoring, 147-152
   Firm performance, 165-167
   Franchisor effort, 152-154
   Multiple Tasks, 161-162
   Outlet size, 144-146
   Product substitutability, 157-158

Franchising and
   Risk, 140-142
   Royalties on Sales, 169-170
   Spillovers, 154-157
   Strategic pricing, 158-161
   Uniform contracts, 167-169
Grossman, S. Hart, O., and Moore, J.,
   Incomplete Contracts, 282-283
Kanodia, C. and Mukherji, A.,
   Take-it-or-leave-it audit pricing, 74-82
Kanodia, C., Coordination and budgeting, 63-71
Li, L. and Whang, S.,
   Time-based competition, 97-105
Li, L., Information sharing in oligopoly, 114-121
Market for corporate control, 27-30
Market microstgructure, 32-33
Markowitz, H., Capital asset pricing model, 18-19
Maskin, E. and Riley, J., Bidding in
   asymmetric settings, 311-315
Merton, R., Intertemporal capital
   asset pricing model, 21
Modigliani, F. and Miller, M.,
   Valuing the firm, 21-23
Morgan, J. and Stocken, P.,
   Pricing of audit risk, 82-85
Multi-Object Auctions and:
   Complementary values, 321-329
   Substitute values, 317-321
Porteous, E. and Whang, S., Coordinating
   marketing and manufacturing, 109-113
Reputation and signalling with
   Endogenous prices, 259-267
   Exogenous prices, 251-259
   Private Information, 267-269
Rubinstein, Alternating offer model, 276-279
Stuart, H., Spatial competition, 205-209
Stuart, H., Supplier-firm-buyer game, 193-199
Vaysman, I., Transfer pricing, 72-73
Vickrey auctions, 317-320, 327-329
Vickrey, W., Auctions and
   revenue equivalence, 300-305, 344-345
Winner's Curse, 305-311

HB144 .G372 2001

Game theory and business
 applications /
    c2001.